Transportation Engineering and Systems

Transportation Engineering and Systems

Edited by
Samuel Morgan

Willford Press
www.willfordpress.com

Published by Willford Press,
118-35 Queens Blvd., Suite 400,
Forest Hills, NY 11375, USA

ISBN: 978-1-68285-337-5

Cataloging-in-publication Data

Transportation engineering and systems / edited by Samuel Morgan.
p. cm.
Includes bibliographical references and index.
ISBN 978-1-68285-337-5
1. Transportation engineering. 2. Railroads. 3. Traffic engineering. I. Morgan, Samuel.
TA1145 .T73 2017
629.04--dc23

For information on all Willford Press publications
visit our website at www.willfordpress.com

WILLFORD PRESS

Printed in China.

Contents

Preface

Transportation engineering is an evolving field of study that focuses on the construction and alteration of the urban environment for the smooth and efficient passage of vehicular traffic. This is achieved through the study of conceptual systems that can simulate, predict and make patterns out of data of existing vehicular movement. This book on transportation engineering and systems discusses topics related to the role of technology in the control and navigation of traffic. The book uses the latest data and research to merge common as well as complex theories of systems control and transportation engineering. It is a compilation of various case-studies that have been gathered from many parts of the world. This book includes contributions of experts and scientists which will provide innovative insights into this field. Students, researchers, experts and all associated with the discipline of transportation engineering will benefit alike from this book.

The main aim of this book is to educate learners and enhance their research focus by presenting diverse topics covering this vast field. This is an advanced book which compiles significant studies by distinguished experts. This book addresses successive solutions to the challenges arising in the area of application, along with it; the book provides scope for future developments.

It was a great honour to edit this book, though there were challenges, as it involved a lot of communication and networking between me and the editorial team. However, the end result was this all-inclusive book covering diverse themes in the field.

Finally, it is important to acknowledge the efforts of the contributors for their excellent chapters, through which a wide variety of issues have been addressed. I would also like to thank my colleagues for their valuable feedback during the making of this book.

Editor

1

Calibration of a rule-based intelligent network simulation model

A. A. Memon[1] · M. Meng[2] · Y. D. Wong[3] · S. H. Lam[4]

Abstract This paper is focused on calibration of an intelligent network simulation model (INSIM) with real-life transportation network to analyse the INSIM's feasibility in simulating commuters' travel choice behaviour under the influence of real-time integrated multimodal traveller information (IMTI). A transportation network model for the central and western areas of Singapore was simulated in PARAMICS and integrated with INSIM expert system by means of an application programming interface to form the INSIM. Upon calibration, INSIM was able to realistically present complicated scenarios in which real-time IMTI was provided to commuters and the network performance measures being recorded.

Keywords Traffic simulation · Integrated traveller information · Calibration · Mode choice

✉ M. Meng
mengm@ntu.edu.sg

A. A. Memon
ab_ahad@emirates.net.ae

Y. D. Wong
cydwong@ntu.edu.sg

S. H. Lam
soihoi.lam@gmail.com

1 IVES (International Vision for Engineering Solutions), Abu Dhabi, United Arab Emirates

2 TUM CREATE, 1 CREATE Way #10-02 CREATE Tower, Singapore 138602, Singapore

3 School of Civil and Environmental Engineering, Centre for Infrastructure Systems, Nanyang Technological University, Singapore 639798, Singapore

4 Transportation Infrastructure Office, Macao S.A.R., China

1 Introduction

Empirical findings of travel behaviour studies have provided evidence that commuters presented with integrated multimodal traveller information (IMTI) in congested travel environment, exhibited a certain degree of mode switching [1–4]. However, from the transportation system's perspective, the potential efficiency and success of various policy measures for eliminating or reducing traffic problems in metropolitan areas largely depend on the level of public acceptance of the traveller information [5–7]. A variety of travel behaviour models based on revealed and stated preference surveys have been developed to study the nature and dynamics of traveller's choice behaviour [8–11]. Despite the obvious need for assessing user acceptance and the potential impacts of these systems in terms of improving traffic conditions for individual drivers and the overall transportation system, there is still a lack of models to evaluate their full impacts. The idealised analytical models are not appropriate for observing commuters' behaviour dynamics as they lack the ability of representing decision-making processes occurring in a complex dynamic environment. On the other hand, treating travel behaviour within the framework of discrete choice models may not be that effective because, with the spatial and temporal dimensions, the choice set becomes astronomically large if one wishes to gain adequate levels of precision in forecasts. Furthermore, these tests and experiments are very expensive to conduct and do not allow for effective evaluation of different alternatives. Realising the limitations of other approaches with respect to the dynamics of travellers' behaviour under the influence of real-time information, several researches have invested considerable effort in laboratory experiments being conducted with the support of the simulation models [12–16].

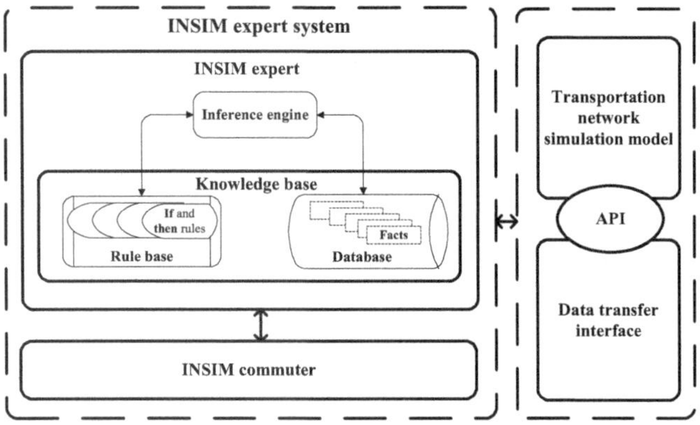

Fig. 1 Architecture of the INSIM

Currently, the simulation models are not sufficiently capable such that all the existing ITS components can be evaluated. Furthermore, in many modelling applications a more disaggregated approach to model socio-economic processes is generally desirable to reduce the model aggregation [17–21]. These two observations together have led inevitably to the adoption of a sophisticated micro-simulation platform to model the ITS systems in an intelligent environment, where the simulation is closer to reality based on the production-rule system approach. The premise underlying a production-rule system is that choice behaviour in a certain context can be better described by a series of "if-then" rules. If the condition in the "if" part is true, then the action of the "then" part follows. This approach utilises an external element (agent) that is integrated with the system to rationalise the "if-then" process within the system. Such an approach is known as the agent-based simulation and has been shown to have the potential to model complicated cognitive processes. This is the approach taken in order to fill the gap of existing simulation research into the impact of IMTI system, the authors have developed a rule-based mode choice model, the intelligent network simulation model (INSIM), as reported in an earlier study [22]. The INSIM possesses the sophistication to imitate the commuters' mode choice behaviour in an information-rich environment.

INSIM has four main components: INSIM expert system (IES), transportation network simulation model (based on PARAMICS), application programming interface (API), and data transfer interface (DTI), as shown in Fig. 1. The IES is responsible for generating mode choice decision. The transportation network simulation model simulates the traffic and generates travel statistics as output files. These output text files are collected by the API, and transformed to Microsoft Access Database (MDB) files and exported to DTI, which is a database that stores the PARAMICS output files in MDB format while retaining intact the original time-dependent updated origin and destination (OD) matrix file. The API takes in the information in the OD matrix file and releases the vehicles in the simulated network according to the pattern specified in the file. The IES is designed in a way to provide the flexibility to adopt an alternative mode choice model that can give prediction of commuters' mode choice. Therefore, the impacts of real-time IMTI on the overall performance of the transportation network can be analysed.

In most current research on intelligent network simulation model, a risk strategy is typically adopted to handle the risk of network uncertainty [23–28]. In the similar manner, the most important characteristics of the IES are its high-quality performance and adaptive learning. The high-quality performance assures system reliability and timeliness. The adaptive learning allows the IES to update dynamically and learn on its own through its experience. The reliability of the IES depends upon its decision-making process, which can be observed by two specific features, i.e. the capability to explain about the decision taken, and to deal with uncertainty. The IES is capable to explain how and why a decision was reached. The execution of rules tending towards certain decisions can be explained by the inference chain or the rule flow. Furthermore, the rule of inference has also been designed to deal with knowledge that is incomplete or not completely certain. A heuristic algorithm is developed to solve the uncertainty problem and generate a suitable decision for the provided facts. This ability of the IES allows it to provide a solution even when the provided information is not 100 % accurate or is incomplete. In this regard, it is also necessary to analyse the level of accuracy of the provided solution. To check the accuracy of the designed algorithm, an experiment was conducted and its level of mode prediction accuracy was analysed and has been reported in an earlier study by the Authors [22].

Accurate simulation results are highly dependent on the selection of model parameters and the calibration methodology [29, 30]. Calibration studies have been performed for various traffic simulation tools [31–33]. The typical calibration method is to minimise the differences between the observed data and simulation output during a certain time period on a typical day. Before application of INSIM, each module in INSIM is to be well calibrated with real traffic data to ensure realistic simulation results. This paper covers the important task involved in the calibration of the INSIM, especially IES, and is organised as follows. The study network and network coding are presented in Sect. 2. Detailed discussions regarding the calibration of INSIM are then provided in Sect. 3 and followed by conclusions in Sect. 4.

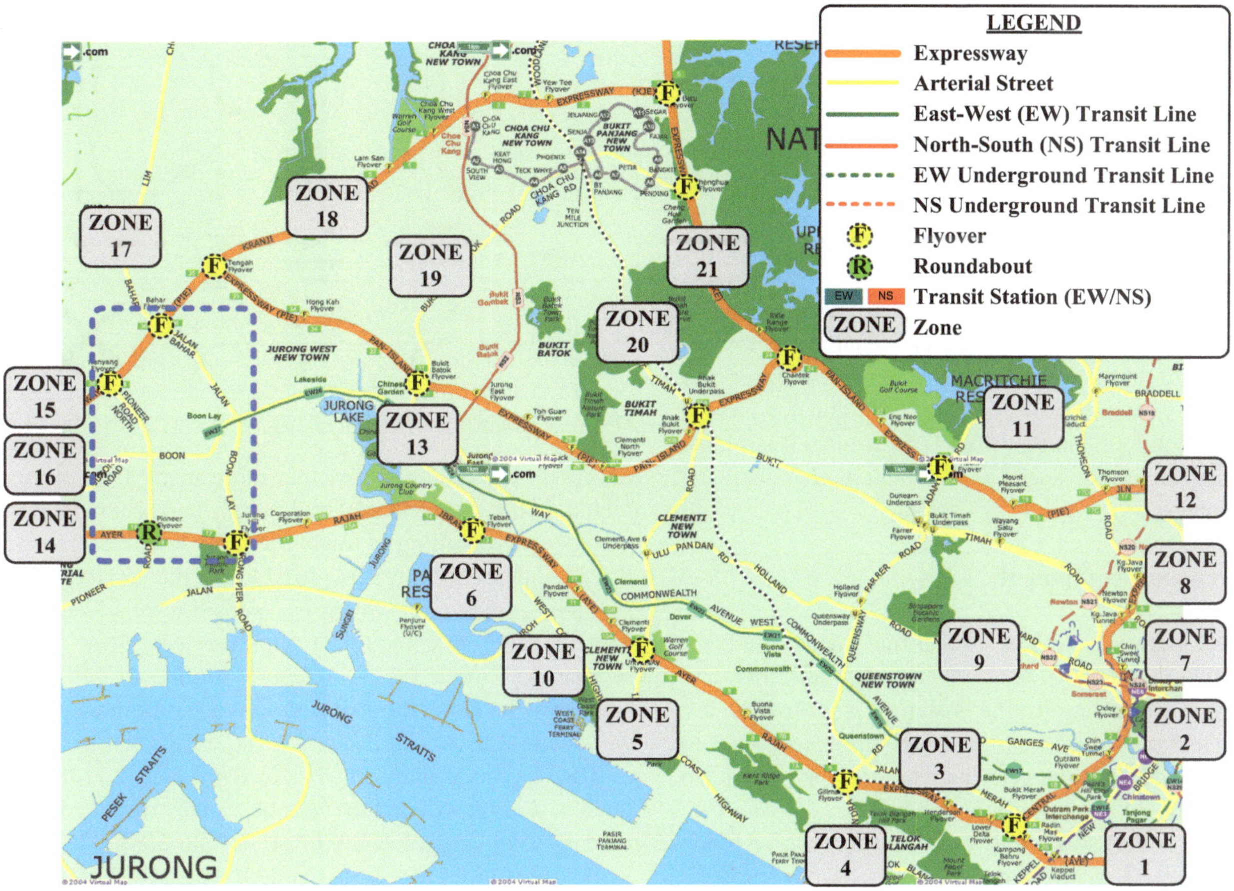

Fig. 2 Overall view of the simulated transportation network (source: Onemap, Singapore)

2 Study network

To assess the capabilities of the integrated IES with the transportation network simulation model, the transportation network for the central and western part of Singapore was selected for the simulation. The simulated network covers the area bounded by 5 major expressways and numerous arterial streets. The expressways are Pan-Island Expressway (18 km), Ayer Rajah Expressway (19 km), Central Expressway (9 km), Bukit Timah Expressway (5 km), and Kranji Expressway (9 km). In the multimodal network established for this study, there are a total of 21 residential and commercial zones, 2 Mass Rapid Transit (MRT) Lines, 23 MRT stations, 2 MRT interchanges, 38 selected signalised road junctions, 14 flyovers, and 1 roundabout. Every MRT station has feeder bus services, which traverse between nearby zones and MRT stations. Figure 2 shows the overall view of the simulated multimodal transportation network.

The physical data of the selected transportation network were collected through a survey of the study area, and also from several reliable web-based sources. All the selected expressways as well as two arterial streets (i.e. Jalan Boon Lay and Pioneer North Road) were covered. The transport network design criteria provided by the Land Transport Authority (LTA) of Singapore [34] was also consulted. The road geometrics and details of the signalised junctions were observed and sketched during site surveys. The details of public mode of transport were collected from Singapore Mass Rapid Transit's website [35].

2.1 Geometric design

The imported map of the study area provided all the essential details regarding the road network layout, location of flyovers, roundabouts and road junctions, MRT network, MRT stations, and bus interchanges. It also gave the details about residential, recreational, and commercial zones. This map was loaded and displayed as the background, over which nodes, links, bus stops, and MRT stations were placed.

The expressways are mostly 3-lane dual carriageways, except at certain locations where they are upgraded to 4-lane dual carriageways. The expressways have on-ramps and slip lanes that allow the vehicles to enter or exit, respectively. All the arterial roads are 2-lane dual carriageways. The lane widths are 3.7 m for expressways and 3.4 m for other roads. The speed limits vary from 80 to 90 km/h for expressways and 60 to 70 km/h for arterials roads.

The road network in PARAMICS was coded according to the road geometrics, including the junctions, roundabouts, and flyovers. The type of links for expressways and arterial streets were specifically designed and provided in the "category" file, where details about number of lanes and speed limits were defined. The speed limit for expressways was standardised at 85 km/h and for arterial streets at 55 km/h, given that PARAMICS models simulate the average travelling speed at 10 km/h higher than the defined values.

2.2 Surveillance equipment

In the real network, traffic streams are being routinely observed via video cameras as well as loop detectors, and the statistics regarding traffic volumes and speeds are estimated and provided to certain users through web-based services. The data for the selected network were extracted from the archived database of such records.

To collect the simulated traffic statistics, 18 loop detectors were coded at points where surveillance cameras and loop detectors were present in the physical network. The loop detectors provided information about traffic counts and average travelling speeds on specific links. The gathered traffic data were also utilised in the calibration process of the simulated transportation network model.

2.3 Signalised junctions

A total of 38 signalised junctions were selected for the simulation model. These junctions were either cross-junctions or T-junctions, with slip lanes for left turning vehicles. The signal phasing patterns are fixed with varying signal cycle lengths. All the signalised junctions have pedestrian cross-walks. The signal phasing pattern is such that there are 3 s amber time and 2 s all-red time. The maximum cycle time is 120 s and maximum green time of any approach is 50 s. The details regarding the coded phasing pattern schemes are shown in Fig. 3. The un-signalised junctions are as follows: no control (57 junctions), yield sign control (27 junctions), and stop sign control (33 junctions).

2.4 Public mode of transport

The public mode of transport consists of two modes: one is the bus service and the other is the MRT. It was found from the travel behaviour survey that more than 90 % of commuters in the sample reported that they had access to MRT, and they were willing to travel by public mode of transport if it provided better service [36]. One of the service criteria was the journey time, which is assumed to be lower in the case of MRT as compared to bus service. Thus, the bus service was modelled as a feeder service to MRT, and MRT was taken as the major mode of public transport for work/school trips.

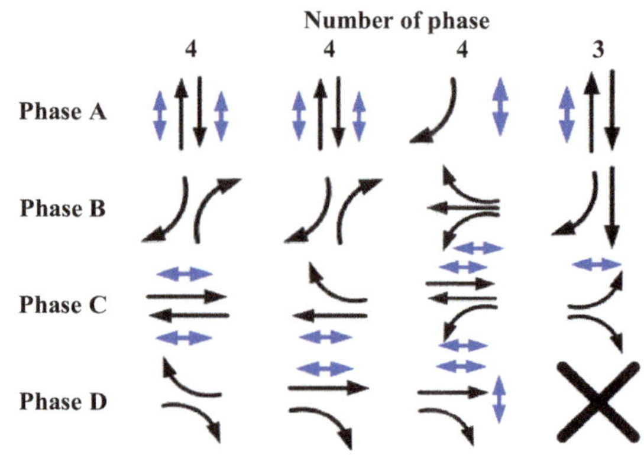

Fig. 3 Coded phasing pattern schemes

The bus routes that served as feeder service to MRT and the bus stops associated to these bus routes were coded and modelled by geographical location. Similarly, the MRT lines and stations were also coded. The MRT lines were modelled as single-lane dual carriageways, with rights-of-way dedicated to MRT services only. Each MRT station was designated a zone, and explicitly simulated in such a way that the residential and commercial zones were connected to the MRT zones (stations) by feeder bus services. The simulated buses were 12 m long with a seating capacity of 47 passengers and a top speed of 20 km/h. The simulated trains had 6 cars with a total length of 138 m and a top speed of 45 km/h. Each train had a seating capacity of 372 passengers.

2.5 Traffic information and OD matrix

It is assumed that travellers commute in an information-rich environment, where pre-trip IMTI and en-route traffic information were available. All the driver vehicle units (DVU) were coded as controlled by familiar drivers and dynamic feedback assignment was adopted. The PARAMICS model disseminated interval traffic information to all the familiar drivers and information was updated at every minute.

An OD matrix was estimated based on the traffic counts for calibrating the developed transportation network. A traffic survey was conducted and the statistics such as traffic counts, average travelling speed, and average travel time were gathered for the calibration.

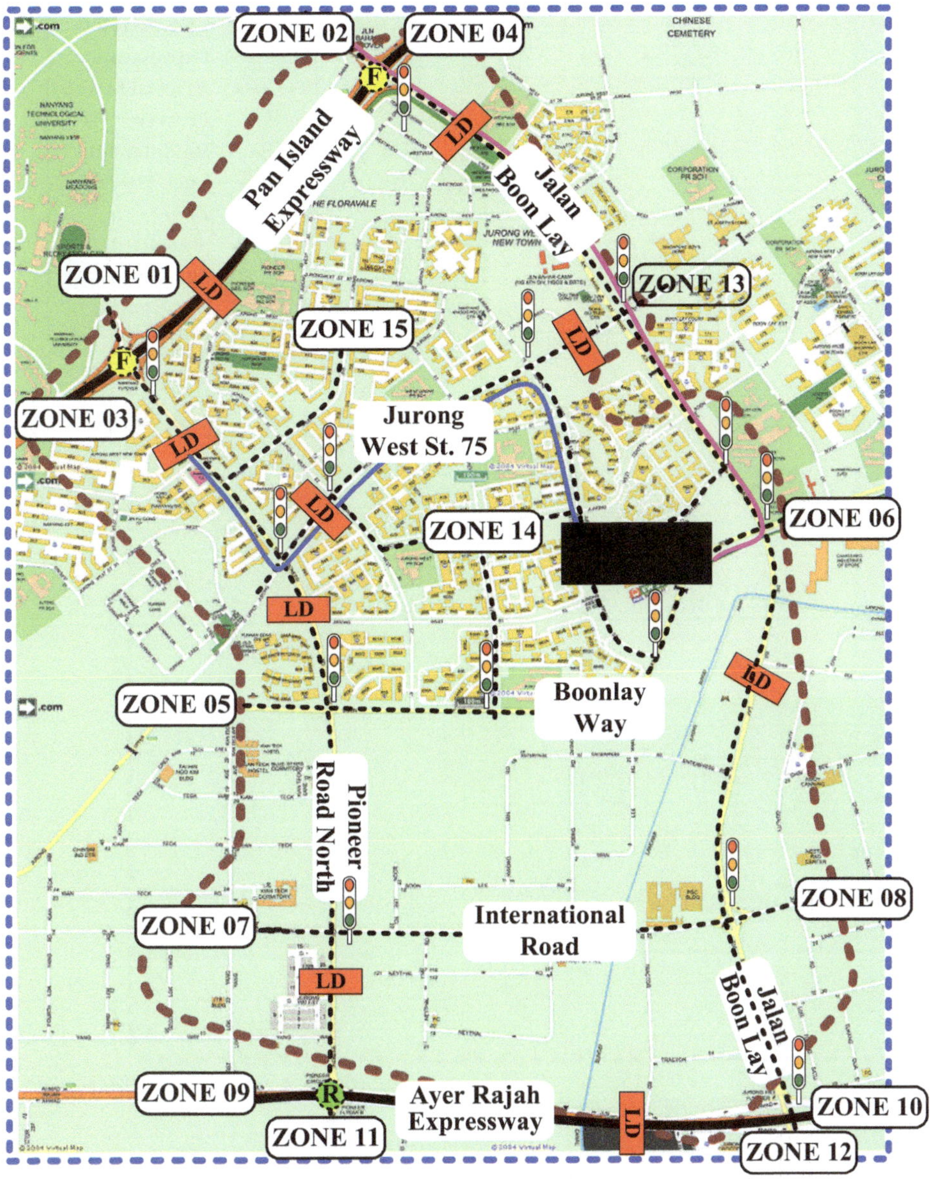

Fig. 4 Detailed map and overview of the simulated network

3 Calibration of the INSIM network

The calibration of the simulated network was done from two different perspectives. The first one was to adjust the traffic counts and the OD matrix in the simulation as compared with the existing situations, such that the simulated network within the complete simulation framework can be close to the real traffic conditions. The second perspective was the calibration of behavioural parameters of the PARAMICS simulation model, such that they imitated the local commuters' travel behaviour. The calibration of the behavioural parameters was necessary because the IES was developed based on the commuters' travel behaviour as observed/estimated from the travel behaviour surveys conducted in Singapore. To integrate IES with PARAMICS to simulate the local traffic environment in accordance with IES, both models should have the same behavioural assumptions, so that the behavioural parameters that govern the PARAMICS simulation model can realistically represent the behaviour of local commuters.

3.1 Study site and basic input data

The network chosen for the simulation model is shown in Fig. 4. To calibrate the behavioural parameters, a portion of the selected study network (marked with blue broken line) in Fig. 4, was considered. This portion of the network was selected because it covers an area bounded by two

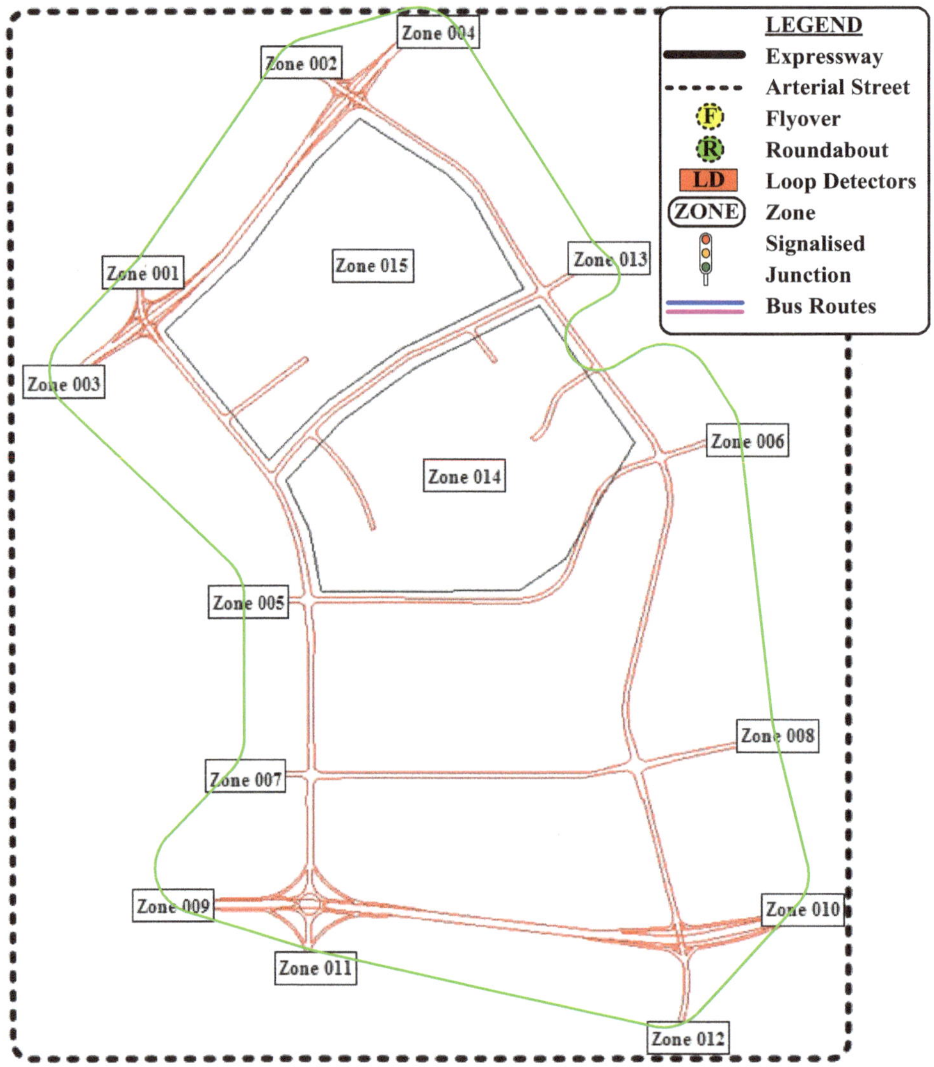

Fig. 5 Detailed simulated network showing zones, links and nodes

expressways (i.e. Pan-Island Expressway and Ayer Rajah Expressway), four arterials (i.e. Jalan Boon Lay, Boon Lay Way, International Road, Pioneer Road North), and one collector (i.e. Jurong West Street 75).

There is a major bus interchange integrated with the MRT station. Such an arrangement facilitated possible transfer between feeder buses and the transit mode. Thus, this portion of the network provided the opportunity to collect travel data for both the public and private modes, for similar trips within the same corridor. The details and overview of the simulated network are presented in Figs. 4 and 5.

The basic input data include network geometry, driver behaviour, vehicle characteristics, transportation zones, travel demands, traffic control systems, traffic detection systems, bus routes, and bus stops. The road geometry and infrastructure locations were obtained from field surveys. The transportation zones were located at the entrance and exit nodes of all the major roads.

3.2 Required data for calibration

The calibration involved the checking of model results against observed data and adjusting the values of the parameters until the model results converged within an acceptable error threshold. The collected data for model calibration included traffic volume, average travelling speed, and travel time data. The traffic volume and average travelling speed data on expressways and arterial streets were obtained from an archived database. The travel time data were obtained by floating car method for Pioneer North Road and Boon Lay Way.

To construct the real-world traffic variations in the simulation, a typical traffic variation can be represented by the traffic conditions of a typical day, of which the traffic data were the target of calibration process. The selection of a typical day can be implemented based on the comparison of traffic volume (at any selected station) of a candidate day with the average traffic volume of all candidate days using the GEH statistic, used by British engineers:

$$\text{GEH} = \sqrt{\frac{(E-V)^2}{(E+V)/2}}, \tag{1}$$

where E is the candidate data and V is the average data. If the GEH values for more than 85 % of the selected stations are less than 5, the traffic condition and the demand pattern of the candidate day are considered typical. There were 35 stations for which data were available. These stations were located on the two expressways i.e. PIE and AYE, and various arterial streets within the network. The data were collected for 1 h during the morning peak period from 8 to 9 AM. It was found that traffic condition on Wednesday was typical with respect to all weekdays. Therefore, the traffic volume and average travelling speed data on a certain Wednesday were chosen for calibration purpose. It was observed from the extracted data that the aggregated average traffic volume on PIE was 3600 vehicles per hour with an average travelling speed of 94 km/h, whereas on AYE the aggregated traffic volume was 5300 vehicles per hour with an average travelling speed of 87 km/h.

The travel time data were collected by floating car method. On the typical day (Wednesday) during the peak period (8–9 AM) several runs were made between Nanyang Technological University (NTU) and Boon Lay Interchange (BLI). The length of the selected route was 4.2 km. The number of floating car runs was obtained from the following equation:

$$n = \left(\frac{ts}{\mu\varepsilon} \times 100\right)^2, \tag{2}$$

where n is the required number of runs, μ is the mean travel time in the runs, s is the standard deviation, ε is the desired margin of error (percent of μ), and t (Student's t-statistic) is the confidence coefficient.

The standard deviation was unknown prior to runs being conducted. A total of 12 runs were performed. It was observed from the floating car runs that the average travel time was 314 s and the standard deviation was 22.3 s. The estimated number of runs with margin of error at 5 % was 9.75 ≈ 10. The 95 % confidence interval at 12 runs was estimated to be ±14.2 s, and it was inferred that the true mean lies between 300 and 328 s. The corresponding average travelling speed on this route was estimated to be between 46.1 and 50.4 km/h. The average travelling speed on the typical day for this route was observed to be 50.4 km/h with standard deviation of 14 km/h. These results provided evidence that the estimated travel time statistics by floating car method gave the details regarding the basic travelling conditions.

Similarly, the travel time data for the public mode of transport were collected. Two bus routes were considered: the first one (SBS 199) served between NTU and Boonlay Interchange (BLI), and the other (SBS 242) served between Jurong West area and BLI. The route length of SBS 199 is 4.2 km, and it is associated with 9 bus stops out of which 3 have bus bays. The route length of SBS 242 is 3 km with 13 bus stops, out of which 3 have bus bays. The frequency of SBS 199 is 4 min during peak hours and 7 min during off-peak hours and for SBS 242 is 3 min during peak hours and 5 min during off-peak hours on week-days, respectively. On each route, there are 4 signalised junctions and 2 mid-block pedestrian signals.

Initially, 30 runs were conducted on each bus route, and the travel time data were collected. For SBS 199 route, it was found that the average travelling time was 752 s with a standard deviation of 51.4 s. With an error margin of 5 %, the estimated number of runs was 7.8 ≈ 8. The resulting 95 % confidence interval was ±19.5 s for 30 runs, which means that the true mean travel time can be between 732 and 772 s. For SBS 242 route, the average travelling time was 556 s and the standard deviation was 46.2 s. With an error margin of 5 %, the estimated number of runs was 11.6 ≈ 12. The resulting 95 % confidence interval was estimated to be 17.5 s, which means that the true mean travel time can be in-between 538 and 574 s. The average travelling speeds based on the observed travel time values on routes SBS 199 and SBS 242 were 23.9 and 32.4 km/h, respectively.

The PARAMICS simulation model needs an OD matrix as a starting point of the calibration process. The time-dependent OD matrix was generated by IES. The OD estimation process was dependent on commuters release profile and zone-based trip distribution. It is necessary that the generated OD matrix should match with some reference OD matrix, so that the PARAMICS model can simulate realistic traffic conditions. The reference OD matrix was estimated from the traffic count data obtained for the typical day.

The cordon points of the network were taken as zones, and 15-min interval traffic counts at all cordon points were gathered from the data. The total traffic attractions and generations of each zone were then assumed to be the distribution of traffic with respect to the OD matrix. The available loop detector data were then compared for checking the network-wide equilibrium of traffic volumes, such that the traffic entering the network should be equal to the traffic leaving the network. To estimate the OD matrix, the observed traffic counts at the cordon points were taken into consideration. These traffic counts were assumed to be the total attraction and generation from the respective zones corresponding to the cordon points. Once the traffic volume being generated and attracted to each zone was estimated, it was then compared with a reference OD matrix. This reference OD matrix was obtained from the study done by Ang [37], which provided the details regarding the traffic volumes being generated during the morning peak period.

The data gathered from the traffic survey provided details about the traffic volumes for 7 zones. The traffic volume data for these zones were compared with those provided in the reference OD matrix and the growth factor was estimated. Based on the estimated growth factor, which was found to be 5.69 % per annum, the total productions and attractions for each zone of the reference OD matrix were revised. The adopted expansion procedure was

(a) the total production/attraction for each zone of the reference OD was expanded, excluding the total productions/attractions for the zones that were obtained from the traffic survey;
(b) all the expanded total productions/attractions were summed up, and compared with the expanded total traffic volume, to estimate the overall difference,
(c) the estimated overall difference was distributed among all the zone-based total productions/attractions depending upon their share of traffic volume aggregated in the total traffic volume;
(d) each zone-based total productions/attractions were then adjusted according to the allocated share of the estimated difference;
(e) the adjusted zone-based total productions/attractions were again summed up, such that their sum was now equal to the expanded total traffic volume; and
(f) the finalised zone-based total productions/attractions were taken as the target values to be used in the final version of the estimated OD matrix.

Each cell value of the reference OD matrix was also expanded, corresponding to the expanded total traffic volume. The zone-based total productions/attractions for these cell values were estimated and taken as the base case values. The Furness technique was then used for balancing the base case values with the target values of the zone-based total productions/attractions. Only the expanded values of the zone-based total productions/attractions were changed, and the total productions/attractions estimated from the traffic survey were unchanged. Based on the estimated OD matrix, the percentages of trips being productions and attractions for each zone were calculated. The zone-based distribution of trips was estimated and is presented in Table 1. The zone-based trip distribution was utilised by IES to generate the time-dependent OD matrix.

3.3 Preliminary testing

In INSIM, three types of demand-related data are necessary: zone-based percentile distribution of trips, percentage of total traffic flow to be released onto the network in each time interval, and the total number of trips to be generated. Field data were processed to gather the demand-related data along with the required performance measures. The transportation network model was simulated based on the demand-related data, and was calibrated by comparing the simulated output results with the gathered performance measures. It was observed from the field data that on a typical day the peak period was from 8 to 9 AM, and 33,120 private mode trips were generated within the physical geography of the simulated network during this period. The peak period was defined by observing the traffic volumes on a typical day from 7 to 10 AM. It was found that the traffic volumes were highest during the time period of 8–9 AM. Available information in recent years [38] revealed that roughly 40 % commuters took private mode for their work trip and 50 % took public mode, and 10 % did not require any motorised transport. Based on these statistics, a total of 79,877 trips were simulated for a 1-h time period. The estimated zone-based percentile trip distribution (Table 1), and a uniform trip release profile was also defined. Default values for vehicle population composition, vehicle characteristics, and driver aggressiveness and awareness distributions were not changed. The dynamic feedback assignment model provided by PARAMICS was adopted to replicate route choice behaviour.

Initially, 10 simulation runs were conducted in order to identify critical variables that may significantly influence the performance of PARAMICS. Variables related to driver behaviour (e.g. the distributions of aggressiveness and awareness), compositions of vehicles, driver's familiarity, etc. were examined. The number of simulation runs was based on number of vehicles (NV). The average NV value was 32,580 and the standard deviation was 335. The estimated number of runs with a margin of error of 1 % was $5.12 \approx 6$. The required simulation runs were conducted with different seed numbers. The numerical outputs of the median simulation run were then compared to corresponding real data.

Outputs used for comparison were related to the total demand generated and the total flow reaching the destination. The outputs were aggregated and analysed using Microsoft Excel. Outputs aggregated in this step for further comparison included total generations (total vehicles released in 1 h), and total attractions (total number of

Table 1 Zone-based trip distribution (%)

O/D	1	2	3	4	5	6	7	8	9	10	11	12	21	22	23	24	25	26
1		0.1	1	0.2	0.3		0.2		0.1	0.4	0	0.5	0.2	0.2	0.6		0.1	0.7
2	0		0.3	0.1	0.2		0		0	0.1		0.1	0.1	0.1	0.1		0.1	
3	0.6	0.3		0.8	2.1	0.2	0.3	0.8	0.6	0.7	0.2	0.5	0.2	0.5	0.8		0.3	0.1
4	0.1	0.1	0.7		0.3		0	0.2	0.1	0.3	0.1	0.1	0.1	0.1	0.2		0.1	0.2
5	0.2	0.3	2	0.3		0.1	0.1	0.7	0.4	1.1	0.1	0.4	0.8	0.6	1.4	0	0.1	0.2
6			0.2		0.1		0		0	0.2	0	0.2	0.2	0	0.2		0.1	0.1
7	0.1		0.4	0	0.1	0		0.3	0.1	0.2	0	0.4	0.1	0.1	0.2		0.1	0.3
8	0.1		1.5	0.3	0.4		0.5		0.1	1.4	0.1	2.1	0.8	0.3	1.4		0.2	
9	0.1	0	0.8	0.1	0.5	0	0	0.1		0.8	0.1	0.6	0.4	0.2	0.9		0.2	0.6
10	0.2	0.2	0.7	0.3	1	0.2	0.2	0.6	0.8		0.2	0.4	0.7	0.3	0.5		0.1	0.1
11	0		0.1	0.1	0	0	0	0.3	0.1	0.2		0.1	0.1	0	0.1		0	0.2
12	0.4	0.2	0.7	0.3	0.9	0.2	0.5	3	0.9	0.6	0.2		0.2	0.3	0.6		0.6	0.8
21	0.1	0	0.1	0.1	0.8	0	0.1	0.7	0.3	0.6	0.1	0.1		0.1	0.5		0	0.1
22	0.3	0.2	1.9	0.3	2.3	0.1	0.1	0.7	0.7	1	0.1	0.8	0.5		3.6	0.1	0.8	0.3
23	0.3	0.2	0.9	0.2	1.5	0.2	0.2	1.2	0.7	0.5	0.2	0.2	0.6	1.2		0.1	0.9	0.2
24					0									0.1	0.1			
25	0.1	0.1	0.4	0	0.2	0.1	0.1		0.2	0.3	0	0.5	0.1	0.3	1.2			0.2
26	1		0.2	0.3	0.5	0.6	0.9		0.7		0.4	1.2		0.2	0.4		0.6	

Empty cells denote ZERO value

Table 2 Output data comparison with the un-calibrated model

	Simulated	Actual	Error (%)
Total demand (production)	28,782	33,120	−13.09
Total outflow (attraction)	25,288	33,120	−23.64

vehicles completing their journey in 1 h). Results in Table 2 indicate a shortage in the number of vehicles being released into the network, and the number of vehicles reaching their destinations. The difference in productions and attractions is due to vehicles which were released in the network but have not reached their destinations. Thus, the difference between the productions and attractions is equal to the total number of vehicles which are still in the network. Therefore, parameter values other than the default needed to be explored. In order to improve the accuracy of the simulation at this stage, the types of vehicles using the network and their respective percentages in the vehicle population were adjusted. There was no systematic way for adjusting the percentage of different types of vehicles except to experiment different combinations so as to get acceptable results. After several combinations the final percentages of different types of vehicles used in the simulation are presented in Table 3.

Table 3 Calibrated percentages of different types of vehicles

Type		Percentage
Serial number	Shape	
1	Private car	50
2	Taxi	15
3	LGV	9
4	OGV 1	8
5	OGV 2	8
6	Mini bus	7
7	Coach	3
Total		100

It was observed during the simulation that the DVU traversing on expressways did not use the outermost lanes, specifically at the merging points at on-ramps along expressway lanes. At such locations, the vehicles slowed down for longer time periods, and such behaviour resulted in heavy congestion. To analyse such issue, three parameters concerned with driver behaviour were investigated. Two parameter sets in particular, aggressiveness and awareness, were examined—since these two significantly influence the behaviour of drivers. There are four default types of distributions associated with each of these two parameters. Numerous runs with different combinations of distributions were conducted until the outputs of lane use were improved further. The third parameter was the ramp awareness distance. It was observed from the PARAMICS graphical interface of the simulated network that increasing the length of ramp awareness distance increased the flow on expressways and improved the outermost lane utilisation. The ramp awareness distance was adjusted one by one for all the ramps, and the minimum value was adjusted to 150 m. Using combination of the above adjustments, 10 simulation runs were conducted and the improved results are summarised in Table 4.

Theoretically, it is a dynamic OD demand estimation problem. To date, there are some OD estimation tools e.g. Estimator of PARAMICS, but their capabilities and potentials for a corridor network are still unclear [39]. The time-dependent OD matrix was estimated on the basis of zone-based percentile trip distribution. The proposed dynamic OD estimation process could be regarded as a process that assigns the total OD to a series of consecutive time stamps. The time-dependent OD demand for each time stamp was then extracted by reconstructing the dynamic OD matrix based on a set of demand profiles for every time stamp, e.g. 1 or 5 min. INSIM allows the user to define the time-dependent releases of demand (trips) and the zone-based trip distribution.

The field data provided 15-min interval counts for all the cordon points in the network. Based on these data, the profile of vehicle production from any origin zone and that of vehicle attraction to any destination zone were thus estimated. Four different trip release profiles for different time stamps were developed as shown in Fig. 6, such that:

(a) A uniform distribution was assumed for the demand release profile, i.e. RP1. The total demand was uniformly spread over 60 min of the simulation period with a time stamp value of 5 min. For the first four time stamps, 9 % of the total volume was released, and in the remaining 8 time stamps 8 % were released.
(b) A nonuniform distribution was assumed for the demand profile RP2. The 15-min counts were segregated into 5-min counts, by uniformly distributing the 15-min counts over 3 periods of 5-min duration, such that 7 % of demand was released in each 5-min time stamp for the first 15-min time interval, 9 % of demand was released in each 5-min time stamp for second 15-min time interval, and so on.
(c) A normal distribution was assumed for the demand profile RP3, such that the maximum release of trips occurred during the 2nd and 3rd quarters of the simulation period. In this profile the demand increased uniformly and arrived at the peak discharge of 12 %, after which it decreased uniformly.
(d) A normal distribution skewed towards the right was assumed for the demand profile RP4. In this profile the peak discharge of 12 % occurred in the 3rd quarter of the simulation period. This profile imitated the real traffic flow condition.

Table 4 Output data comparison based on the calibrated model

	Simulated	Actual	Error (%)
Total demand (generation)	32,428	33,120	−2.09
Total outflow (attraction)	31,914	33,120	−3.64

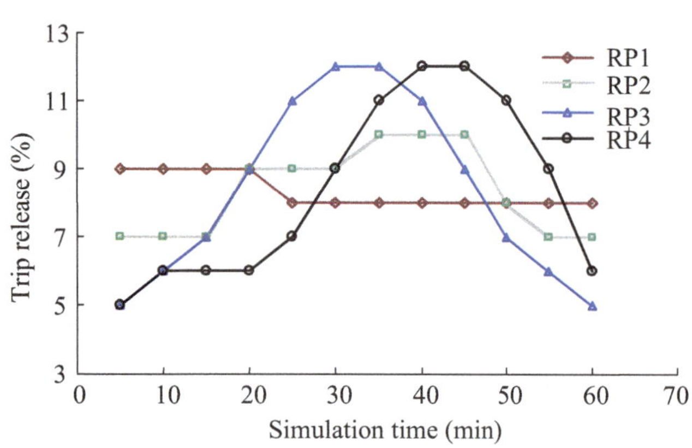

Fig. 6 Different trips release profiles

3.4 Calibration of route choice model

In the OD estimation process, the network, traffic and route choice behaviour parameters were to be fixed initially, as this process is based on the traffic assignment matrix that is affected by any change in the simulation input parameters.

Due to the existence of expressways and parallel streets in the study network, the routing algorithm adopted in the PARAMICS simulation was important. The network was calibrated using the dynamic feedback assignment model provided by PARAMICS. Dynamic feedback assignment in PARAMICS assumes that different drivers perceive different costs from a decision node to the destination. The perceived cost was calculated, and the perceived shortest route was chosen at the decision node. At this stage the parameter to be calibrated for the route choice model is the number of drivers who are familiar with the road network. Since there was no data to calibrate it, it was assumed that most drivers in the morning peak period were familiar drivers who had the knowledge of road network and traffic conditions. Therefore, it was assumed that 95 % of drivers were familiar drivers, who could choose their route from the available options.

3.5 Secondary testing

Multiple simulation runs for each parameter combined with the each trip release profile were also performed. During these simulations, certain changes in the cell values (i.e. specific origin to destination traffic volumes) of the

Table 5 Traffic count calibration results for release profile RP4

ID	Traffic count (0800–0815)			Traffic count (0815–0830)			Traffic count (0830–0845)			Traffic count (0845–0900)		
	Act	Sim	GEH	Act	Sim	GEH	Act	Sim	GEH	Act	Sim	GEH
02851	793	778	0.54	947	875	2.39	961	1037	2.4	853	870	0.58
06958	161	158	0.24	192	193	0.07	195	231	2.47	173	200	1.98
06966	68	57	1.39	81	71	1.15	82	68	1.62	73	82	1.02
06967	122	96	2.49	146	115	2.71	148	118	2.6	131	137	0.52
06975	575	564	0.46	687	674	0.5	697	701	0.15	619	581	1.55
07818	219	195	1.67	262	268	0.37	265	250	0.93	236	176	4.18
07819	194	149	3.44	232	183	3.4	235	184	3.52	209	252	2.83
07822	192	144	3.7	229	276	2.96	233	203	2.03	207	210	0.21
07823	140	145	0.42	167	117	4.2	170	124	3.79	151	181	2.33
07842	234	244	0.65	279	330	2.92	284	292	0.47	252	266	0.87
07843	192	156	2.73	229	252	1.48	233	323	5.4	207	283	4.86
10570	73	64	1.09	87	93	0.63	88	101	1.34	79	82	0.33
10571	219	173	3.29	262	287	1.51	265	306	2.43	236	187	3.37
10574	170	97	6.32	203	214	0.76	206	255	3.23	183	183	0
10578	176	131	3.63	210	162	3.52	213	236	1.54	189	193	0.29
10579	315	252	3.74	376	359	0.89	382	395	0.66	339	323	0.88
10586	648	592	2.25	774	797	0.82	785	849	2.24	697	745	1.79
10587	352	321	1.69	420	475	2.6	427	550	5.57	379	386	0.36

Act actual, *Sim* simulation

estimated OD matrix were incorporated. After multiple iterations, the calibration criteria were satisfied. The highest GEH value was obtained by adopting RP4 trip release profiles with 95 % of familiar drivers. The calibration results of this step are shown in Table 5, which presents the 15-min intervals traffic counts at the selected measurement stations. It shows that for more than 87.5 % of all measurement locations, their GEH values were lower than 5, which satisfied the calibration acceptance criterion of this step.

3.6 Calibration of driving behaviour model

The last step was to calibrate the driver behaviour model to reflect local driver characteristics. The local driver characteristics can be examined through the comparison of the simulated and observed volume-occupancy curves drawn based on aggregated point detector data, and the point-to-point travel time measurements. Thus, the network was further calibrated under the following conditions: the network has been coded and partially calibrated for the parameters in above section; there was no data that can support the calibration of route choice model; the dynamic assignment model in the microscopic simulator can be accepted; and the zone-based percentile trip distribution and trip release profile have been obtained.

The objective function is to minimise the deviations among the observed and the simulated volume-occupancy curves, and the point-to-point travel time measurements, subject to the condition that the traffic counts, which have already been calibrated, do not change. But if the traffic counts do change, then they must satisfy the GEH calibration acceptance criteria. The point-to-point travel time match was performed only for one route between NTU and Boon Lay Interchange (BLI). The driver behaviour model can be calibrated by adjusting the car-following and lane-changing models. The global parameters for the car-following and lane-changing models are mean headway and driver reaction time, which can drastically influence the simulated driver behaviour. These two parameters were fine-tuned by trial-and-error. The purpose was to reconstruct traffic variations and match the congestion pattern of the study network, so that the resulting simulated travel time measurement matched with the real life estimated travel time values. Five different combinations of mean headway and driver reaction time were analysed. For each combination 5 simulation runs were performed and the median NV value was taken into consideration. It was observed that higher values of mean headway and reaction time resulted in lower travelling speeds and vice versa.

The final calibrated mean target headway and driver reaction time were 0.85 and 0.75, respectively. The calibrated demand profile for the simulated peak period is shown in Fig. 7.

The root mean square percent error (RMSP) was estimated to examine the error between the simulated traffic demand and the actual observed traffic demand, and it was found to be 8.29 %. Figure 8 shows the comparison of observed and simulated point-to-point travel time for the selected route, which has an RMSP value of 2.49 %. Based on the lower value of RMSP, it can be stated that calibration process has been effective as indicated by the test statistics.

3.7 Calibration of public mode of transport

The last stage of calibration process was to calibrate the public mode of transport. Two public modes of transport i.e. train and bus services were simulated. As the bus shares the right-of-way on the surface street network, its journey time can be affected by the congestion occurring in the network. The bus journey time is dependent on two parameters namely the average travelling speed of buses and the number of passengers waiting at bus stops.

The number of waiting passengers can influence the travel time by affecting the bus dwell time at the bus stops, which is an integral part of travel time. Increasing the number of passengers can increase the boarding passengers' queue length resulting in longer bus dwell time at the bus stops and vice versa. Thus, the bus dwell time can be a function of number of passengers arriving at the bus stop. The public mode of transport was calibrated by adjusting the bus travel time. The selected parameters were the average travelling speed by bus and the arrival rate of passengers at the bus stop. The objective function was to minimise the difference between the simulated and observed bus travel times. The field data were available for two bus routes, i.e. SBS 242 and SBS 199. Along with other bus services these two routes were simulated.

Initially the bus travel time was calibrated by adjusting the travelling speed. Three different speeds, i.e. 15, 20, and 25 km/h, were simulated. Each speed was simulated for 5 times with different seed number. Then the bus travel time

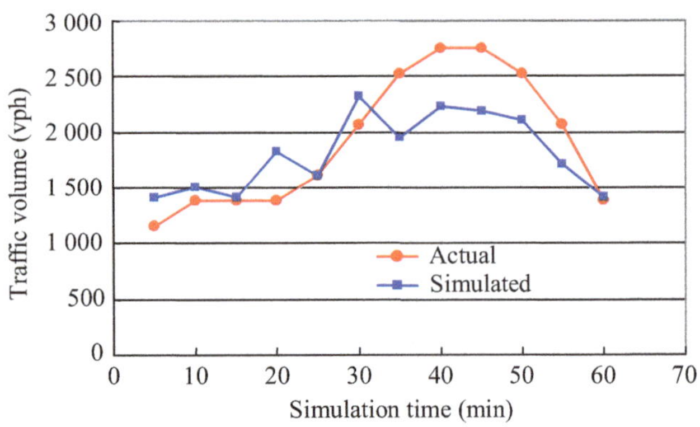

Fig. 7 Calibrated demand profile

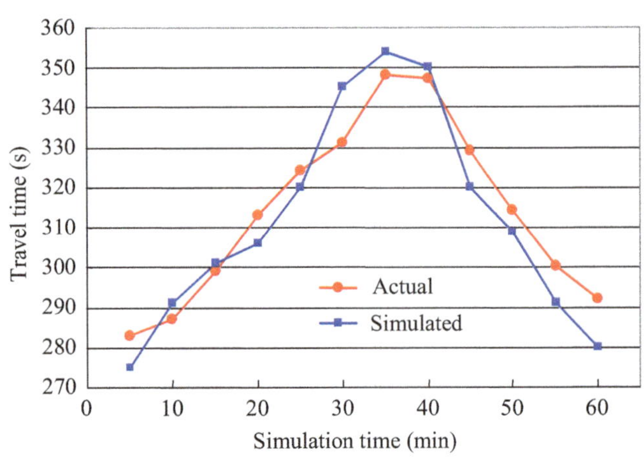

Fig. 8 Observed and simulated point-to-point travel time

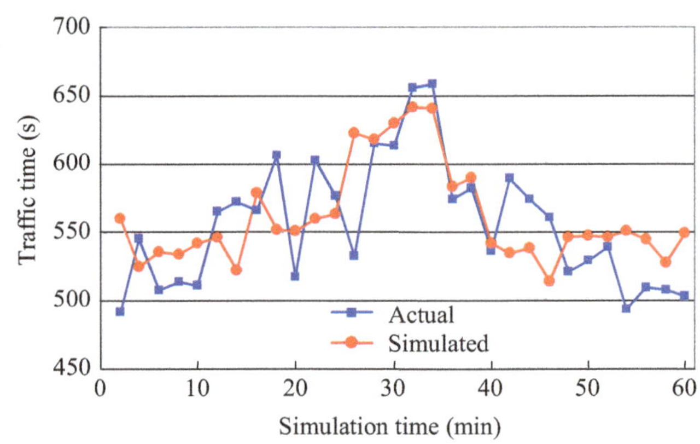

Fig. 9 Observed and simulated travel time by bus route SBS 242

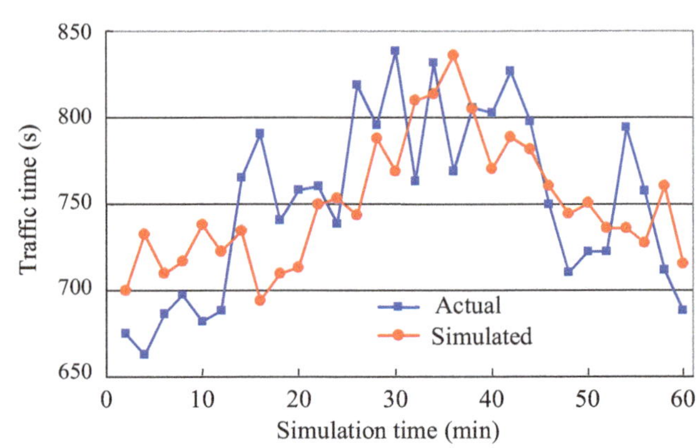

Fig. 10 Observed and simulated travel time by bus route SBS 199

was fine-tuned by adjusting the arrival rate of passengers at each bus stop. After multiple simulation runs, the bus journey time was found to be acceptable. The average travel time measured for buses from simulation were compared with the average travel time for buses measured in the field. The comparison of observed and simulated point-to-point travel times on bus route SBS 242 and SBS 199 are presented in Figs. 9 and 10, respectively. The RMSP was estimated to be 9.54 % for SBS 242, and 9.01 % for SBS 199 route.

4 Conclusion

This paper calibrated a self-developed traffic simulation model, INSIM, with the observed field data from the central and western part of Singapore. Results reflected that the improved model and the generated outputs are acceptable. The network coding errors were corrected and a reliable time-dependent OD matrix was estimated. The INSIM was adjusted, such that the commuter release profile and the percentile trip distribution resulted in simulated traffic counts that were acceptable according to the GEH statistics. The PARAMICS route choice model was adopted and 95 % of the driver population was assumed to be familiar with the road network. The driver behaviour model was calibrated, and it reflects the local driver behaviour. The mean headway and driver reaction time were adjusted to achieve the calibration of the driver behaviour model. Finally, the public mode of transport was calibrated by adjusting the average travelling speed of buses and the arrival rate of passengers at the bus stops.

Model calibration is a necessary process for various kinds of traffic simulation model, especially for newly developed rudiment before application. This paper provides the whole calibration procedure using local travel network and data. It can simulate the travel time and travelling speed parameters, which shall be utilised by the IES to generate commuters' travel mode choice under the influence of integrated traveller information, within the statistically acceptable limits, as compared to the observed data. Future tasks include applying the calibrated INSIM in various kinds of real networks to achieve its functions.

Acknowledgments This work was conducted under the PhD research programme of the first author (Dr Abdul Ahad Memon) at the Nanyang Technological University.

References

1. Chorus CG, Molin EJ, Van Wee B (2003) Use and effects of advanced traveller information services (ATIS): a review of the literature. Transp Rev 26(2):127–149
2. Tsirimpa A, Polydoropoulou A, Antoniou C (2007) Development of a mixed multinomial logit model to capture the impact of information systems on travelers' switching behavior. J Intell Transp Syst 11(2):79–89
3. Meng M, Shao CF, Zeng JJ, Dong CJ (2014) A simulation-based dynamic traffic assignment model with combined modes. PROMET-Traffic & Transp 26(1):65–73
4. Ben-Elia E, Shiftan Y (2010) Which road do I take? A learning-based model of route-choice behavior with real-time information. Transp Res Part A 44(4):249–264
5. Emmerink P, Nijkamp P, Rietveld P (1995) Is congestion pricing a first-best strategy in transport policy? A critical review of arguments. Environ Plan B 22(5):581–602
6. Garling T, Jakobsson C, Loukopoulos P, Fujii S (2004) Adaptation of private car use in response to travel demand management measures: potential roles of intelligent transportation systems. J Intell Transp Syst 8(4):189–194
7. Lam SH, Toan T (2006) Land transport policy and public transport in Singapore. Transportation 33(2):171–188
8. Srinivasan KK, Mahmassani HS (2003) Analyzing heterogeneity and unobserved structural effects in route-switching behaviour under ATIS: a dynamic kernel logit formulation. Transp Res Part B 37(9):793–814
9. Hensher DA, Green WH (2003) The mixed logit model: the state of practice. Transportation 30(2):133–176
10. Beesley ME (1965) The value of time spent in travelling: some new evidence. Economica 32(126):174–185
11. Adler JL, Blue VJ (1988) Towards the design of intelligent traveller information systems. Transp Res Part C 6(3):157–172
12. Chen PS, Mahmassani HS (1993) Dynamic interactive simulator for studying commuter under real time information supply strategies. J Transp Res Rec 1413:12–21
13. Adler JL, Recker WW, McNally MG (1993) A systematic evaluation of the impacts of real-time traffic condition information on traffic flow. PATH Research Report, UCB-ITS-PRR-93-6, California Department of Transportation
14. Koutsopoulos HN, Lotan T, Yang Q (1994) A driving simulator and its applications for modelling route choice in the presence of information. Transp Res Part C 2(2):91–107
15. Miller J (2006) Integrated urban models: progress to date; future prospects. In: 11th international association for travel behaviour research conference, Kyoto
16. Srinivasan K (2000) Dynamic decision and adjustment processes in commuter behaviour under real time information. PhD Dissertation. The University of Texas Austin, pp. 21–26
17. Rakha HA, Van Aerde MW (1996) Comparison of simulation modules of TRANSYT and INTEGRATION Models. J Transp Res Rec 1566:1–7
18. Barceló J, Delgado M, Funes G, García D, Perarnau J, Torday A (2007) Online microscopic traffic simulation supports real-time traffic-management strategies. Proceeding of the 6th ITS European Congress, paper no. 2607
19. Jayakrishnan R, Jun-Seok O, Sahraoui A (2001) Calibration and path dynamics issues in microscopic simulation for advanced traffic management and information systems. J Transp Res Rec 1771:9–17
20. PTV AG (2001) VISSIM 3.5 User Manual. Karlsruhe
21. Wang PTR, Glassco RA (1995) Enhanced Thoreau traffic simulation for intelligent transportation systems (ITS). In: Proceedings

of winter simulation conference, IEEE computer society, pp. 1110–1115
22. Memon AA, Meng M, Wong YD, Lam SH (2015) Rule-based mode choice model—INSIM Expert System. J Transp Eng ASCE 141(4):04014088
23. Bell MGH, Cassir C (2002) Risk-averse user equilibrium traffic assignment: an application of game theory. Transp Res Part B 36(8):671–681
24. Chen A, Zhou Z (2010) The α-reliable mean-excess traffic equilibrium model with stochastic travel times. Transp Res Part B 44(4):493–513
25. Lam WHK, Shao H, Sumalee A (2008) Modeling impacts of adverse weather conditions on a road network with uncertainties in demand and supply. Transp Res Part B 42(10):890–910
26. Siu B, Lo HK (2006) Doubly uncertain transport network: degradable link capacity and perception variations in traffic conditions. J Transp Res Rec 1964:59–69
27. Siu BWY, Lo HK (2008) Travel time budget in doubly uncertain transport network: degradable capacity and stochastic demand. Eur J Oper Res 191(1):166–181
28. Zhang C, Chen X, Sumalee A (2011) Robust Wardrop's user equilibrium assignment under stochastic demand and supply: expected residual minimization approach. Transp Res Part B 45(3):534–552
29. Lee JB, Ozbay K (2009) New calibration methodology for microscopic traffic simulation using enhanced simultaneous perturbation stochastic approximation approach. J Transp Res Board 2124:233–240
30. Chen XM, Wei ZH, Gao L (2011) Simulation research of traffic behavior of road intersection pedestrian. J Cent South Univ 42(s1):1037–1041
31. Balakrishna R, Antoniou C, Ben-Akiva M, Koutsopoulos HN, Wen Y (2007) Calibration of microscopic traffic simulation models: methods and application. J Transp Res Board 1999(1):198–207
32. Appiah J, Rilett LR (2010) Joint estimation of dynamic origin-destination matrices and calibration of micro-simulation models using aggregate intersection turn count data. In: Transportation research board 89th annual meeting, No. 10-2764
33. Flötteröd G, Bierlaire M, Nagel K (2011) Bayesian demand calibration for dynamic traffic simulations. Trans Sci 45(4):541–561
34. Land Transport Authority. Civil design criteria for road and rail transit systems (2010). Available via fromhttp://www.lta.gov.sg/content/dam/ltaweb/corp/Industry/files/DC_EGD09106A1_Overall.pdf
35. SMRT (2010) Network map. Available via http://www.smrt.com.sg/Trains/NetworkMap.aspx
36. Meng M, Memon AA, Wong YD, Lam SH (2015) Dynamic interactions between commuters' mode choice behaviour and integrated traveller information. PROMET-Traffic & Transp 26(6):485–495
37. Ang SZ (2003) Surface street network simulation studies using TRANSYT and PARAMICS. M. Sc. Dissertation, School of Civil & Environmental Engineering, Nanyang Technological University, Singapore
38. Singstat (2010) Transport characteristics. Available via http://www.singstat.gov.sg/publications/publications_and_papers/cop2010/census_2010_release3/cop2010sr3.pdf
39. Chu KC, Yang L, Saigal R, Saitou K (2011) Validation of stochastic traffic flow model with microscopic traffic simulation. In: Automation science and engineering (CASE), 2011 IEEE conference on, IEEE, pp. 672–677

2

A multidimensional examination of performances of HSR (High-Speed Rail) systems

Milan Janić[1]

Abstract This paper deals with a multidimensional examination of the infrastructural, technical/technological, operational, economic, social, and environmental performances of high-speed rail (HSR) systems, including their overview, analysis of some real-life cases, and limited (analytical) modeling. The infrastructural performances reflect design and geometrical characteristics of the HSR lines and stations. The technical/technological performances relate to the characteristics of rolling stock, i.e., high-speed trains, and supportive facilities and equipment, i.e., the power supply, signaling, and traffic control and management system(s). The operational performances include the capacity and productivity of HSR lines and rolling stock, and quality of services. The economic performances refer to the HSR systems' costs, revenues, and their relationship. The social performances relate to the impacts of HSR systems on the society such as congestion, noise, and safety, and their externalities, and the effects in terms of contribution to the local and global/country social-economic development. Finally, the environmental performances of the HSR systems reflect their energy consumption and related emissions of green house gases, land use, and corresponding externalities.

Keywords HSR (high-speed rail) systems · Performances · Multidimensional examination · Overview · Analysis · Modeling

✉ Milan Janić
M.Janic@tudelft.nl

[1] Faculty of Civil Engineering and Geosciences, Department of Transport & Planning, Delft University of Technology, Stevinweg 1, 2628 CN Delft, P.O. Box 5048, 2600, GA, Delft, The Netherlands

1 Introduction

The high-speed rail (HSR) systems as the rather innovative systems within the railway transport mode, particularly as compared to its conventional (rail) passenger counterpart, have been developing worldwide (Europe, Far East-Asia, and United States of America (USA)). Despite the common name, different definitions of these systems have been used as follows:

- *Japan* The HSR system called 'Shinkansen' (i.e., 'new trunk line') is defined as the main line along almost its entire length (i.e., route) where trains can run at the speed of at least and above 200 km/h. The 'Shinkansen' system's network has been built with the specific technical standards (i.e., dedicated tracks without the level crossings and the standardized and special loading gauge). This HSR system represents a part of the overall Japanese Shinkansen transportation system [1].
- *Europe* The definition of HSR system includes (a) infrastructure, (b) rolling stock, and (c) compatibility of the infrastructure and rolling stock [2].
  - *Infrastructure* Infrastructure of the trans-European HSR system is considered a part of the Trans-European rail transport system/network. It is specially built and/or upgraded for the high-speed (HS) travel. This may include connecting lines and junctions of the new lines upgraded for the HS, and the stations located on them, where the train speeds must take into account the local conditions. The HSR lines include those specially built for the speeds equal to or greater than 250 km/h (*Category I*), those specially upgraded for the speeds of the order of 200 km/h (*Category II*), and those upgraded with the particular features resulting from

the topographical relief or the town-planning constraints (*Category III*). Therefore, the *Category I* lines are exclusively considered as the real HSR lines.

- *Rolling stock* The HS trains are designed to guarantee safe and uninterrupted travel at the speed of at least 250 and 300 km/h under the appropriate circumstances on the *Category I* lines, about 200 km/h on the specially upgraded *Category II* lines, and at the highest possible speed on the other *Category III* lines.
- *Compatibility of the infrastructure and rolling stock* The HS trains are designed to be fully compatible with the characteristics of infrastructure, and vice versa, which influences the performances in terms of safety, quality, and cost of services.

- *China* According to Order No. 34, 2013 from China's Ministry of Railways, the HSR system refers to the newly built passenger-dedicated lines with (actual or reserved) speed equal and/or greater than 250 km/h. Its specific acronym is China railway high-speed (CRH). In addition, a number of new 200 km/h express passenger and 200 km/h mixed (passenger and freight) lines have been building as the components of the country's entire HSR network [3].
- *USA* (*United States of America*) The HSR system is defined as that providing the frequent express services between the major population centers on the distances from 200 to 600 mile (mi) with a few intermediate stops, at the speeds of at least 150 mph (mi/h) on the completely grade-separated, dedicated rights-of way lines (1 mi = 1.609 km). It is also considered as the system providing regional, relatively frequent services operated at the speeds from 110 to 150 mph between the major and moderate population centers on the distances between 100 and 500 miles with some intermediate stops, grade separated with some dedicated and shared tracks using the positive train control technology [4, 5]. In both cases, the HSR system has been expected to relieve congestion at the highways and airports, in the latter case particularly by competing with the short- to medium-haul airline flights.

This paper deals with a multidimensional examination of the infrastructural, technical/technological, operational, economic, social, and environmental performances of the above-mentioned HSR systems by providing their overview, analysis, and limited (analytical) modeling. In addition to the introductory section, this paper consists of eight other sections. Section 2 introduces the concept of performances of the HSR system(s). Section 3 analyzes the systems' infrastructural, Sect. 4 technical/technological, Sect. 5 operational, Sect. 6 economic, Sect. 7 social, and Sect. 8 environmental performances. The last section summarizes some conclusions.

2 A concept of performances of HSR systems

The performances of transport systems can be defined as their ability to fulfill the needs and expectations of particular actors/stakeholders involved, which are usually users/passengers, rail operators, and the third parties. For the HSR systems, these performances can generally be classified as infrastructural, technical/technological, operational, economic, social, and environmental [6].

- *Infrastructural and technical/technological performances* imply the system's physical, constructive, technological, and technical characteristics of infrastructure, vehicles, i.e., HS trains, and supporting facilities and equipment, i.e., the power supply, signaling, and traffic control/management system(s);
- *Operational performances* reflect the system's capabilities to serve the specified volumes of user/passenger demand under given conditions;
- *Economic performances* express the costs and revenues, the latter based on the charges (prices) to users/passengers, and their relationship(s); and
- *Social and environmental performances* reflect the scale of the system's effects and impacts on the society and environment, the later usually expressed in the monetary terms as the external costs, i.e., externalities, if internalized by the related policies.

The above-mentioned performances of the HSR systems are frequently considered individually although being inherently dependent and influential on each other as shown in Fig. 1.

As can be seen, according to the 'top-down' approach, the infrastructural performances directly influence the technical/technological performances, thus causing their mutual influence as well as the influence between them and all other performances. According to the 'bottom-up' approach, the social/policy performances can directly influence the infrastructural and technical/technological performances, thus creating the mutual influence of these and all other performances.

3 Infrastructural performances of HSR systems

The infrastructure of HSR systems consists of lines with the rail tracks connecting the stations/stops along them and the end stations/terminuses, both considered exclusively as the above-mentioned *Category I* of the HSR lines. The lines and stations constitute the HSR network spreading

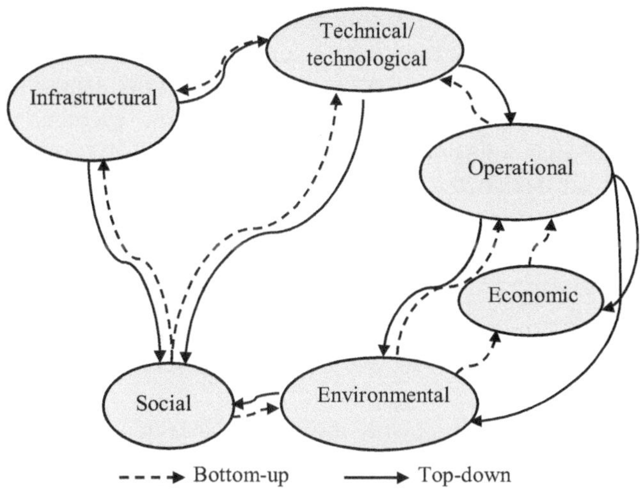

Fig. 1 A potential relationship of the performances of HSR systems [6]

over a given region, country, and/or a continent. Table 1 provides an illustration of the progress so far in developing the HSR networks at particular continents.

As can be seen, the longest HSR network currently operating and being under construction is in Asia, mainly thanks to the fast developments in China, followed by that in Europe. The last are those in both Americas and Africa.

3.1 Lines

The lines as links connecting particular stations as the nodes of HSR network are mainly characterized by their three-dimensional layout and geometry of tracks. The most relevant parameters of geometry of tracks are the distance between their centers, gauge, the maximum axle load, gradient, the minimum horizontal and vertical radius of curvature, the maximum cant and the maximum cant gradient, and the length of transition curves corresponding to the minimum curve radius. For example, in Europe, except track gauge (1,435 mm), all other parameters are dependent on the maximum design speed. In addition, the HSR tracks can be broadly ballasted and ballast less [7, 8]. The former are present at the most already built HSR lines, while the latter have been considered particularly for the lines with long segments of tunnels and/or bridges such as those in Japan.[1] In addition, they have been expected to increase the capacity of HSR lines, operating speed, reduce the maintenance costs through reducing the frequency of maintenance operations, and consequently increase the level of safety.

Table 1 Development of the HSR networks at particular continents [14, 50]

Status	Continent			World
	Europe	Asia	Others[a]	
In operation (km)	7,351	15,241	362	22,954
Under construction (km)	2,929	9,625	200	12,754
Total (km)	10,280	24,866	562	35,708

[a] Latin America, USA, Africa

3.2 Stations

The HSR stations mainly characterized by location and design enable facilitation of the HSR system with its users/passengers. The main aspect of location as the nodes of corresponding HSR network is their number along particular lines. Then, it is their micro-location in urban areas/cities and often at airports, which should enable safe, efficient, and effective accessibility by individual (car) and mass urban public transit systems (bus, tram, light rail, metro, and regional rail).[2] Furthermore, it is their functional design, which includes (i) the track and platform technical aspects (number, arrangement, dimension, safety, and electrical, signaling, and communication systems); (ii) the user/passenger service and comfort aspects (accessibility, inter-modal transfer, security, ticketing and travel information, station facilities, etc.); and (iii) the environmental aspects (choice of building/construction materials and protection of the local environment from noise) [11, 12].

In particular, an additional important aspect of design of the HSR stations is the arrangement of tracks and platforms for users/passengers. In general, two main concepts have been used: the side platforms facing the track(s) by one side and the island platforms facing the tracks by both sides, as shown in Fig. 2a, b, respectively [11].

The safety aspect of design of the HSR stations is important for users–passengers and accompanies standing on the platforms in cases when the non-stopping trains are passing by at relatively high speeds. These people could be affected (sucked toward a passing train if standing too close to the platform edge) by air streams generated by the HS trains. For example, some research indicated that the people standing on the platforms at the distance of 2 m from the HS train passing by at the speed of 240 km/h could be under a real risk [13].

[1] In the year 1972, the ballastless 'slab track' had been developed and applied to the Sanyo Shinkansen line; in the year 2007, the 'slab track' was used for 1244-km-long line, which shared about 57 % of the Shinkansen network [9]. In China, both ballast and ballastless slab tracks have been used [10].

[2] For example, the new CRH South Guangzhou station on the Hangzhou–Shenzhen line (China) has 15 platforms with 28 tracks and is the largest in Asia at the moment [10].

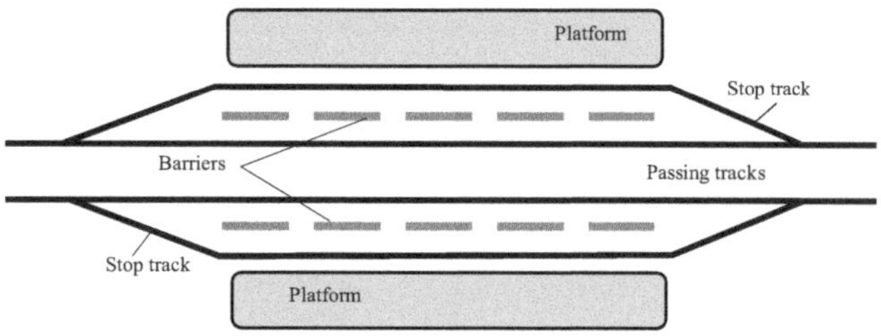

(a)

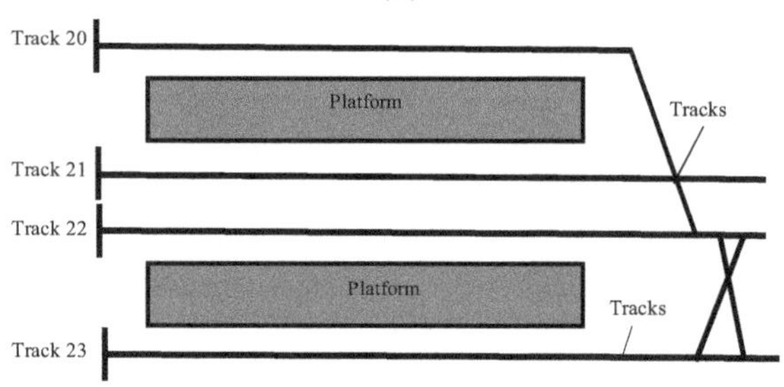

(b)

Fig. 2 Simplified schemes of arrangements of the platforms and tracks at the HSR station. **a** Line station with the side platforms and two passing and two stopping tracks [11]. **b** Begin/end station/terminus with the inland platforms (Tokyo Shinkansen) [24]

3.3 Network

The above-mentioned lines and stations constitute the HSR infrastructure network, which spreads over a given country (http://www.johomaps.com/eu/europehighspeed). Table 2 shows some characteristics of the main grid (eight national backbone lines) of the HSR network in China.

The specificity of this (Chinese) compared to the other HSR rail networks worldwide, particularly those in Europe, is the length of lines between the end stations/terminuses, which varies from 1,000 to 2,400 km. In Europe, these lengths are much shorter and vary, for example, from 280 km between Berlin and Hamburg (Germany) to 770 km between Paris and Marseille (France) [14]. However, the experience so far has shown that the average travel distances on some of these long Chinese lines have been about 560–620 km, which appears comparable to some of their (long) European counterparts [15].

4 Technical/technological performances of HSR systems

The technical/technological performances of HSR systems relate to their rolling stock, i.e., high-speed trains (HSTs) and supportive facilities and equipment, i.e., power supply, signaling, and traffic control/management system(s).

4.1 Rolling stock

The HSR rolling stock, i.e., trains, are characterized by an optimized aerodynamic shape; fixed composition and bi-directional set; self-propelling, concentrated, or distributed power; interior signaling system(s); several braking systems; power electronic equipment; control circuits; computer network; automatic diagnostic system; particularly high level of reliability, availability, maintainability, and safety (RAMS); maintenance by inspection in fixed time intervals and preventively; and compatibility with infrastructure (track and loading gauge, platforms, catenary, etc.) [1]. Table 3 provides the selected technical/technological specifications for different HS trains. As can be seen, the maximum design speed varies from 250 to 350 km/h. The locomotives are powered by the electric energy. These are the so-called multi-system locomotives interoperable for at

Table 2 Some characteristics of the main grid of China rail high-speed (CRH) network [10, 15] (https://en.wikipedia.org/wiki/High-speed_rail_in_China/)

Relation	Orientation	Length of line (km)	Design speed (km/h)
Beijing–Harbin	N–S	1,800	350
Beijing–Shanghai	N–S	1,318	350
Beijing–Hong Kong	N–S	2,383	350
Hangzhou–Shenzhen	N–S	1,499	250/350
Sub-length		7,000	
Qingdao–Taiyuan	E–W	940	200/250
Xuzhou–Lanzhou	E–W	1,434	250/350
Chengdu–Shanghai	E–W	2,066	200/250
Kunming–Shanghai	E–W	2,056	350
Sub-length		6,496	
Total length		13,469	

[a] North–South (N–S); East–West (E–W)

Table 3 Technical/technological characteristics of different HS trains [51] (http://en.wikipedia.org/wiki/ICE_3; http://en.wikipedia.org/wiki/Siemens_Velaro; http://www.trainweb.org/tgvpages/tgvindex.html; http://en.wikipedia.org/wiki/New_Pendolino)

Type of HS train	Build date (year)	Max. speed[a] (km/h)	Supply voltage[b]	Traction (kW, kV)	Length/weight (m, ton)	Configuration[c] (–, seats)	Performance metrics (kW/seat)
TGV PSE (Paris Sud-East)	1978–1985	300/270	25 kV 50 Hz AC 1.5 kV DC	6,450, 25 3,100, 1.5	200, 385	1 + 8 + 1, 385	18.34
TGV LA Poste	1981–1984	270	25 kV 50 Hz AC 1.5 kV D	6,450, 25 3,100, 1.5	200, 345	–	–
TGV Atlantique	1989–1992	300	25 kV 50 Hz AC 1.5 kV DC	8,800, 25	238, 484	1 + 10 + 1, 485	18.14
TGV Resau	1992–1996	300	25 kV 50 Hz AC 1.5 kV DC	8,800, 25	200, 386	1 + 8 + 1, 377	23.34
Eurostar	1993–1995	300	25 kV 50 Hz AC 3 kV DC 1.5 kV DC, 750 V DC	12,200, 25	394, 752	1 + 18 + 1, 794	15.90
TGV Duplex	1995–1997	320	25 kV 50 Hz AC 1.5 kV DC	8,800, 25	200, 380	1 + 8 + 1, 545	16.15
Thalys	1995–1998	300	25 kV 50 Hz AC 1.5 kV DC	8,800, 25	200, 385	1 + 8 + 1, 377	23.14
ICE 3	1998–1999	330	15 kV 16.7 Hz AC 1.5 kV DC	8,000, 1.5	201, 435	1 + 8 + 1, 441	18.14
ICE 3 M	2000	330	25 kV 50 Hz AC 3 kV DC	8,000, 25	201, 435	1 + 8 + 1, 430	18.60
ICE Velaro CN[d]	2004	350	25 kV 50 Hz	9,200, 25	200, 447	1 + 8 + 1, 610	15.31
AVE	1991–1992	300	25 kV 50 Hz AC 3 kV DC	8,800, 25	200, 392	1 + 8 + 1, 320	26.75
KTX (TGV Korea)	1997–2002	300	25 kV 50 Hz AC	13,200, 25	381, 701	1 + 18 + 1, 935	14.12
N700-I (Japan)	2007	330	25 kV 60 Hz	9,760, 25	204.7, 365	8, 636	15.35
ETR 600[d]	2008	250	25 kV 50 Hz AC 3 kV DC	5,500, 25	187, 387	2 + 3 + 2, 430	12.79

[a] Design speed

[b] AC: alternating current, DC: direct current

[c] Power car(s)–trailers–power car(s)

[d] Operating in China—version CRH2C

least two different electric power supply systems. The traction power varies from 5,500 to 13,200 kW/train set. The length of a train set is predominantly about 200 m, and the corresponding weight is between 350 and 450 tons. Typical configuration of an HS train set is 1 power car + 8 trailers + 1 power car. The performance metrics vary across the considered set of HS trains from 12 to 23 kW/seat.

In addition, Fig. 3 shows the relationship between the performance metrics and the seat capacity of the selected HS trains.

As can be seen, the performance metrics expressed by the installed traction per seat (kW/seat) decreases more than proportionally with the increase of the number of seats, thus indicating economies of the train size in terms of the installed (and required) traction. This indicates that the HS trains with higher seating capacity do not need to have the proportionally stronger traction.

As well, Fig. 4 shows the relationship between the maximum designed and operating speed of the HS trains [16].

As can be seen, the speeds ranging from 200 to 320 km/h coincide with each other for many HS trains. Nevertheless, generally, with the increase of the maximum design speed, the positive difference between this and the maximum operating speed tends to increase. This particularly happens for the speeds ranging between 270 and 380 km/h. Consequently, at particularly high maximum design speeds (above 300 km/h), it is likely to expect the lower maximum operating speeds for about 10 %–20 %, as shown in this case.

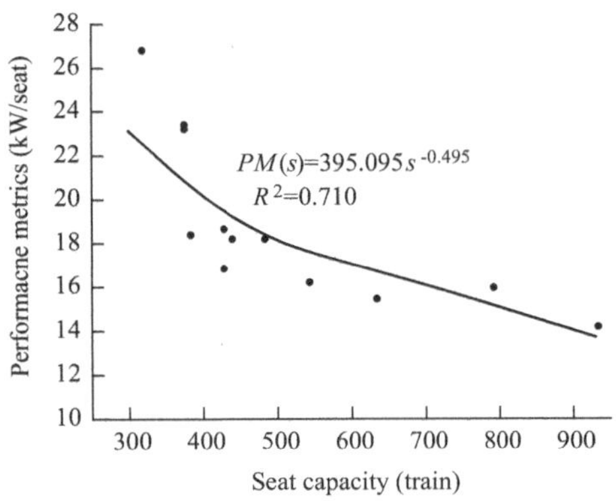

Fig. 3 Relationship between the performance metrics and the seat capacity of HS trains (Table 3)

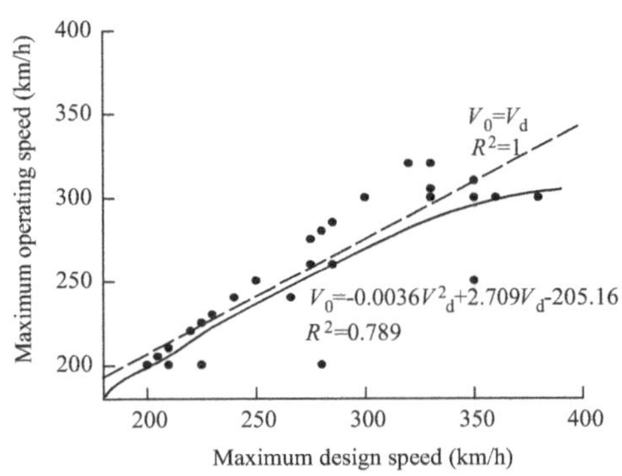

Fig. 4 Relationship between the maximum design and maximum operating speed of the selected HS trains (http://en.wikipedia.org/wiki/List_of_high-speed_trains)

4.2 Supportive facilities and equipment

The main supportive facilities and equipment of the HSR system in the given context are power supply, signaling, and traffic control/management system.

4.2.1 Power supply system

The power supply system is an integrated system including the high-voltage electric power lines, substations, contact line, HS trains, and the remote command and control system ensuring efficient, reliable, and safe supply of electric power to the HSR lines and trains, and consequently operations. The electrified networks for the HSR lines generally use the alternate current (AC) or direct current (DC). As given in Table 3, the typical voltage and frequencies are 25 kV 50 Hz AC, 1.5 kV DC, and 15 kV 16.7 Hz AC. The latest has been installed in Germany and supplied from the dedicated high-voltage network called the 'Railway Frequency.' The above-mentioned general system components can further be divided into two main components: the HSR electrical infrastructure and the HS rolling stock traction equipment [17].

4.2.2 Signaling systems

The different HSR signaling systems have been applied in different countries. For example, each European country has its own HSR signaling systems: in France it is Transmission Vole Machine (TVM), in Germany LinienZugBeeinflussung (LZB), in Spain German's LZB (for speeds up to 300 km/h) and Electrique Bureau CABine (EBICAB) (for speeds up to 220 km/h), and in Italy Blocco Automatico a Correnti Codificate (BACC) (for speeds up to 250 km/h). In addition, the European rail traffic management system (ERTMS—Level 1 and/or 2) has been introduced on the particular lines in different countries as an alternative and/or a complement to the existing national systems [17].

The type of signaling system influences the length of a block of the track, which can be occupied exclusively by a single HS train. The number of such successive empty blocks determining the (breaking) distance between any pair of HS trains moving in the same direction depends of their maximum operating cruising speed and the breaking/deceleration rate(s).

4.2.3 Traffic control/management system

In general, at the HSR rail lines/networks the rail traffic control/management systems is fully computer supported and can include the following main components: TOC—train operation controller; PC—power controller; STC—signal and telecommunication controller; CCC—crew and car utilization controller; PSC—passenger service controller; and TSMC—track and structure maintenance controller. These components are usually accommodated in the same room with the corresponding staff [18].

5 Operational performances of HSR systems

The main operational performances of HSR systems are demand, capacity, and quality of services, the latest as an outcome from the dynamic interaction between the former two. These performances can be considered for an individual line/route and/or for the entire network serving a given region, i.e., country.

5.1 Demand

The demand for HSR services consists of the self-generated demand and the demand attracted from other transport

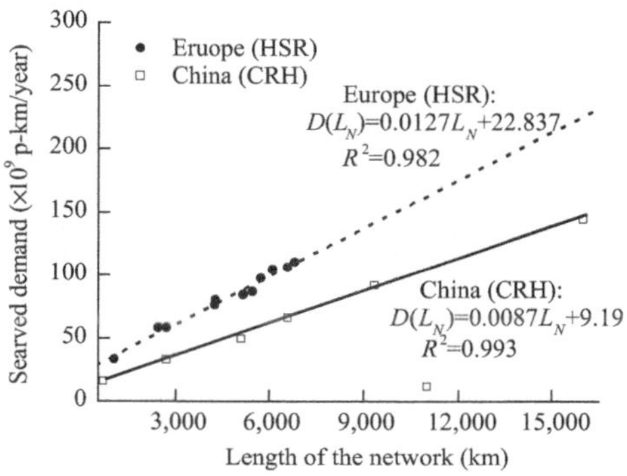

Fig. 5 Relationship between the annual passenger demand and the length of HSR networks in Europe and China (period 1995–2014) [3, 50, 52]

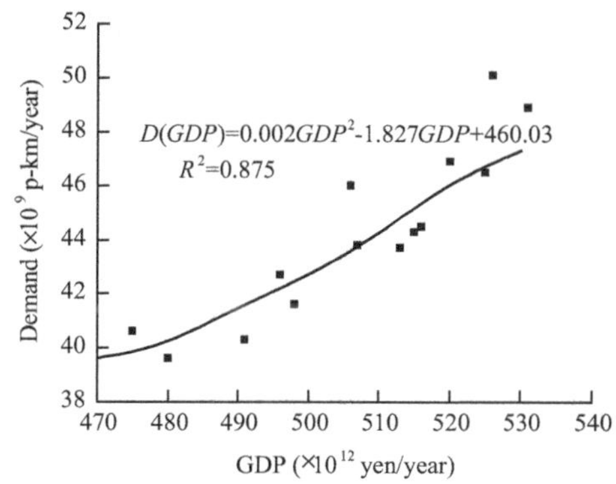

Fig. 7 Relationship between the satisfied passenger demand by Japanese Tokaido Shinkansen HSR system and the national gross domestic product (GDP) (period 2001–2015) [19]

modes on the competitive routes such as individual car, conventional railways, and air passenger transport (APT).

In general, the self-generated demand for HSR services has been stimulated by expansion of the HSR network and increase of the welfare in terms of the national gross domestic product (GDP). Figure 5 shows the relationship between the served passenger demand and the length of HSR network in Europe and China.

In both regions, the served passenger demand has grown linearly with the increase of the length of HSR networks. In terms of absolute values, the served passenger demand in China has exceeded that in Europe during the relatively short period of time (7 years), which has indicated the very strong user/passenger preference to the new CRH speed system as shown in Fig. 6.

In Europe, the served passenger demand has continuously been growing during the specified period of time. In China, since the start of implementing the CRH speed network, the corresponding passenger demand has been growing tremendously and very quickly exceeded that in Europe. In both cases, this has been possible primarily thanks to expanding the HSR network as shown in Fig. 6

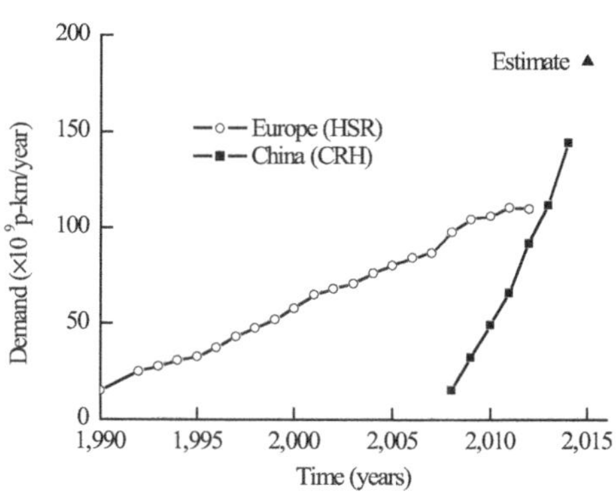

Fig. 6 Development of the satisfied passenger demand in the European HSR and Chinese CRH speed network (period 1990–2014) [50, 52]

and the other above-mentioned demand-stimulating factors. Figure 7 shows the relationship between GDP and the satisfied HSR passenger demand in Japan during the observed period [19].

As can be seen, the served passenger demand has increased more than proportionally with rising of GDP, thus indicating that GDP has generally been, is, and will continue to be a strong generator of demand in the given context.

The attracted and satisfied HSR passenger demand from other transport modes on the competitive routes has resulted from their competition. Figure 8 shows the passenger market share of HSR compared to that of APT dependence on the line travel time.

As can be seen, the relative market share of HSR (that of APT is complement to 100 %) has decreased linearly (Europe, Japan) and more than linearly (China) with the increase of the line/route travel time within the given range.

5.2 Capacity

The capacity of HSR systems can generally be calculated for their components of infrastructure—stations, lines/routes, and rolling stock. In general, for the infrastructure

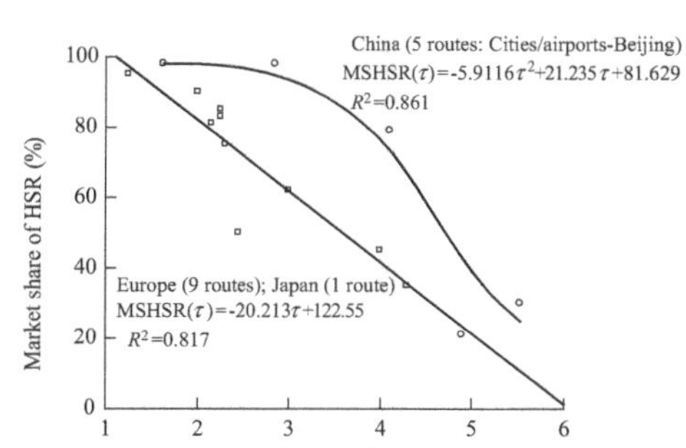

Fig. 8 Relationship between the market shares of HSR and APT, and the line/route travel time [40, 53–55]

components, the 'ultimate' and 'practical' capacity can be considered. Both are dependent on the operational rules and procedures providing a safe separation of trains while operating along the lines and at the stations in the same and/or different directions. These rules specify the minimum time separation between occupying the same section of the line(s) and/or of the station(s), which mainly influences their corresponding capacities. In addition, the transport work and productivity can be considered as the measures integrating in some way the capacity of infrastructure components and that of operations of the rolling stock.

5.2.1 Infrastructure components

5.2.1.1 'Ultimate' capacity

- *Line* The 'ultimate' capacity of a given HSR line/route is defined by the maximum number of trains, which can pass safely through the selected 'reference location' on the line where it is counted under given conditions, i.e., usually constant demand for service. This capacity can be estimated as follows [20, 21]:

$$\mu_l(T) = \frac{T}{t_{ij/\min}}, \qquad (1a)$$

where *i, j* are the leading and trailing trains in the sequence of two successive trains (*ij*) passing through the 'reference location' for their counting, respectively, which can be any location along the open line/route; $t_{ij/\min}$ the minimum time interval at which the successive trains (*i*) and (*j*) moving in the same direction pass through this 'reference location' (min); and *T* is the period of time for calculating the ultimate capacity of particular infrastructure component (h).
This minimum time interval ($t_{ij/\min}$) in Eq. 1a is mainly influenced by the HS train's maximum operating speed, acceleration and deceleration/braking performances, length, the way of its control, and also the spacing and design of the stations/terminuses, gradients along the line/route, and type of traffic control (signaling) system. In general, this time can be estimated as follows [22]:

$$t_{ij/\min} = \frac{V_j}{a_j^-(V_j)} + \frac{S_{b/j} + L_i}{V_j}, \qquad (1b)$$

where *i, j* are the leading and trailing HS trains, respectively, of the pair of successive trains (*in*); V_j is the maximum operating speed of the trailing train (*j*) (km/h); $a^-(V_j)$ is the average deceleration rate of the trailing train (*j*) at the maximal braking rate (m/s°); $S_{b/j}$ is the "buffer" zone for the trailing train (*j*) (m); and L_i is the length of the leading train (*i*) (m).
The maximum operating speeds of HS trains are usually about 250–350 km/h. The deceleration[3] rate a^- varies, i.e., it generally increases with the decrease of speed during the breaking phase of trip. The buffer zone (typically of the length of $S_{b/j} = 100$ m) is the distance added to the braking distance of trailing HS train to allow a margin for its safe separation from the leading train (*i*) [22, 23]. The train length is typically $L = 200$ or 400 m. The latter is the length of Eurostar and 2-unit German-designed Velaro train operating in China (Table 3).

- *Station along the line/route* If the leading train (*i*) is to stop and the trailing train (*j*) is to pass through a station along the line, the 'reference location' for counting trains, i.e., calculating the capacity, can be the exit signal of the station. The 'ultimate' capacity of the station in this case can be estimated as follows [6, 20]:

$$\mu_{s/l}(T) = \frac{T}{t_{ij/s/\min}}, \qquad (1c)$$

where $t_{ij/s/min}$ is the minimum time interval at which the successive trains (*i*) and (*j*) pass in the same direction through the station (min).
The minimum time ($t_{ij/s/\min}$) in Eq. 1c can be estimated as follows: the leading train (*i*) after being dispatched from the station should be at least at the minimum breaking distance of the trailing train (*j*) at the moment when this arrives at the exit signal of the station, which in this case will allow it to proceed. In such case, the time ($t_{i\,j/\min}$) in Eq. 1b can generally be extended by the dwell time of the train (*i*) at the station as follows:

$$t_{ij/s/\min} = \tau_i + \left[\frac{S_{b/j} + L_i}{2a_j^-(V_j)}\right]^{1/2} + \frac{V_j}{a_j^-(V_j)}, \qquad (1d)$$

where τ_i is the dwell time of the leading train (*i*) at the station (min).
The other symbols are analogous to those in the previous equations.
At most HSR systems, the dwell time is typically $\tau = 2$–3 min at the stations located along the lines/routes and $\tau = 5$ min for those located at airports, the latter mainly due to enabling users/passengers to handle

[3] For example, it can be $a^- = 0.30$ m/s^2 for the speeds between $V = 350$ and 300 km/h (first 1,000 m of breaking distance), $a^- = 0.35$ m/s^2 for the speeds $V = 300$–230 km/h (second 1,000 m of breaking distance), and $a^- = 0.6$ m/s^2 for the speeds $V = 230$–0 km/h (the rest of 6,000–7,000 m of breaking distance). Consequently, the average deceleration rate of $a^- = 0.5$ m/s^2 is usually used in these calculations [22].

their baggage. This time also includes the time for closing the doors, setting up the conflict-free exit path, and dispatching the leading train (*i*).

- *End terminus/station* The 'ultimate' capacity in this case can be estimated as follows [6, 20]:

$$\mu_{s/\text{arr}}(T) = \frac{T}{t_{ij/\text{min/arr}}}, \tag{1e}$$

where $t_{ij/\text{min/arr}}$ is the minimum time interval at which the successive trains (*i*) and (*j*) arrive at the entry signal of the given end station/terminus (min) as the 'reference location' for their counting.
The time ($t_{ij/\text{min/arr}}$) in Eq. 1e can be estimated as follows:

$$t_{ij/\text{min/arr}} = \frac{V_j}{a_j^-(V_j)} + \left[\frac{S_{b/j} + L_i}{2a_j^-(V_j)}\right]^{1/2} + \tau_{ij} + \tau_b, \tag{1f}$$

where τ_{ij} is the time for changing the route of trains (*i*) and (*j*) arriving at the end station/terminus of the given line/route (typically 10 s); and τ_b is the time of blocking the entrance of the end station/terminus by other trains(s) (typically 25 s).
The other symbols are analogous to those in the previous equations.

- *Begin terminus/station* The 'ultimate' capacity in this case can be estimated as follows [6, 20]:

$$\mu_{s/\text{dep}}(T) = \frac{T}{t_{ij/\text{min/dep}}}, \tag{1g}$$

where $t_{ij/\text{min/dep}}$ is the minimum time interval at which the successive trains (*i*) and (*j*) pass the exit signal of the given station/terminus as the 'reference location' for their counting (min).
The time ($t_{ij/\text{min/dep}}$) in Eq. (1g) can be estimated as follows:

$$t_{ij/\text{min/dep}} = \max\left\{\left[\frac{S_{b/i} + L_i}{2a_i^+(V_i)}\right]^{1/2} + \frac{V_i}{a_i^+(V_i)}; \tau_{j/r} + \tau_{j/gl} + \tau_{j/cf} + \tau_{j/d}\right\}, \tag{1h}$$

where $\tau_{j/r}$ is the time for setting the exit path for the trailing train (*j*) in a given departing sequence (*ij*) (usually 10 s); $\tau_{j/gl}$ is the time for setting the green light for trailing train (*j*) in a given departing sequence (*ij*) (usually 25 s); $\tau_{j/cf}$ is the time of blocking exit of the station/terminus for departing trailing train (*j*) by other incoming and outgoing trains (usually 60–75 s); and $\tau_{j/d}$ is the dispatching time of the trailing train (*j*) in a given departing sequence (*ij*) (usually 30 s).

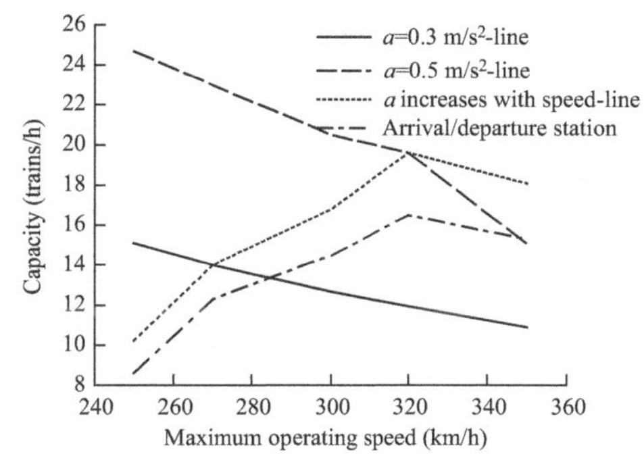

Fig. 9 Relationships between the ultimate capacity of the HSR line/route, begin/end station/terminus, and the maximum train operating speed

The other symbols are analogous to those in the previous equations.
Equation 1h indicates that the minimum time between departures of the successive trains (*i*) and (*j*) from the begin station/terminus should be set up as the maximum of two time periods: the time the leading train (*i*) needs to reach the minimum breaking distance from the trailing train (*j*) and the time for setting up a safe departure path for this trailing train (*j*).

Figure 9 shows examples of the above-mentioned 'ultimate' capacities of the HSR line/route and begin/end station/terminus dependent on the train's maximum operating speed calculated by Eq. 1.

As can be seen, the line/route capacity decreases with the increase of speed if the same average deceleration/acceleration rate is applied ($a = 0.5$ m/s^2 for the speeds of $V = 250$–350 km/h). However, if this rate increases with the increase of speed ($a = 0.5$ m/s^2 for the speed of $V = 250$ km/h, $a = 0.3$ m/s^2 for the speed of $V = 270$ km/h, $a = 0.4$ m/s^2 for the speed of $V = 300$ km/h, and $a = 0.5$ m/s^2 for the speeds of $V = 320$ and 350 km/h), the capacity generally tends to increase. In the latest case, the capacity again decreases due to applying the same deceleration/acceleration rate to the increasing speed. Similar happens with the arrival and departure capacities of begin/end station/terminus, respectively. In all cases, the train length is assumed to be $L = 400$ m and the buffer distance $S_b = 100$ m [22]. Consequently, the line/route capacity can be estimated as the minimum of the above-mentioned four 'ultimate' capacities. In practice, the 'ultimate' capacity of the HSR lines/routes and stations is typically $\mu = 13$–15 trains/h.[4] In addition, the required number of tracks at the end/begin station/terminuses can be determined as the product of the above-mentioned

[4] The number of the Shinkansen "Nozomi" services has been scheduled to be 10 dep/h during the peak hours [19].

'ultimate' capacities and the HS train's dwell time at the stations—stop time at the line and the turnaround time[5] at the begin/end station(s).

5.2.1.2 'Practical' capacity The 'practical' capacity of a given HSR line/route is defined as the maximum number of HS trains, which can be accommodated during the specified period of time under conditions when each of them is imposed an average delay [25]. However, in this case, the mutual interferences between the HSR services of equal priority operating on the above-mentioned *Category I* lines causing their delays are prevented by the stability of timetable. This implies that the maximum permissible delay of leading train in the sequence of two trains is defined in a way not to cause an additional delay of the following train. As such, this delay indicates some kind of the system's margin allowing delays of the HS trains anyway. The longer delays causing disruption of the timetable occur generally due to other causes.

5.2.2 Rolling stock

The capacity of HSR rolling stock reflects its size expressed by the number of trains of a given seating capacity required to operate under the conditions specified in the timetable. These conditions are usually characterized by the service frequency during the given period of time (h, day) and the train's turnaround time along the given line/route. Consequently, the required number of rolling stocks/trains to carry out at the specified service frequency on a given line, $m_{rs}\,[T; f(T)]$, can be estimated as follows [6, 26]:

$$m_{rs}[T;f(T)] = f(T)\tau, \tag{2a}$$

where $f(T)$ is the train service frequency on a given line during time (T) (trains/h; trains/day) and τ is the average turnaround time of a train along a given line (h).

The service frequency $f(T)$ in Eq. 2a can be either considered to be equal to the line/route 'ultimate' capacity determined by Eq. 1 or set up to satisfy the expected demand as follows [26]:

$$f(T) = \min\left[\mu_l(T); \mu_{s/l}(T); \mu_{s/arr}(T); \mu_{s/\mathrm{dep}}(T); \frac{D(T)}{\rho(T)s}\right], \tag{2b}$$

where $D(T)$ is the expected user/passenger demand on a given HSR line during time (T)(pax); $\rho(T)$ is the average load factor on a given line during time (T) ($\rho(T) \leq 1.0$); and s is the seat capacity of a train operating on a given line (seats/train).

The other symbols are analogous to those in the previous equations.

The train's turnaround time (τ) increases with the increase of the operating time along the line/route (the ratio between the length of line/route and the operating speed), the number and duration of intermediate stops, all in both directions, including those at the beginning and end station/terminus, and vice versa. The train's seat capacity is usually constant per service frequency indicating the above-mentioned homogeneous HS train fleet on a given line/route. For example, if the given line/route operates at the service frequency of $f(T) = 15$ trains/h, and if the average turnaround time per train is $\tau_l = 4$ h, the required number of trains will be $m_{rs}(T) = 15 \times 4 = 60$. In addition, if the average train's seat capacity is $s = 485$ (TGV Atlantique, see Table 3), the total number of required seats will be $m_s(T) = 29{,}100$.

5.3 Transport work and productivity

The transport work and productivity of a given HSR line/route can be calculated for the supply and demand sides. On the supply side, it counts the total offered number of seats during a given period of time. On the demand side, it counts the total number of used seats under the same conditions.

Based on Eq. 2, the transport work on a given line for the supply (s-km) (seat-kilometers) and demand (p-km) sides, respectively, can be calculated as follows [26]:

$$\mathrm{TWS}(T) = f(T) \times s \times d \quad \text{and} \quad \mathrm{TWD}(T) = f(T) \times s \times \rho(T) \times d. \tag{3a}$$

Similarly, the productivity of both supply and demand sides of a given line expressed as the volumes of seat-km/h and pax-km/h, respectively, can be calculated as follows:

$$\mathrm{TPS}(T) = f(T) \times s \times v \quad \text{and} \quad \mathrm{TPD}(T) = f(T) \times s \times \rho(T) \times V, \tag{3b}$$

where d is the length of a given line (km) and V is the operating speed of HS trains on a given line (km/h).

The other symbols are analogous to those in the previous equations.

As can be seen, the transport work increases with the increase of the length of line, service frequency, seat capacity per frequency, and load factor. The productivity increases with the increase of the service frequency, seat capacity, load factor per frequency, and the average train operating speed, and vice versa. For example, for the HS trains, each with the seat capacity of $s = 485$ seats and the average load factor $\rho = 0.80$, operating on the line of length of $d = 500$ km at the operating speed of

[5] This time is used for disembarking the incoming passengers and their baggage, cleaning the interior of the train, replenishing water, restock, king victuals, changing the crew, and embarking the outgoing passengers and their baggage. It is typically about 20 min at most HSR systems. In Japanese HSR system (Shinkansen), it is about 12 min [24].

$V = 300$ km/h and the service frequency in the single direction of $f(T) = 15$ trains/h, the transport work on the line's demand and supply sides during the period of 1 h will be TWS = 15 × 500 × 485 = 3,637,500 (seat-km) and TWD = 15 × 500 × 485 × 0.80 = 2,910,000 (pax-km), respectively. The corresponding productivity under the same conditions will be TPS = 15 × 485 × 300 = 2,182,500 (seat-km/h) and TPS = 15 × 485 × 0.80 × 300 = 1,746,000, respectively (pax-km/h).

5.4 Quality of service

The quality of service provided by HSR systems can be expressed by the attributes such as schedule delay, trip time in combination with the reliability and punctuality of services, the comfort on board HS trains, and accessibility of the HSR stations [27].

5.4.1 Schedule delay

The schedule delay is defined as the difference between the desired and the available time of boarding a chosen HSR service. Under an assumption that the users/passengers familiar with the timetable arrive uniformly during the time between any two successive HS trains' departures on the same line/route/direction, this delay can be roughly estimated as follows [28, 29]:

$$SD(T) = \frac{T}{4f(T)}, \tag{4a}$$

where all symbols are analogous to those in the previous equations. For example, for the service frequency of $f(T) = 1$ train/h, the schedule delay will be $SD(T) = 15$ min; for the service frequency of $f(T) = 15$ trains/h, the schedule delay will be SD = 1 min ($T = 1$ h or 60 min).

5.4.2 Trip time, reliability, and punctuality

- *Trip time* by HSR systems is much shorter than that of their conventional counterparts at the same lines/routes. The potential time savings on a given route can be estimated as follows:

  $$\Delta = d/(1/V_{CON} - 1/V_{HSR}), \tag{4b}$$

  where d is the length of a given line/route (km); V_{CON} is the operating speed of the conventional passenger train (km/h); and V_{HSR} is the operating speed of the HS train (km/h). Figure 10 shows an example for this in Italy.

  As can be seen, the difference in trip time by the conventional and HSR trains increases with the increase of the line/route length, which in the given case amounts to 33 %–42 %.

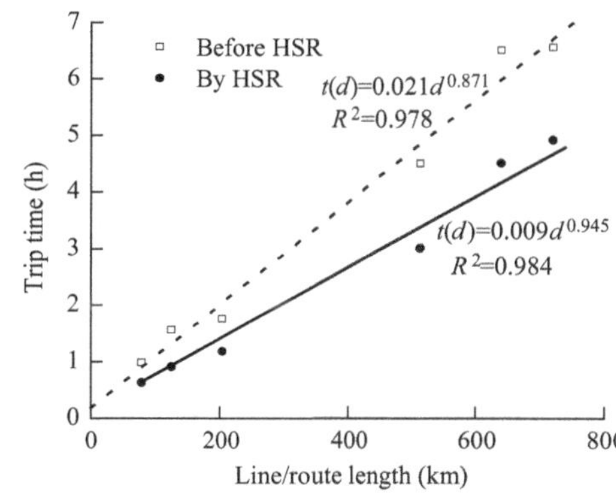

Fig. 10 Relationship between the trip time by the HS and conventional rail, and length of line/route in Italy [56]

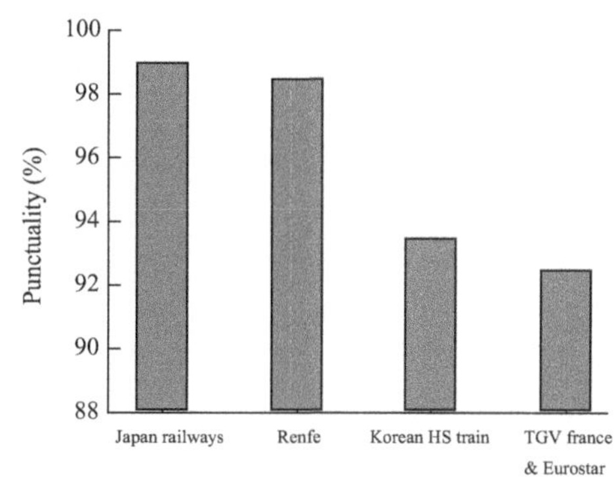

Fig. 11 Punctuality of services—the ratio—of the selected HSR systems [30]

- *Punctuality* of the HSR services can be expressed by two attributes: (i) the ratio of the number of transport services carried out on time, i.e., according to the timetable, or with the specified maximum or average delays, and the total number of services carried out, and (ii) the average delay per delayed service. Both attributes are recorded during a given period of time (day, month, year) under given conditions. The experience so far has shown that these services in general and on the particular lines/routes have been highly punctual as shown in Fig. 11 [30].

  As can be seen, the Japanese HSR system has generally been the most and the UK's the least punctual. In addition, Fig. 12 shows an example of the punctuality of the Japanese HSR system expressed by the average delay per service.

  As can be seen, the average delay per HSR service has varied from 0.3 to 0.5 min. In addition, the average delay of the Shinkansen HSR system has been about 0.6 min per service over the last decade [24, 31].

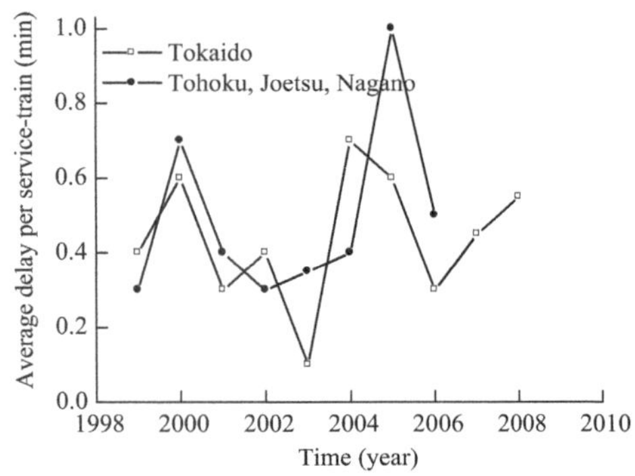

Fig. 12 Punctuality of services—the average delay—of the Japanese HRS system (period 1999–2008) [24, 57]

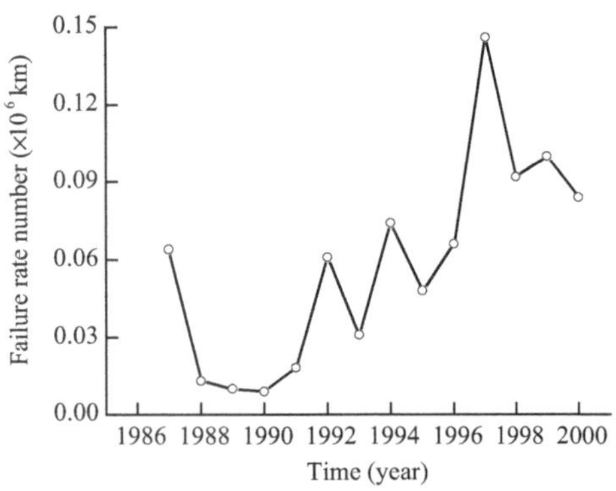

Fig. 13 Reliability of the HSR rolling stock (East Japan Railways—period 1987–2000) [24]

As can be seen, the average delay per HSR service has varied from 0.3 to 0.5 min. In addition, the average delay of the Shinkansen HSR system has been about 0.6 min per service over the last decade [24, 31].

- *Reliability* of the HSR services can be expressed as the ratio between the realized and planned transport services during a given period of time (day, month, and year) under given conditions. This is dependent on the rate of failure of rolling stock due to any system's internal and/or external reasons causing cancelation or long delays of the affected services. Figure 13 shows an example of the Japanese HSR system.

As can be seen, this rather very low failure rate has fluctuated during the observed period with an average of 0.084 failures/10^6 km.[6]

[6] This has been achieved by maintaining the rolling stock at four levels: (i) daily inspection (every 2 days), i.e., inspection of the wear parts (pantograph strip, refreshing water/waste); (ii) regular inspection (every 30 days or 30,000 km) (test of conditions and function, inspection of the important parts/components without decomposition); (iii) inspection of bogie (every 1.5 year or 600,000 km) (bogie parts by decomposition); and (iv) the overall inspection (every 3 years or 1,200,000 km) (inspection of the overall rolling stock by decomposition) [32].

5.4.3 Accessibility

Accessibility of stations is an important attribute of the overall quality of services provided by the HSR systems. In most cases, the new dedicated HSR stations are usually located and designed to fit as good as possible within the surrounding urban and/or sub-urban layout on one hand and enable the satisfactory quality of accessibility on the other. In some other cases, the parts of conventional railway stations have been appropriately upgraded and adapted to serve the HSR services. In both cases, the quality of accessibility needs is expected to be efficient, effective, and safe. This implies a reasonable (acceptable) time and costs from/to the doors of users/passengers by a variety of urban and sub-urban transit modes (car, taxi, and frequent, punctual, reliable, and safe, i.e., without incidents/accidents due to known reasons, bus, tram, metro, regional rail, etc.), respectively.

5.4.4 Comfort on board the HS trains

The comfort offered to their users/passengers on board of the HS trains usually includes the booked seats and the very limited number of stops along the lines/routes compared to those at the conventional train counterparts. As far as the comparison with the ATP system as the main competitor on the short- and medium-haul liens/routes is concerned, the attributes for comparison have usually been the distance between seats and internal mobility, diversity and type of services, noise on board, and the potential impact on health. Table 4 summarizes these for both systems/modes.

As can be seen, the HS trains have generally possessed higher comfort on board than their aircraft counterparts.

6 Economic performances of HSR systems

The economic performances of HSR systems include their costs and revenues. The costs are imposed by implementation and operation of the systems. The revenues obtained mainly by charging users/passengers cover the costs and provide some funds for updating the system and the profits for particular stakeholders involved. In any case, both revenues and costs need to be balanced in order to guarantee the economic and financial stability of the system.

6.1 Costs

The total costs of a given HSR system generally consist of the infrastructure and operating costs. The infrastructure costs include: (i) planning the system and acquisition and preparing the land; (ii) building the lines and stations

Table 4 Some attributes of the comfort on board of the HS trains and commercial passenger aircraft [30]

Attribute	System/mode	
	HS train	Aircraft
Distance between seats (cm)	87–97	78–85
Internal mobility (general)	Higher	Lower
Services (food, drink, internet, etc.) (general)	Similar	
Noise (dBA)	62–69	70–82
Impacts on health (general)	Lower	Higher

including tunnels and bridges, and the supportive facilities and equipment including the signaling systems, catenaries and electrification mechanisms, and communications and safety installations; and (iii) maintenance of the entire infrastructure and supporting facilities and equipment [33]. The operating costs include acquiring, operating, and maintaining the rolling stock, selling services, and administration. The costs of labor, material, and energy have the largest share in the total costs [33].

Table 5 gives an indication of the average infrastructure cost of the already built and planned HSR lines, which do not include the cost of planning, and acquisition and preparation of the land.

As can be seen, the average infrastructure cost for both already built and under-construction HSR lines has significantly varied in both European and non-European, i.e., two Asian countries. In Europe, the lowest cost has been in France and Spain, and much higher in Italy, Germany, and Belgium. It can be shown that the average infrastructure cost has been 18 million€/km. In addition, the average cost of building the new HSR lines in Asian countries (Japan, South Korea, except China) has been slightly higher than those in particular European countries [34, 35]. As well, the average maintenance cost per unit of length of the HSR system infrastructure has also highly varied, mainly depending on the length of lines. Some estimates indicate that the average maintenance cost in European countries has amounted from about 13–72 thousands/year [35, 36].

The average cost of operating the HSR services has also differed throughout the European counties and rest of the world as well. This cost has been mainly influenced by the local pricing of the particular above-mentioned inputs and type of the HS trains. Some estimates indicate that this average operating cost for 12 types of the HS trains operating in the corresponding European countries has been: $\overline{C} = 0.14626$ €/seat-km. In this total, the cost of maintenance of the rolling stock has shared about 8.5 %. Under an assumption that the average load factor was: $\theta = 0.8$ (i.e., 80 %), the total average operating costs of the HSR services throughout Europe would be: $\overline{C} = 0.183$ €/p-km [34, 35].

Table 5 The average infrastructure cost of the HSR lines worldwide [3, 35, 42]

Country	Cost (million €/km)	
	Built (in service) lines	Under construction lines
Austria	–	18.5–39.6
Belgium	16.1	15.0
France	4.7–18.8	10.0–23.0
Germany	15.0–28.8	21.0–33.0
Italy	25.0	14.0–65.8
Netherlands	–	43.7
Spain	7.8–20.0	8.9–17.5
Japan	20.0–30.0	25.0–40.0
South Korea	–	34.2
China[a]	8.4–12.3[b]/11.3–22.0[c]	–

[a] 27 HSR lines in operation

[b] Design speed: 250 km/h

[c] Design speed: 350 km/h; 1 RMB (Chinese Yuan) ≈ 0.12 €

6.2 Revenues

The HSR systems obtain revenues from different sources such as the transport-based charging users/passengers, merchandise, and others [37]. In particular, the prices for users/passengers are set up to cover the systems' total operating cost in cases of the lack of subsidies. The latter can be used as an element for enabling stronger competition with the other transport modes such as the conventional rail and particularly APT, both on the above-mentioned competitive lines/routes. Figure 14 shows relationship between the annual revenues and the annual satisfied demand of the HSR systems in different countries [19].

As can be seen, the revenues have generally linearly increased with increasing of the volumes of satisfied demand at an average of 17.44 ¢US$/p-km, which is in line with the above-mentioned corresponding costs.

6.3 Balancing revenues and costs

The HSR systems intend to operate in the profitable way, i.e., to cover their costs by revenues. Figure 15 shows an example of the profitability of the Japanese HSR operating both HSR and conventional rail services.

As can be seen, despite a relatively high variations the profitability has generally increased with increasing of the volume of the company's output during the given period of time. This case could be used as an example how the HSR system can be profitable in the medium- to long-term period—by careful balancing the revenues and costs while at the same time increasing the scale of operations to satisfy the growing user/passenger demand.

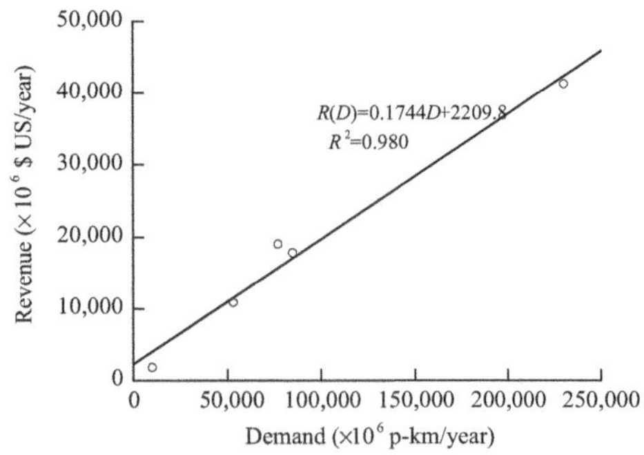

Fig. 14 Relationship between the annual revenues and the satisfied passenger demand of particular HSR systems—Japan, France, Germany, UK, USA (period 2012–2015) [19]

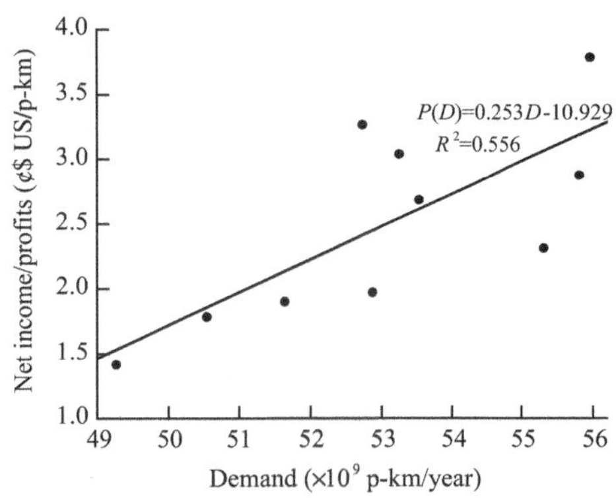

Fig. 15 Relationship between the annual demand and the net income/profits—Central Japan Company (period 2004–2013) [37]

7 Social performances of HSR systems

The social performances of HSR systems include the impacts and effects. The impacts embrace noise, congestion, and safety, i.e., traffic incidents and accidents. The effects generally refer to the system's overall welfare expressed by savings of the user/passenger time, relieving congestion from roads, and contribution to the regional GDP through direct and indirect employment.

7.1 Impacts

The HSR system generally impacts the society/people by noise, congestion, and safety, i.e., traffic incidents and accidents.

7.1.1 Noise

The HS trains generate noise while operating at the high speed(s), which comprises rolling, aerodynamic, equipment, and propulsion sound. This noise mainly depends on its level generated by the source, i.e., moving HS train(s), and its distance from an exposed observer(s). Figure 16 shows a scheme of changing the distance and time of exposure to noise by an HS train of an observer.

The shadow polygon represents an HS train of length (L) passing by an observer (small triangle at the bottom) at the speed (V). He/she starts to consider noise of an approaching train when it is at distance (β) from the point along the line, which is at the closest right angle distance (γ) from him/her. The consideration stops after the train moves behind the above-mentioned closest point again for the distance (β). Under such circumstances, the distance between the observer and the passing-by HS train changes over time as follows:

$$\rho^2(t) = (L/2 + \beta - V \times t)^2 + \gamma^2 \quad \text{for } 0 < t < = (L + 2 \times \beta)/V, \tag{5a}$$

where the last term represents duration of the noise event, i.e., the time needed for a train to pass by the observer (The length of HS trains is given in Table 3). If the level of noise received from the train passing by an observer with the speed (V) at the shortest distance (γ) is $L_{eq}(\gamma, V)$, the level of noise at any time (t) can be estimated as follows:

$$L_e[\rho(t), V] = L_{eq}(\gamma, V) - 8.6562 \ln[\rho(t)/\gamma]. \tag{5b}$$

The second term in Eq. 5b represents the noise attenuation with distance over the area free of barriers. The total noise exposure of the observer from $f(T)$ successive trains passing by during the period (T) can be estimated as follows:

$$L_{eq}[f(T)] = 10 \log \sum_{k=1}^{f(T)} 10^{\frac{Le[k,\rho(t),V]}{10}}. \tag{5c}$$

As a standard approach, the noise from HS trains is measured at the right angle distance of $\gamma = 25$ m from the track(s). Figure 17 shows the results of some such measurements across Europe depending on maximum operating speed of the HS trains.

As can be seen, the noise has generally linearly increased with increasing of the train's operating speed: at the lower rate for the speeds up to about 300 km/h, and at the higher rate for the speeds above $V = 300$ km/h. The variation of noise level at the given speed has been about 3–4 dBA. This noise has included the train's rolling (wheel), pantograph/overhead, and aerodynamic noise. Some additional measurements have shown that the rolling and pantograph/overhead noise has predominated and increased with increasing of the HS train's speed approximately at the rate of $30\lg V$ up to the speed(s) of about 300 km/h (some data have shown that this is 370 km/h). The aerodynamic noise depending on the HS train's (aerodynamic) design has also increased, equalized with the rolling noise at the above-mentioned (transition) speed(s), started predominating and further increasing at an approximate rate of $80\lg V$ [38]. In addition, in cases when

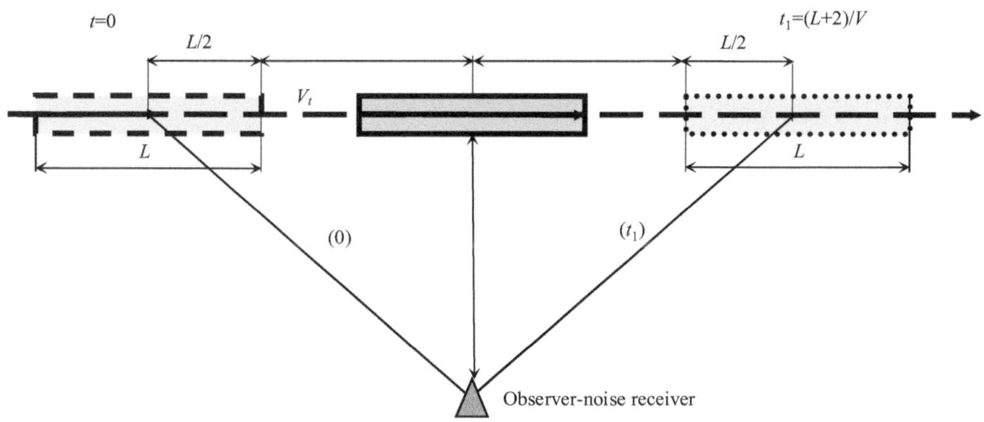

Fig. 16 Scheme for determining the noise exposure of an observer by passing by HS train [58]

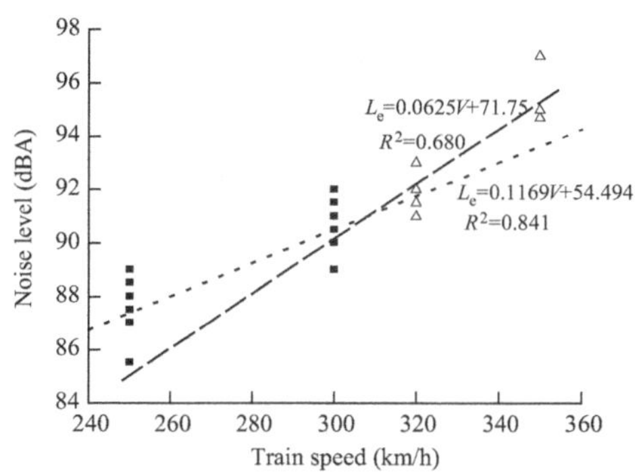

Fig. 17 Relationship between the noise and the maximum operating speed of the passing-by HS train(s) measured at the right angle distance of 25 m (Belgium, France, Germany, Spain, Italy) [59]

the frequent HSR services are carried out along the particular lines/routes, their noise becomes persistent over time and can be estimated from Eq. 5c. As well, the time of exposure of an observer to noise by a passing by HS train can be estimated from Eq. 5a. If $\beta = 0$ m, $L = 200$ m, and $v = 250$ km/h, this exposure time to the maximum noise will be about $t_1 = 3$ s; if $V = 350$ km/h, this time will be about $t1 = 2$ s.

Last but not least, while considering the actual exposure of the population located close to the HSR lines to noise by the passing-by HS trains, it is necessary to take into account the noise-mitigating barriers protecting the particular land use activities, i.e., a quiet land with intended outdoor use, a land with the residence buildings objects, and a land with the daytime activities (businesses, schools, libraries, etc.), all by absorbing the maximum noise levels for about 20 dB(A) (single barrier) and 25 dB(A) (double barrier).

7.1.2 Congestion

Thanks to applying the above-mentioned separation rules in addition to designing timetable(s) on particular lines/routes and the entire HSR network accordingly, the HSR systems are free of congestion and consequent delays due to the direct mutual influence of trains on each other while 'competing' to use the same segment of given lines/routes at the same time. However, the substantive delays due to some other reasons can propagate (if impossible to absorb and neutralize them) through the affected HS trains itineraries as well as along the dense lines/routes also affecting the other otherwise non-affected services. Under such conditions, the severely affected services are usually canceled in order to prevent further increase and propagation of their delays. On the one hand, this contributes to maintaining the punctuality but on the other, it compromises the reliability of the overall services (as mentioned above). Nevertheless, the already mentioned figures indicate that both reliability and punctuality of the HSR system services worldwide have been very and in some cases extremely high (The latter is the example of Japanese HSR system).

7.1.3 Safety, i.e., traffic incidents/accidents

Experience so far has indicated that the HSR and APT system have been the safest transport systems/modes in which traffic incidents/accidents have rarely occurred, usually due to the previously unknown reasons. This means that the number of traffic incidents/accidents and related person injuries, deaths, and the scale and cost of damaged properties both of the systems and the third parties per, for example, 10^9 s-km and/or p-km carried out over a given period of time, have been extremely low. In particular, high safety of the HSR systems has been provided also a prior by designing completely the grade-separated lines and the other supportive built-in safety features at both infrastructure and rolling stock. This implies that the safety has been achieved on the account of increased investments and maintenance cost. As well, the HSR operators and

Table 6 Characteristics of the HSR fatal accidents [60–62]

Country/system/number of trains	Date	Cause	Passengers on board	Fatalities	Injuries
Germany/ICE/1	03/06/1998	Wheel disintegration	287	101	88
China/2	23/07/2011	Railway signal failure	1,630	40	>210
Spain/Alvia/1	24/07/2013	Derailment due to excessive speed	222	>79	139

infrastructure managers have continuously practiced a risk management and training approach aiming at maintaining a high level of safety and particularly with increasing of the maximum speeds. Nevertheless, the HSR systems in different countries have not been completely free from traffic incidents/accidents. For example, some relevant statistics for the TGV system in France indicate that there have not been accidents with the fatalities (deaths) and severe injuries of the users/passengers, staff, and/or third parties since starting the HSR services started in the year 1981 despite the trains have been carrying out annually about 10×10^6 p-km. In addition, some incidents happened on the HSR lines/routes such as broken windows, opening of the passenger doors during operating at the cruising speed, couple of fires on board, collision with animals and concrete block on the tracks, and the terrorist attempts to bomb the tracks. The incidents and accidents of TGV trains operated on the conventional tracks have been more frequent with fatalities, injuries, and damages of properties but all at the relatively low scale. In these cases, the HS trains have been exposed to the external risk similarly to their conventional counterparts (http://www.railfaneurope.net/tgv/wrecks.html). Similarly, since started in 1960s, the Japan's Tokaido Shinkansen HS services[7] have also been free of accidents causing the user/passenger and staff fatalities and injuries due to the derailments and collisions of trains. This has been achieved despite the services have been exposed to the permanent threat of the relatively frequent (and sometimes strong) earthquakes.

Nevertheless, the fatal accidents with deaths and injuries of the users/passengers and staff happened at the HSR systems in Germany, Spain, and China (one in each country). Table 6 gives the main characteristics of these three accidents.

7.1.4 Cost of the social impacts—externalities

Quantifying the social impacts of HSR systems in the monetary terms as externalities has usually represented an ambiguous and often politically challenging task. Nevertheless, some estimates of these externalities for the HSR systems and other transport modes in Europe have been carried out. They have indicated that the total social externalities of HSR systems have amounted 22.9 €/10^3 p-km. In this total, the noise and traffic incidents/accidents externalities have shared about 22 % and 2 %, respectively. Since the HSR systems are free of congestion, the corresponding externality has not been considered. On the other hand, for comparison, the total externalities of APT have estimated to be 52.5€/10^3 p-km, of which the noise and traffic incidents/accidents externalities shared about 4 % and 3 %, respectively [39, 40].

7.2 Effects

The effects of HSR systems have consisted of contribution to the direct and indirect employment and consequently the economic-social development and welfare, both at a global-country and the local–regional scale.

7.2.1 Direct employment

The direct employment relates to manufacturing, building, and maintaining the infrastructure and manufacturing, operating, and maintaining the rolling stock and supporting facilities and equipment, i.e., the main system's components, of the HSR systems. For example, the number of employees operating the HSR services in particular countries is strongly dependent on the length of HSR networks as shown on Fig. 18.

A can be seen, in the considered countries, the number of employees increases linearly with increasing of the length of HSR network with an average of 7.3 employees/km.

7.2.2 Indirect employment

The indirect employment relates to the non-rail staff supplying the HSR system(s) with different kinds of daily consuming material and energy on the one hand and that generated just thanks to existing of the system on the other. These latter are the non-rail related economic activities around and at the HSR stations such as: business services (banking, insurance, and advertising), information and

[7] The Tokaido Shinkansen line/route of the length of 552.6 km connects Tokyo and Shin Osaka station is free of the level crossings. The trains operate at the maximum speed of 270 km/h covering the line/route in 2 h and 25 min. The route/line capacity is: $\mu_l = 13$ trains/h/direction. The number of passengers carried is about 386 thousand/day and 141 million/year (2011) [31].

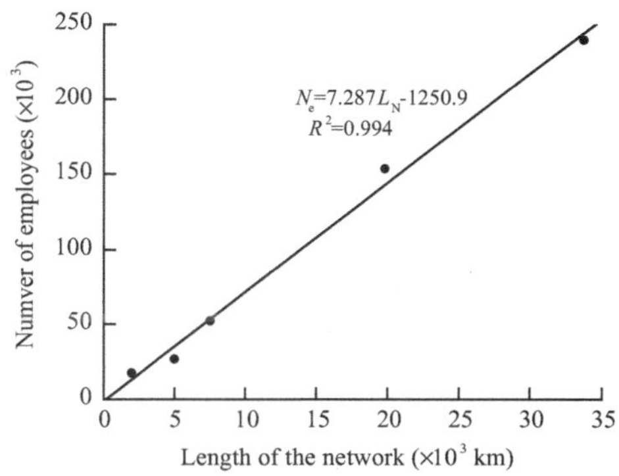

Fig. 18 Relationship between the number of employees and the length of HSR network—Japan (Central, East, West), SNCF (France), DB AG (Germany) (period 2014) [19]

retail services, research and development, higher education, tourism, and political institutions [30]. At the larger scale, these businesses have created urban (both business and housing) agglomerations around the HSR stations, which themselves have induced additional demand for the HSR services. Such development has been taking place mainly at the HSR stations already located in the larger urban agglomerations connected by the HSR lines/routes, but also within them. For example, inclusion of the city of Lille (France) in the HSR line/route Paris-Brussels has brought an enormous economic development of the city itself and its region in terms of increasing of business and touristic activities and related employment. In the UK, the substantial economic activities have been created in the cities 2 h from London area just thanks to the HSR [41].

7.2.3 Contribution to the local and global economy and welfare

In general, the above-mentioned employment has contributed to the economic-social development and welfare, both at a global-country and local–regional scale. For example, at the global-country scale, the direct effects have been contribution of the investments in HSR systems to the national GDP, which in Europe has estimated to be about 0.25 % of the national GDPs. At the regional scale, this contribution has been about 3 % of the regional GDP [42, 43]. This contribution has been much higher in the cities with the primarily service-oriented than in those with the primarily manufacturing-oriented economy [44]. In addition, the German regions with the cities of Montabaur and Limburg, with populations of 12,500 and 34,000 respectively, have recorded growth of GDP of about 2.7 % just due to increase in their market accessibility to the larger cities Frankfurt and Cologne thanks to the HSR services [45]. In Japan, the HSR has generated growth of population in the cities of about 1.6 % compared to those being bypassed where this growth has been for about 1 %. This growth has taken place primarily in the cities with the information industry and higher education [44].

8 Environmental performances of HSR systems

The environmental performances of the HSR systems generally include the energy consumption and related emissions of GHG, the area of land used for settling down the system's infrastructure, and the related costs considered, if internalized, as externalities. For the given HSR system, these performances can be considered at different time and spatial scale. In the former case, this could be the instant, short, medium, and/or life cycle assessment (LCA). In the latter case, in combination with the former, these performances can be considered for the particular HSR lines and/or the entire network [46].

8.1 Energy consumption and emissions of green house gases (GHG)

Energy consumption and related emissions of GHG are considered exclusively from operations of the HSR systems, which excludes those from building the infrastructure (lines) and manufacturing the supporting facilities and equipment and rolling stock (trains) [47].

In general, the HS trains consume electric energy primarily for accelerating up to the operating/cruising speed and then for overcoming rolling/mechanical and aerodynamic resistance to motion at that speed. This also includes the energy for overcoming resistance of grades and curvatures of tracks along the given line/route. As well, the energy is consumed for powering the equipment on board the trains. In particular, during the acceleration phase of a trip the electric energy is converted into kinetic energy at an amount proportional to the product of the train's mass and the square of its speed(s). A part of this energy recovers by regenerative breaking during deceleration phase before the train stops. During cruising phase of a trip, the energy is mainly consumed to overcome the rolling/mechanical and the aerodynamic resistance, which for a given type of HS train can be expressed as follows [48]:

$$R = R_M + R_A = (a + bV)W + cV^2, \tag{6a}$$

where R_M, R_A are the rolling/mechanical and aerodynamic resistance, respectively (N); W is the weight of a train (tons); V is the operating/cruising speed of a train (km/h); and a, b, c are the experimentally estimated coefficients.

Equation 6a essentially reflects the Davis's equation with the corresponding coefficients. It indicates that the aerodynamic resistance generally increases with the square of operating/cruising speed. The rolling mechanical resistance increases linearly with the increase of this speed and weight of the HS train. Some experiments carried out for Shinkansen Series 100 HS trains estimated the total resistance depending on the cruising/operating speed as follows: $R(V) = 8.202 + 0.10656\ V + 0.00116232\ V^2$ ($R(V)$ in kN and v in m/s) [40, 48]. The above-mentioned relationship emphasizes the importance of reducing both the weight of train and its aerodynamic resistance in order to achieve savings in the energy consumption during the longest phase of trip—cruising at high speed.

Estimates of the energy consumption by different types of HS trains including acceleration/deceleration/cruising phase of a trip have differed and changed over time, just thanks to the above-mentioned permanent improvements of their both characteristics (aerodynamic, weight) and operations. Table 7 provides some recent estimates of this energy efficiency for different types of the HS trains.

As can be seen, the Japanese Shinkansen is the most and the Eurostar the least energy efficient trains. One of the reasons is the relatively large difference in the seat capacity between them. As an indication, at present, the average energy efficiency of an HS train is assumed to be about EC = 0.033 kWh/s-km. Considering this and taking into account the emission rates of the primary sources for producing electricity in Japan, the average rate of emissions of GHG by Shinkansen trains is EMR = 42 gCO_2/s-km [19]. Under the analogous conditions, in Europe, this average rate is EMR = 21 gCO_2/s-km with an ambition to be reduced to EMR = 5.9 gCO_2/s-km by the year 2025, 1.5 gCO_2/s-km by the year 2040, and 0.9 gCO_2/s-km by the year 2055. This reduction is expected to be achieved through further improvement of the energy efficiency of HS trains and their operations on one side and by changing type and composition of the primary sources for producing electric energy on the other. In the latter case, the aim is to produce as much as possible electric energy from the renewable decarbonized primary sources [30, 47].

Table 7 Energy efficiency of different types of HS trains [49, 51]

Train type	Operating speed (km/h)	Seat capacity (seats)	EC (energy efficiency) (kWh/s-km)
Shinkansen Series 700	300	1,323	0.029
AVG	300	650	0.033
TGV Reseau	300	377	0.031
TGV Duplex	300	545	0.032
Pendolino Class 300	300	439	0.033
Eurostar Class 323	300	750	0.041
Velaro D	320	601	0.030

For some comparison, the emission rate of an average passenger car is around EMR = 140 gCO_2/km. This is likely to decrease to about EMR = 130 gCO_2/km by the year 2020. However, the new cars to be launched in the meantime are expected to have the emission rate of about EMR = 120 gCO_2/km, which is just according to the EU proposals. In addition, this could be reduced to about EMR = 80 gCO_2/km mainly thanks to more massive introduction of hybrid cars by the year 2030, and to about EMR = 57 gCO_2/km during the period between the years 2040 and 2055 when the electric or fuel-cell cars are supposed to only really contribute to the more significant reduction of the above-mentioned emission rates. Similarly to the HS trains, this will be carried out in parallel to the changing the structure of the primary sources for producing electric energy. In addition, the fuel efficiency and related emissions of CO_2 and other GHG by APT competing with the HSR on the short- to medium-haul lines/routes will also be improved in the forthcoming decades. For example, the emission rate of CO_2 is expected to decrease from today's average of EMR = 97–62 gCO_2/s-km by the year 2025 to EMR = 47 and 41 gCO_2/s-km by the years 2040 and 2055, respectively (the emission conversion factor is 1 g of Jet A fuel = 3.18 gCO_2/s-km; the aircraft types considered are similar to today's A319 and B737-800 models). The mentioned improvements are expected to be achieved by improving the aircraft airframe and engine efficiency. Beyond the year 2050, further improvements may be expected means by introducing the alternative fuels such as, for example, liquid hydrogen [6, 49]. Nevertheless, the above-mentioned figures indicate that the HSR systems will remain superior in terms of energy efficiency and related emissions of GHG (CO_2) as compared to its competitors—passenger cars and the short- to medium-haul commercial aircraft.

8.2 Land use

The HSR infrastructure directly occupies much smaller area of land than its road–highway counterpart. For example, if the width of an HSR line is (w) and the length (d), the total occupied land can be estimated as follows:

$$A = wd. \qquad (6b)$$

For example, if $w = 25$ and $d = 1$ km line, the total area of directly taken land will be $A = 2.5$ ha (ha—hectare) (the average gross area of taken land is 3.2 ha). For a highway with three lanes in both directions whose width is $w = 75$ m and length $d = 1$ km, the directly taken land is $A = 7.5$ ha (the average gross area of the taken land is about 9.3 ha, i.e., three times greater than that of the HSR line). In addition, utilization of the taken land by both modes is quite different. The capacity of HSR line/route in both directions is two times of 12–14 trains/h, i.e., 24–28 trains/h. If each train carries about 600 passengers, the intensity of land use will be 24–28 × 600/2.5 = 5,760–6,720 pax/h/ha. In case of the above-mentioned highway with the capacity of 4,500 veh/h and the occupancy rate of 1.7 pax/car, the intensity of land use will be 1,020 pax/h/ha, which is for about 6–7 times lower than that of HSR [40].

8.3 Externalities

The energy consumption and related emissions of GHG and land use by the HSR systems have also been considered as externalities. Similarly to the case of social externalities, the HSR systems have been shown to be rather superior compared to the other competing transport modes such as road passenger cars and APT. Some estimates have indicated that the air pollution associated with the climate change shares about 26 % and the land use about 30 % in the total HSR system externalities of 0.00229€/10^3 p-km. After including the above-mentioned share of the social externalities, the rest to 100 % is the share of up- and downstream, and urban externalities. The corresponding figures for APT are 86 % for the emissions of GHG and 2 % for the land use. After including the share of social externalities, the rest to 100 % is the share of urban, and up- and downstream externalities in the total of about 0.00525€/10^3 p-km [39, 40].

9 Conclusions

This paper has dealt with the multidimensional examination of infrastructural, technical/technological, operational, economic, social, and environmental performances of the HSR systems. The infrastructural performances have been related to the geometrical characteristics and design of the HSR lines and stations. The operational performances have included demand, capacity, and their dynamic relationship reflected through the quality of transport services provided to the users–passengers. The economic performances included the cost and revenues of setting up and operating the HSR system(s) and the revenues gained from charging users–passengers. The social/policy performances have included the impacts and effects of the HSR systems on the society. The former have included noise, congestion, and traffic incidents/accidents, i.e., safety, and their externalities. The latter have included global and local direct and indirect contributions (benefits) to the economy in the widest sense. The environmental performances have embraced the energy consumption and related emissions of GHG (Green House Gases), and land use with both direct and indirect impacts on the environment, and associated externalities.

The particular performances have been elaborated in both descriptive and analytical ways dependent on the most influential factors. In the latter case, some analytical models of particular performances have been presented. In addition, where considered appropriate, a comparison of the performances of HSR systems with those of the competing systems operated by other transport modes has been carried out.

Finally, the HSR systems have been shown to be the mass high-speed inter-urban transport systems serving the user/passenger demand generally efficiently, effectively, and safely through competition and/or cooperation with its conventional rail counterpart, car, and APT, where and if appropriate.

References

1. UIC (2010) Necessities for future high speed rolling stock. Report January 2010, International Union of Railways, Paris
2. EC (1996) Interoperability of the trans-European high speed rail system, Directive 96/48/EC. European Commission, Brussels
3. Ollivier G, Sondhi J, Zhou N (2014) High-speed railways in China: a look at construction costs. China Transport Topics No. 9, World Bank Office, Beijing, pp 1–8
4. USDOT (2009) Vision for high-speed rail in America. U.S. Department of Transportation, Federal Railroad Administration, Washington DC
5. Wendell C, Vranich J (2008) The California high speed rail proposal: a due diligence report. Reason Foundation (with Howard Jarvis Taxpayers Association and Citizens Against Government Waste), Los Angeles, California
6. Janić M (2014) Advanced transport systems: analysis, modelling, and evaluation of performances. Springer-Verlag, New York
7. UIC (2002) Feasibility of "Ballastless" track. UIC Report, Infrastructure Commission—Civil Engineering Support Group, International Union of Railways, Paris
8. UIC (2004) UIC Code 406. International Union of Railways, Paris
9. Takai H (2013) 40 Years experiences of the slab track on Japanese high speed lines. Presentation, Railway Technical Research Institute, Tokyo
10. Takagi K (2011) Expansion of high speed rail services: development of high-speed railways in China. Jpn Railw Transp Rev 57:36–41
11. Anderson T, Lindvert D (2013) Station design on high speed railway in Scandinavia: a study of how track and platform technical design aspects are affected by high speed railway concepts planned for the Oslo Göteborg line. MSc Thesis, Chalmers University of Technology, Göteborg
12. Kido ML (2005) Aesthetic aspects of railway stations in Japan and Europe, as a part of context sensitive design for railways. J East Asia Soc Transp Stud 6:4381–4396

13. USDOT (1999) Assessment of potential aerodynamic effects on personnel and equipment in proximity to high-speed train operations. U.S. Department of Transportation, Federal Railroad Administration, DOT/FRA/ORD-99/11, DOT-VNTSC-FRA-98-3, Washington DC
14. UIC (2014) High Speed Lines in the World. Updated 1st September 2014, UIC High Speed Department, International Union of Railways, Paris
15. Fu J, Nie L, Meng L, Sperry RB, He Z (2015) A hierarchical line planning approach for a large-scale high speed rail network: the China case. Transp Res Part A 75:61–83
16. Crozet I (2013) High speed rail performance in France: from appraisal methodologies to ex-post evaluation. Discussion Paper No. 2013-26, The Roundtable on Economics of Investments in High Speed Rails, International Transport Forum, 18–19 December 2013, New Delhi
17. ABB (2014) Powering the world's high speed rail networks. ABB ISI Rail, Geneva
18. JR Central (2011) Data Book 2011, Central Japan Railway Company, Nagoya, Japan
19. JR Central (2015) Annual Report 2015. Central Japan Railway Company, Nagoya
20. Janić M (1984) Single track line capacity model. Transp Plan Technol 9:135–151
21. Parkinson T, Fisher I (1996) Rail Transit Capacity Report 13. Transit Cooperative Research Program, Transportation Research Board, National Research Council, National Academy Press, Washington DC
22. Connor P (2011) Rules for high speed line capacity or how to get realistic capacity figure for a high speed rail line. Info Paper No. 3, Rail Technical Webpages, PRC Rail Consulting Ltd., Sutton Bonington, Loughborough
23. Hunyadi B (2011) Capacity Evaluation for ERTMS (European Rail Traffic Management System) Level 2 Operation on HS2. Bombardier Transportation Rail Control Solutions, Bombardier Inc., Montréal, Canada
24. Nishiyama T (2010) High-speed rail operations in Japan, international practicum on implementing high speed rail in the United States. APTA, American Public Transport Association, New York
25. Janić M (1988) A practical capacity model of a single track line. Transp Plan Technol 12:301–318
26. Vuchic RV (2007) Urban transit systems and technology. Wiley, New York
27. Carol DC (2011) High-speed rail is not about trains, NC 1-NETWORK—High Speed Rail 73: 6–7
28. Janić M (2000) Air transport system analysis and modelling. CRC Press (Taylor & Francis Group), Abington
29. Janić M (2003) High-speed rail and air passenger transport: a comparison of the operational environmental performance. Proc Inst Mech Eng Part F 217(4):259–269
30. UIC (2011) High speed rail and sustainability. UIC Publications, International Union of Railways, Paris
31. JR Central (2012) Data book 2011. Central Japan Railway Company, Nagoya
32. Yanase N (2010) High speed rolling stock in Japan: international practicum on implementing high speed rail in the United States. APTA, American Public Transportation Association, Washington DC
33. UIC (2005) High speed rail's leading asset for customers and society. UIC Publications, International Union of Railways, Paris
34. de Rus G (ed) (2009) Economic analysis of high speed rail in Europe, Informes 2009. Economía y Sociedad, Fundación BBVA, Bilbao
35. Pourreza S (2011) Economic analysis of high speed rail. NTNU (Norwegian University of Science and Technology), Trondheim
36. Henn L, Sloan K, Douglas N (2013) European case study on the financing of high speed rail. In: Proceedings of Australasian transport research forum, 2–4 October 2013, Brisbane
37. JR Central (2014) Visitors guide 2014. Central Japan Railway Company, Nagoya
38. Thompson JD, Eduardo LE, Liu X, Zhu J, Hu Z (2015) Recent developments in the prediction and control of aerodynamic noise from high-speed trains. Int J Rail Transp 3(3):119–150
39. INFRAS/WWW (2000) External costs of transport: accident, environmental and congestion costs in Western Europe. INFRAS Consulting Group for Policy Analysis and Implementation, Zürich, WWW University of Karlsruhe, Karlsruhe
40. UIC (2010) High speed rail: fast track to sustainable mobility. International Union of Railways, Paris
41. Baron, T., (2009), High Speed Rail Contribution to Sustainable Mobility, HAL Id, http://dumas.ccsd.cnrs.fr/dumas-00793156
42. de Rus G, Nombela G (2007) Is investing in high-speed rail socially profitable? J Transp Econ Policy 41(Part 1):3–23
43. Preston J (2013) The economic of investment in high speed rail: summary and conclusions. Discussion Paper No. 2013-30, International Transport Forum (ITF/OECD), Paris, p 38
44. Albalate D, Bel G (2010) High speed rail: lesson for policy makers from experiences abroad. Research Institute of Applied Economics, University of Barcelona, Barcelona
45. Boqué RJ (2012) High-speed rail: economic evaluation, decision-making and financing, master-Arbeit, Technische Universität Dresden, Fakultät Verkehrswissenschaften "Friedrich List". Institut für Gestaltung von Bahnanlagen, Dresden
46. Yue Y, Wang T, Liang S, Yang J, Hou P, Qu S, Zhou J, Jia X, Wang H, Xu M (2015) Life cycle assessment of high speed rail in China. Transp Res D 41:367–376
47. UIC (2010) High speed, energy consumption and emissions. UIC Publications, International Union of Railways, Paris
48. Rochard PB, Schmidt F (2000) A review of methods to measure and calculate train resistances. Proc Inst Mech Eng Part F 214:143–185
49. ATOC (2009) Energy consumption and CO_2 impacts of high speed rail. Association of Train Operating Companies Ltd., London
50. CSP (2014) China statistical yearbook 2014. China Statistical Press, National Bureau of Statistics of China, Beijing
51. Siemens (2014) Valero CN high speed trains for China railways. Mobility Division, Siemens AG, Berlin
52. EC (2014) EU transport in figures: statistical pocketbook 2014. European Commission, Publications Office of the European Union, Luxembourg
53. de Rus G (2008) The economic effects of high-speed rail investments. Discussion Paper, 2008-16, OECD-ITF Transport Research Centre
54. IR (2012) The benefits of high-speed rail in comparative perspective, Invensys rail. www.invensysrail.com/whitepapers/hsh-research-report.pdf
55. Wu J (2013) The financial and economic assessment of China's high speed rail investments: a preliminary analysis. Discussion Paper No. 2013/28, Prepared for the Roundtable on The Economics of Investment in High Speed Rail, 18–19 December 2013, New Delhi, International Transport Forum, Paris
56. Cascetta E, Coppola P (2011) High speed rail demand: empirical and modelling evidence from Italy. European Transport Conference, Glasgow, p 18
57. Tomii N (2010) How the punctuality of the Shinkansen has been achieved. WIT Trans Built Environ 114:111–120
58. Janić M, Vleugel J (2012) Estimating potential reductions in externalities from rail-road substitution in trans-European freight transport corridors. Transp Res D 17:154–160

59. Gautier P-E, Letourneaux FP (2010) High speed trains external noise: a review of measurements and source models for the TGV case up to 360 km/h. http://pdf-ebooks.org/ebooks/high-speed-trains-pdf.html
60. NDTnet (2000) ICE Train Accident in Eschede—Recent News Summary. e-J Nondestruct Test Ultrason 5(2). http://www.ndt.net/news/2000/eschedec.htm/
61. Puente F (2014) Driver error "Only Cause" of Santiago accident, says report, IRJ—Int J Railw, Falmouth, Cornwall. http://www.railjournal.com/index.php/high-speed/wenzhou-crash-report-blames-design-flaws-and-poor-management.html/
62. Qiao H (2012) Wenzhou crash report blames design flaws and poor management. IRJ—Int J Railw, Falmouth, Cornwall. http://www.railjournal.com/index.php/high-speed/wenzhou-crash-report-blames-design-flaws-and-poor-management.html/

3

Injury severity analysis: comparison of multilevel logistic regression models and effects of collision data aggregation

Taimur Usman[1] · Liping Fu[1,2] · Luis F. Miranda-Moreno[3]

Abstract This paper describes an empirical study aiming at identifying the main differences between different logistic regression models and collision data aggregation methods that are commonly applied in road safety literature for modeling collision severity. In particular, the research compares three popular multilevel logistic models (i.e., sequential binary logit models, ordered logit models, and multinomial logit models) as well as three data aggregation methods (i.e., occupant based, vehicle based, and collision based). Six years of collision data (2001–2006) from 31 highway routes from across the province of Ontario, Canada were used for this analysis. It was found that a multilevel multinomial logit model has the best fit to the data than the other two models while the results obtained from occupant-based data are more reliable than those from vehicle- and collision-based data. More importantly, while generally consistent in terms of factors that were found to be significant between different models and data aggregation methods, the effect size of each factor differ substantially, which could have significant implications for evaluating the effects of different safety-related policies and countermeasures.

✉ Taimur Usman
taimurus@gmail.com

Liping Fu
lfu@uwaterloo.ca

Luis F. Miranda-Moreno
luis.miranda-moreno@mcgill.ca

1 Department of Civil & Environmental Engineering, University of Waterloo, Waterloo, ON N2L 3G1, Canada

2 School of Transportation and Logistics, Southwest Jiaotong University, Chengdu 610031, China

3 Department of Civil Engineering & Applied Mechanics, McGill University, Montreal, QC H3A 2K6, Canada

Keywords Injury severity · Multilevel logistic regression models · Collision data aggregation

1 Introduction

The outcome of a collision is polytomous in nature such as no injury (NI), minimal injury, minor injury, major (incapacitating) injury, and fatal injury. This type of data is mostly modeled using logistic regression models. Most of the models are extensions of the multinomial logit models based on the assumption of independent severity classes [1–11]. Although different modeling methodologies are available from literature to examine collision severity as related to various influencing factors, little is known on the relative merits of these alternatives. The first objective of this research is therefore to compare three most widely used logistic regression models, namely, sequential binary logit models, ordered logit models, and multinomial logit models in a multilevel framework for injury severity analysis.

Some of the issues related to injury severity analysis are within-crash correlation, hierarchical nature of collision data, misclassification, underreporting, endogeneity, sample size, and spatial correlation [5, 11–21]. While a number of recent studies have been devoted to addressing some of these issues, the issue pertaining to the hierarchical nature of collision data has not been addressed adequately. Collision data is hierarchical in nature with possible correlation at the occupant or vehicle level. Ignoring such correlation (intra-class correlation) could lead to false estimation of standard errors and undermine the true significance of parameter estimates [22]. However, little work has been done to account for the multilevel structure of the collision data. Jones and Jørgensen [17] and Lenguerrand et al. [20] were among the first, as identified in Usman et al. [23], to recognize the need to consider the hierarchical crash-car-occupant structure of collision data for crash severity modeling. They discussed the potential issues of ignoring the clustering nature of data and the correlation within the clusters, such as erroneous estimates of model coefficients and understated standard errors and confidence intervals for the effects. They, however, did not discuss the effects of data aggregation. Their conclusions were similar to those from other disciplines such as epidemiology, social research, and political science [24–27]. The second objective of this research is therefore to evaluate the effects of data aggregation through an empirical investigation using three levels of aggregations, i.e., occupant level, vehicle level, and collision level.

This paper contributes to the literature by generating new knowledge about the implications of different modeling alternatives and data aggregation methods for collision severity analysis. The paper first describes the data used in the empirical investigation, including study sites, data sources, and data processing and integration. The three different data aggregation methods are discussed in details. Then, an overview of the three logistic regression models in the construct of the multilevel framework is provided, followed by a discussion on the model calibration process and the results. Finally, the main findings are summarized, focussing particularly on the differences from different approaches.

2 Data description

This research makes use of a collision database prepared in our previous effort [23, 28, 29]. This dataset is unique in several aspects, including reliable observations on traffic and environmental conditions when the collision occurred, and extensive spatial and temporal coverage, as described in the following section.

2.1 Study sites

A total of 31 patrol routes, each representing a highway section covered by a single maintenance unit (yard), from different regions of Ontario, Canada, were selected for this analysis as shown in Fig. 1. These sites were selected based

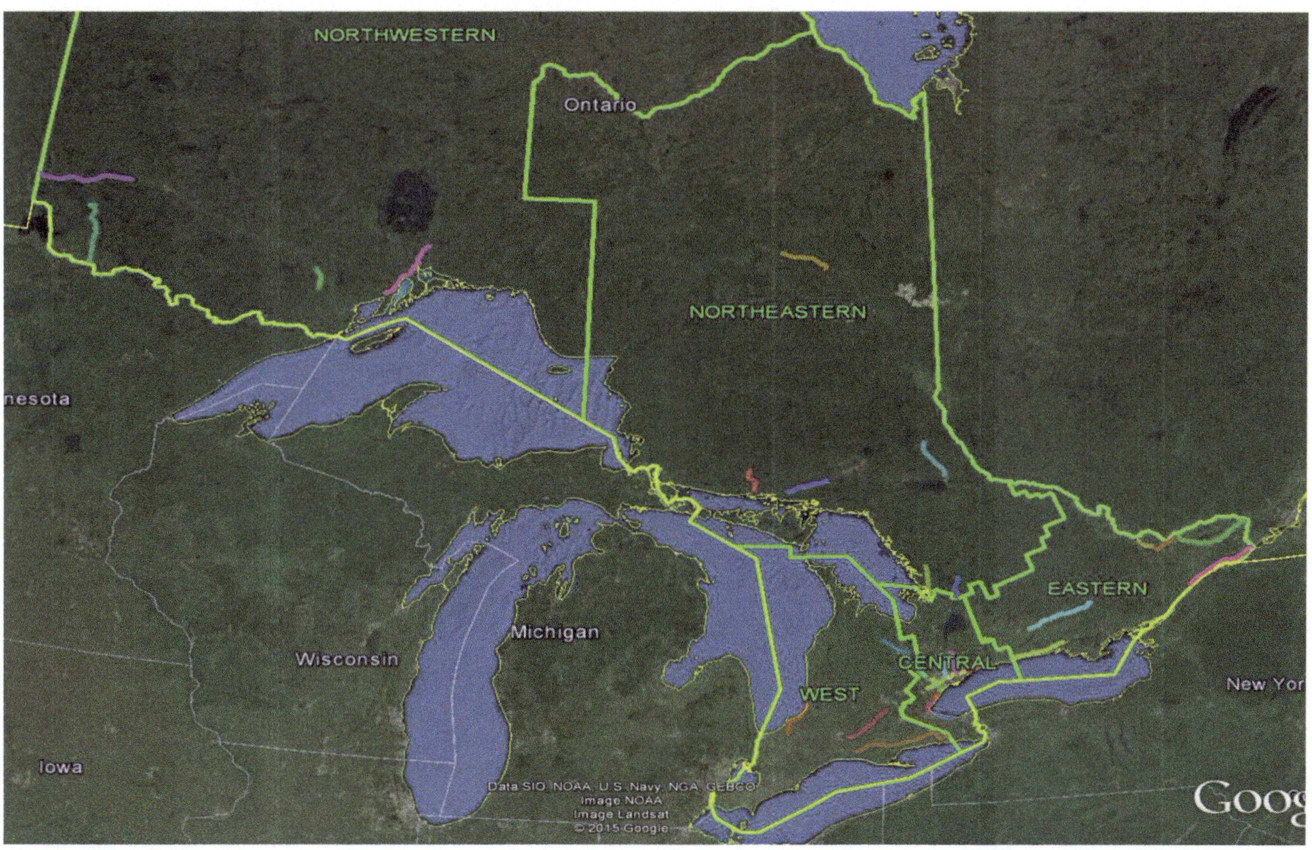

Fig. 1 Study sites

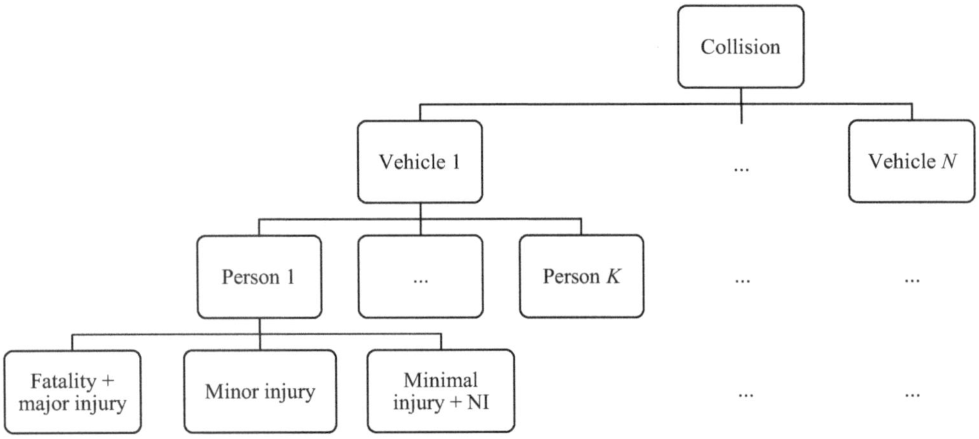

Fig. 2 Hierarchical structure of collision data

on representativeness of different classes of highways, including freeways, four-lane rural highways, and two-lane, two-way highways, and data availability.

2.2 Data sources

Collision data from six winter seasons (2000–2006) were used for collision severity analysis. Detailed description of each data source can be found in Usman et al. [29] and is also given below.

2.2.1 Traffic volume data

Hourly traffic data were obtained from two sources: Ministry of Transportation, Ontario (MTO) COMPASS system and permanent data count stations (PDCS). Both COMPASS and PDCS use loop detectors for collecting traffic data such as volume, speed, and density. The raw data from the sources were screened for any outliers caused by detector malfunction and then merged into hourly traffic volume data. In cases where multiple readings are available for a segment (e.g., from both sources and/or multiple detectors), average values are used.

2.2.2 Traffic collision data

The Ontario Provincial Police (OPP) maintains a database of all of the collisions that have been reported on Ontario highways. A database including all of the collision records for the study routes was obtained from the MTO. The database includes detailed information on each collision, including collision time, location, collision type, impact type, severity level, vehicle information, driver information, etc. One of the important data fields in these data was related to road surface condition. This variable was converted into a continuous variable—road surface index (RSI) as per the criteria set in Usman et al. [28]. This data is person-based data with an inherent multilevel structure where individuals are nested within vehicles and vehicles within collisions. The data used in this research contains 13,775 collisions involving 39,564 people in 19,635 vehicles for the six winter seasons on the selected routes.

2.2.3 Environment Canada (EC) data

Weather data from Environment Canada includes temperature, precipitation type and intensity, visibility, and wind speed. With exception of the precipitation intensity data, all other data are in hourly format. Most of the EC stations have missing data. For this reason, EC data were obtained from 302 stations for the study routes. These data were processed in three steps: In step 1, a 60 km arbitrary buffer zone was assumed around each route and all stations within this boundary were assigned to the particular route. In the next step using *t* test, EC stations were identified, which on average are similar to EC stations near the routes. In the last step, data from different EC stations around a route were converted into a single dataset by taking their arithmetic mean. It was found that arithmetic means provide better results than weighted averages.

2.3 Data processing

As described above, collision data are hierarchical with different outcomes possible for a single collision, as shown in Fig. 2. Collisions are categorized into five distinct injury severity levels as follows:

1. NI, where no injuries were sustained;
2. Minimal injury, where the victim suffered minor abrasions and complained of pain but did not go to the hospital;
3. Minor injury, where the victim was treated in the emergency room but not admitted;

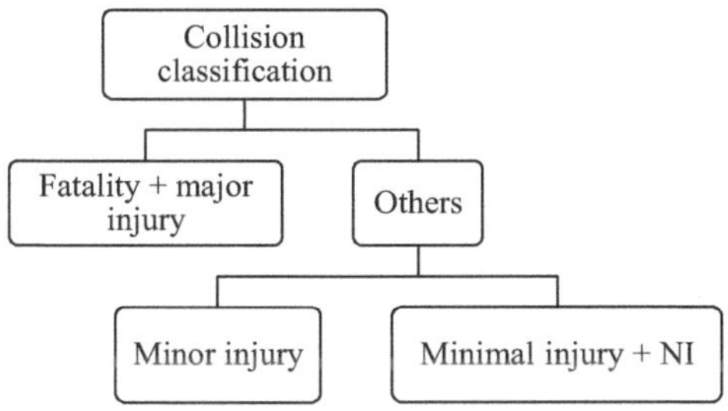

Fig. 3 Data classification scheme (vehicle- and collision-based data)

4. Major injury, where the victim was admitted to the hospital either for treatment or observation;
5. Fatality, where the victim died within 30 days of collision or on site.

Minimal injury and NI collisions were grouped together into one category because they are similar in terms of consequence. Similarly, major injuries and fatalities were also grouped into a single category. This merging of categories will also take care of the possible correlation that could exist between those closely related outcomes of a collision severity [12, 30]. The hierarchic structure of collision data is shown in Fig. 2, which shows that for a given collision, vehicles are nested within the collision and persons are nested within vehicles and each person could have a given level of severity.

Data from other sources such as weather and traffic were merged with the person-based collision data based on date, time, and location for the 31 patrol routes. A stepwise aggregation process was followed to convert the data from occupant-based records to vehicle-based, and finally to collision-based records. Three datasets were thus formed for this analysis: occupant-based dataset with three levels (occupant—vehicle—collision), vehicle-based dataset with two levels (vehicle—collision), and collision-based dataset with a single level. For the vehicle- and collision-based data, collision severity levels were assigned to the respective vehicles and collisions as per the classification scheme shown in Fig. 3. Note that this classification scheme was not used for occupant-based data as each person has a unique injury severity level.

3 Model development

Different approaches can be used for collision severity analysis: (a) incorporating severity into the collision frequency models by modeling collisions classified by severity types [31–34]; and (b) modeling the conditional probability of each severity level for a given collision [14, 15, 17, 35, 36]. In this research, we adopted the second approach for three reasons: (i) different factors could have different effects on collision occurrence and severity (e.g., seat belt use has nothing to do with collision occurrence, but is an important factor in severity analysis); (ii) data that could be used for joint models are limited in nature because most of the data are collected after the collision has happened [12]; and, (iii) consequence outcomes and injury data are at the individual, vehicle, or accident level. Three different model structures were considered for the conditional probability of a collision for each of the three datasets discussed previously.

Multilevel framework is used to account for the correlation between vehicles in a collision or persons in a vehicle. In a multilevel setting, correlation at a sub-level is taken care of by inclusion of random parameters which are constant within the sub-level but are allowed to vary at the upper levels [18, 20, 37].

3.1 Multilevel logistic regression models

The first modeling structure considered is the multilevel multinomial logit model. In this model, a base category is selected out of the different outcomes and other categories are estimated with respect to the base category. Many researchers have used multinomial logit models for accident severity analysis [1–10]. If the three severity levels are represented by 0, 1, and 2 with 0 as the reference or base category then the model structure for a three-level data structure (occupant-based data) is given by Eq. (1). The resulting models are called multilevel multinomial logit models (MML).

$$\left.\begin{aligned} \ln\left[\frac{P(Y=1/X)}{P(Y=0/X)}\right] &= \beta_{10} + \sum_{n=1}^{N} \beta_{1n} X_{1ijkn} + U_{jk} + V_k, \\ \ln\left[\frac{P(Y=2/X)}{P(Y=0/X)}\right] &= \beta_{20} + \sum_{n=1}^{N} \beta_{2n} X_{2ijkn} + U_{jk} + V_k, \end{aligned}\right\} \tag{1}$$

where P represents the probability of severity level (either 0, 1 or 2); i, j, and k represent occupant, vehicle, and collision levels, respectively; U_{jk} and V_k denote the second level (vehicle) and the third level (collision) random effect factors which are assumed to follow a logistic distribution; β is a model coefficient to be estimated; and X_{ijk} represents a set of explanatory variables at the individual level. U_{jk} remains constant for occupants within a vehicle but varies across vehicles and collisions. Similarly, V_k is constant for vehicles in a collision but varies across collisions. U_{jk} and V_k are obtained by considering the intercept as a random parameter.

The second modeling structure is the sequential binary logistic model. Collision data were divided into two mutually exclusive injury outcomes for a given collision at a given level, and binary logit models were specified at

Table 1 List of variables used in the analysis

Category	Variable	Definition
Road characteristics	Road classification	Freeway = 1, kings highway multilane = 2, kings highway two lane[c] = 3
	Road alignment	Straight on level[c] = 1, straight on hill = 2, curve on level = 3, curve on hill = 4
	Number of lanes	Number of lanes
	Collision location	Intersections = 1, segment[c] = 2, bridges/underpasses = 3
	Speed limit	km/h
Weather and environment conditions	Light	Light + dawn = 1, dark + dusk[c] = 2
	Precipitation type	Other[c] = 0, freezing rain/snow = 1
	Hourly precipitation	Precipitation intensity in "cm"
	Temperature	Measured in °C
	Wind speed	km/h
	Visibility	km
	Road surface condition	Road surface condition in winter i.e., icy, snow covered, etc. represented by RSI
	Day	Weekdays = 0, weekends[c] = 1
Vehicle	Vehicle type[b]	SUVs/car/station wagon[c] = 1, van = 2, large trucks etc. = 3
	Vehicle condition[b]	Otherwise = 0, defective[c] = 1
	Vehicle age[b]	In years
Driver/person	Driver age[b]	In years
	Driver sex[b]	Male = 1, female[c] = 2
	Driver condition at time of collision[b]	Otherwise[c] = 0, normal = 1
	Position in vehicle[a]	Front = 1, rear[c] = 2
	Safety equipment used[a]	Used safety device = 0, not used or bad use[c] = 1
Traffic	Hourly traffic volume	ln(Hourly traffic volume)

[a] Used for occupant-based data only

[b] Used for vehicle- and occupant-based data only

[c] Base category

Table 2 Collision count by severity and percent change in collision severity distribution due to data aggregation

All sites	Minimal injury + NI	Minor injury	Fatality + major injury	Total
Collision count by severity				
Occupant-based data	30,246	7,733	1,585	39,564
Vehicle-based data	12,698	5,905	1,032	19,635
Collision-based data	7,349	5,531	895	13,775
Percent change in collision severity distribution due to data aggregation				
Occupant-based data	76.45 %	19.55 %	4.00 %	
Vehicle-based data	64.67 %	30.07 %	5.26 %	
Collision-based data	53.35 %	40.15 %	6.50 %	

each level such as shown in Fig. 2. Many researchers have used binary logit models for accident severity analysis [5, 11, 13–21].

For multilevel data, the resulting model is called the multilevel sequential binary logit model (MBL). The mathematical form of the model for a three-level data structure (occupant-based data) is shown in Eq. (2):

$$\ln\left[\frac{P(Y=1)}{P(Y=0)}\right] = \beta_0 + \sum_{n=1}^{N} \beta_n X_{ijkn} + U_{jk} + V_k, \tag{2}$$

where P represents the probability of severity level (either 0, or 1).

The third modeling structure considered in this research is multilevel ordered logit model. Ordered logit models are extensions of multinomial logit models to account for the inherent ordering of severity levels in collisions, such as, from no injury to injury and to fatal [10, 38–44]. The mathematical form of a multilevel ordered logit model (MOL) for a three-level data structure (occupant-based data) is shown in Eq. (3):

$$\log\left[\frac{S^{s}_{ijk}}{S^{r}_{ijk}}\right] = \beta_0 + \beta_1 X_{ijk} + U_{jk} + V_k, \tag{3}$$

where severity (represented by "S") with superscript "r" represents the base severity against which other severity levels, denoted by superscript "s," are compared at the occupant level. The reference category could be either the least or most severe one. If Y denotes the observed severity level, Y^* the unobserved injury severity level from Eq. (3), and $\mu_1, \mu_2, \ldots, \mu_j$ the cut-off points or threshold values for the injury severity levels, then

$$\begin{aligned} Y &= 1 \text{ if } Y^* \le \mu_1, \\ Y &= 2 \text{ if } \mu_1 < Y^* \le \mu_2, \\ &\vdots \\ Y &= j \text{ if } \mu_{j-1} < Y^*. \end{aligned} \tag{4}$$

The probability of a particular injury severity level $Y = j$ can be estimated using Eq. (5) [45]:

$$\begin{aligned} P(Y=j) &= P(\mu_{j-1} < Y^* < \mu_j) \\ &= \frac{\exp(\mu_j - \sum \beta_k x_k)}{1+\exp(\mu_j - \sum \beta_k x_k)} - \frac{\exp(\mu_{j-1} - \sum \beta_k x_k)}{1+\exp(\mu_{j-1} - \sum \beta_k x_k)}, \end{aligned} \tag{5}$$

where β_k are model coefficients to be estimated and $\{X_1, X_2, \ldots X_k\}$ represents a set of explanatory variables. An important aspect of ordered logit models is the proportional odds (or parallel slopes) assumption, where the variables are assumed to have the same slope across all levels of severity/outcome [46–48] with the exception of the intercept [49]. Results of ordered logit models are therefore unidirectional (show either an increase or decrease in severity) and are thus very easy to interpret. This unidirectional effect can sometimes lead to undesirable effects where a variable could cause the probability of high or low severity collision to increase at the cost of the other [38].

The presence of correlation is confirmed by calculating the intra-class correlation (correlation among observations within the same cluster). Intra-class correlation, denoted by ρ, is a coefficient with values ranging from 0 to 1 and is calculated as the ratio of the variance at the sub-level to the total variance [23, 50, 51] as given in Eq. (6):

$$\rho = \frac{\sigma^2_{\text{sub-level}}}{\sum \sigma^2}. \tag{6}$$

The higher the value of ρ, the greater the correlation is and the higher the consequences of ignoring it will be [30]. For details on how ρ can be calculated, readers are referred to e.g., Jones and Jørgenson [18].

3.2 Exploratory data analysis

There are a large number of factors that influence the severity of collisions under winter conditions [52, 53]. The main factors can be grouped into three categories, namely road driving conditions, vehicle characteristics, and driver attributes. Road driving conditions include road geometry, environment, and pavement surface conditions. The latter are affected by weather and maintenance operations. Different sets of variables were considered in analyzing the three datasets as listed in Table 1.

Table 2 provides a summary of collision counts by severity for the different datasets and the changes in the proportions of different types of injury severity levels due to aggregation at each step.

As shown in Fig. 2, a collision may involve several vehicles and the occupants of an involving vehicle may experience different levels of injury severity. As a result, modeling the collision severity at the collision level will result in a loss of information and misrepresentation of certain severity levels, as show in Table 2. For example, if we aggregate data for a collision with three fatal injuries and two vehicles involved, the fatality count for occupant-, vehicle-, and collision-based datasets will be three (03), two (02), and one (01), respectively.

4 Model calibration and results

MLwin[1] was used to calibrate the three alternative models discussed in Sect. 3. Tables 3 through 5 provide the calibration results for collision-based data, vehicle-based data, and occupant-based data. MLWin uses Quasi-likelihood for models with discrete dependent variables and thus the reported likelihood estimates are only approximate leading to unreliable likelihood ratio tests [54]. A positive sign is used as an indicator of increase in severity level with respect to the associated variable. Results from all the models are consistent in terms of the direction of their effect on severity; however, effect of the size of coefficient varies across different models and aggregation levels. For

[1] Rasbash, J., Charlton, C., Browne, W.J., Healy, M. and Cameron, B. (2005) *MLwin Version 2.22*. Centre for Multilevel Modeling, University of Bristol.

Table 3 Modeling results for collision-based data

Categories	Variable	MBL fatal		MBL minor		MOL		MML—fatality versus NI		MML—minor versus NI	
		Coeff.	Sig.	Coeff.	Sig.	Coeff.	Sig.	Coeff.	Sig.	Coeff.	Sig.
	Constant	−0.856	0.021	0.020	0.917			0.034	0.926	0.016	0.931
Road type	Freeways	−0.114	0.476	0.017	0.875	−0.159	0.101	−0.101	0.523	0.027	0.789
	Multilane kings	0.287	0.089	−0.038	0.772	−0.044	0.704	0.262	0.123	−0.032	0.791
	2 Lane kings	0.000		0.000		0.000		0.000		0.000	
Other	Light/dawn	−0.207	0.004					−0.181	0.009		
	Dark/dusk	0.000						0.000			
	Weekdays	−0.197	0.009	−0.097	0.018	−0.137	0.000	−0.247	0.001	−0.096	0.014
	Weekends	0.000		0.000		0.000		0.000		0.000	
Road related	Road alignment—straight on hill			0.186	0.004	0.180	0.003			0.172	0.004
	Road alignment—curve on level			−0.004	0.955	0.014	0.832			−0.008	0.902
	Road alignment—curve on hill			−0.021	0.795	−0.038	0.617			−0.014	0.852
	Road alignment—straight on level			0.000		0.000				0.000	
	Speed limit	0.013	0.001	0.008	0.000	0.010	0.000	0.014	0.000	0.008	0.000
	Number of lanes	−0.103	0.000	−0.083	0.000	−0.087	0.000	−0.127	0.000	−0.082	0.000
	RSI	0.317	0.018					0.309	0.018		
Weather	Weather—freezing rain, snow					−0.085	0.048				
	Weather—other					0.000					
	Wind speed (km/h)			−0.005	0.012	−0.005	0.012			−0.005	0.012
Traffic	ln(Traffic)	−0.320	0.000	−0.056	0.005	−0.135	0.000	−0.339	0.000	−0.054	0.004
	μ_1					−0.881	0.000				
	μ_2					1.742	0.000				
	−2 × log likelihood (null)	508.443		18,437.2		19,301.4		19,301.4			
	−2 × log likelihood (full)	−3,340.050		17,892.1		15,804		14,944			
	Collision level observations	13,775		12,880		13,775		13,775			

MBL = multilevel binary logit models
MOL = multilevel ordered logit models
MML = multilevel multinomial logit models

Table 4 Modeling results for vehicle-based data

Categories	Variable	MBL fatal		MBL minor		MOL		MML—fatality vserus NI		MML—minor versus NI	
		Coeff.	Sig.	Coeff.	Sig.	Coeff.	Sig.	Coeff.	Sig.	Coeff.	Sig.
	Constant	−0.488	0.260	0.397	0.042			0.487	0.248	0.397	0.037
Road type	Freeways	0.029	0.868	0.165	0.102	−0.026	0.780	0.111	0.519	0.179	0.062
	Multilane kings	0.297	0.112	0.030	0.804	−0.011	0.920	0.318	0.087	0.037	0.746
	2 Lane kings	0.000		0.000		0.000		0.000		0.000	
Other	Light/dawn	−0.175	0.021					−0.184	0.012		
	Dark/dusk	0.000						0.000			
	Weekdays	−0.159	0.047					−0.163	0.034		
	Weekends	0.000						0.000			
	Collision location—intersections	−0.148	0.294			0.134	0.031	−0.227	0.100		
	Collision location—bridges/underpasses	0.665	0.021			−0.240	0.124	0.707	0.009		
	Collision location—segment	0.000				0.000		0.000			
Road related	Road alignment—straight on hill			0.202	0.000	−0.193	0.001			0.198	0.000
	Road alignment—curve on level			0.146	0.032	−0.159	0.013			0.137	0.032
	Road alignment—curve on hill			0.125	0.105	−0.098	0.185			0.129	0.077
	Road alignment—straight on level			0.000		0.000				0.000	
	Speed limit	0.014	0.000	0.008	0.000	−0.008	0.000	0.015	0.000	0.008	0.000
	Number of lanes	−0.109	0.000	−0.080	0.000	0.085	0.000	−0.132	0.000	−0.080	0.000
	RSI			−0.322	0.000	0.266	0.000			−0.312	0.000
Driver	Driver age (years)	0.007	0.020	0.002	0.046	−0.003	0.003	0.008	0.008	0.003	0.003
	Driver—male	0.139	0.078	−0.241	0.000	0.175	0.000			−0.242	0.000
	Driver—female	0.000		0.000		0.000				0.000	
	Driver condition—normal	−0.605	0.000	−0.170	0.000	0.285	0.000	−0.711	0.000	−0.177	0.000
	Driver condition—other (drinking etc.)	0.000		0.000		0.000		0.000		0.000	
Vehicle	Vehicle age (years)			0.013	0.001	−0.012	0.000			0.013	0.000
	Vehicle type—vans	0.192	0.021	0.116	0.003	−0.129	0.000	0.217	0.007	0.112	0.003
	Vehicle type—large trucks etc	0.042	0.771	−0.612	0.000	0.495	0.000	−0.089	0.519	−0.595	0.000
	Vehicle type—car/station wagon	0.000		0.000		0.000		0.000		0.000	
Weather	Wind speed (km/h)			−0.006	0.003	0.006	0.003			-0.006	0.003
Traffic	ln(Traffic)	−0.384	0.000	−0.139	0.000	0.201	0.000	−0.447	0.000	−0.141	0.000
	μ_1					−1.195	0.000				
	μ_2					1.230	0.000				
	−2 × log likelihood (null)	−3,834.8		24,302.0		20,777.2		20,777.2			
	−2 × log likelihood (full)	−11,816.4		22,859.8		13,911.2		11,398.3			
	Variance at collision level	2.352	0.000	0.199	0.000	2.198	0	2.248		0.000	

Table 4 continued

Categories	Variable	MBL fatal		MBL minor		MOL		MML—fatality vserus NI		MML—minor versus NI	
		Coeff.	Sig.	Coeff.	Sig.	Coeff.	Sig.	Coeff.	Sig.	Coeff.	Sig.
	Variance at vehicle level	3.29	0	3.29	0	3.29	0	3.29		0	
	Collision level observations	13,775.0		13,060.0		13,775.0		13,775			
	Vehicle level observations	19,635.0		18,603.0		19,635.0		19,635			

evaluating the effect of individual factors, their elasticities are calculated and given in Table 6. For a continuous variable X_{ki}, elasticity for a particular collision severity outcome "i" is computed as

$$E_{X_{ki}}^{P(i)} = [1 - P(i)]\beta_{ki}X_{ki}, \tag{7}$$

where $P(i)$ is the probability of collision severity outcome "i," and β_{ki} is the coefficient associated with variable X_{ki}. For categorical variables elasticity is calculated as $E = (\exp\beta - 1)/\exp\beta$ [3, 7, 8]. Table 7 gives values predicted from the models and the observed severity ratios.

4.1 Comparison of quality of fitting

As explained in the previous section, likelihood estimates from MLWin are approximate and the usual goodness of fit criterion such as Akaike Information Criterion (AIC) and Bayesian Information Criterion (BIC) could not be applied [54]. AIC [55], defined as $-2LL + 2p$, is a test statistics used to identify the best fit model from a set of models. The term LL is the log likelihood of a fitted model and p the number of parameters, which is included to penalize models with higher number of parameters. A model with smaller AIC value represents a better overall fit. Similarly, Bayesian Information Criterion (BIC) [56], defined as $-2LL + p\ln(n)$, which is another test statistics and a variation of AIC, is used to identify the best fit model from a set of models. The term "n" represents the number of observations used to calibrate the model. A model with smaller BIC value represents a better overall fit. Alternatively, results from the models were compared to the actual observations and it was found MML models have a better prediction performance compared to MOL models except for collision-based fatalities where MOL has a slightly better prediction. Similarly, MML models have better prediction results compared to MBL models for occupant- and vehicle-based data. For collision-based data, MBL results are slightly better for NI + minimal injury and minor injury collisions, whereas for fatality collisions, MML results are closer to the observed severity ratios. Based on the discussion in this section, MML is found to perform better as a whole than MBL and MOL.

4.2 Effects of data aggregation and correlation

If the collision data are used at a disaggregated level of analysis such as occupant based or vehicle based, then efforts should be made to account for the correlation that exists between occupants in a vehicle or vehicles in a collision such as shown by the variance terms in Tables 4 and 5. Occupant-based results (Table 5) show that around 79 % of the variation ($\rho = 0.79$) is accounted for at the

Table 5 Modeling results for occupant-based data

Categories	Variable	MBL fatal		MBL minor		MOL		MML—fatality versus NI		MML—minor versus NI	
		Coeff.	Sig.	Coeff.	Sig.	Coeff.	Sig.	Coeff.	Sig.	Coeff.	Sig.
	Constant	0.341	0.504	0.559	0.003			1.349	0.009	0.561	0.002
Road type	Freeways	0.054	0.791	0.105	0.259	−0.012	0.892	0.048	0.812	0.130	0.149
	Multilane kings	0.340	0.134	0.103	0.353	−0.085	0.409	0.297	0.187	0.094	0.380
	2 Lane kings	0.000		0.000		0.000		0.000		0.000	
Other	Light/dawn	−0.159	0.058			0.057	0.075	−0.166	0.043		
	Dark/dusk	0.000				0.000		0.000			
	Collision location—intersections	−0.151	0.336			0.133	0.024	−0.189	0.223		
	Collision location—bridges/underpasses	0.823	0.015			−0.247	0.100	0.873	0.007		
	Collision location—segment	0.000				0.000		0.000			
Road related	Road alignment—straight on hill			0.131	0.017	−0.131	0.013			0.133	0.012
	Road alignment—curve on level			0.180	0.004	−0.193	0.002			0.165	0.007
	Road alignment—curve on hill			0.198	0.007	−0.168	0.018			0.190	0.007
	Road alignment—straight on level			0.000		0.000				0.000	
	Speed limit	0.016	0.001	0.009	0.000	−0.008	0.000	0.018	0.000	0.009	0.000
	Number of lanes	−0.122	0.000	−0.079	0.000	0.084	0.000	−0.141	0.000	−0.078	0.000
	RSI			−0.295	0.000	0.248	0.000			−0.306	0.000
Driver	Driver age (years)	0.007	0.000	0.004	0.000	−0.004	0.000	0.008	0.000	0.004	0.000
	Driver—male			−0.377	0.000	0.329	0.000			−0.381	0.000
	Driver—female			0.000		0.000				0.000	
	Driver condition—normal	−0.561	0.000			0.147	0.000	−0.590	0.000		
	Driver condition—other (drinking etc.)	0.000				0.000		0.000		0.000	
Vehicle	Vehicle age (years)			0.018	0.000	−0.016	0.000			0.017	0.000
	Vehicle type—vans			−0.287	0.000	0.263	0.000	−0.190	0.037	−0.288	0.000
	Vehicle type—large trucks etc			−0.910	0.000	0.790	0.000	−0.442	0.001	−0.886	0.000
	Vehicle type—car/station wagon			0.000		0.000		0.000		0.000	
	Vehicle condition—non-defective	−0.411	0.031			0.156	0.063	−0.419	0.024		
	Vehicle condition—defective	0.000				0.000		0.000			
Person	Position in vehicle—front			0.241	0.000	−0.215	0.000	0.166	0.062	0.244	0.000
	Position in vehicle—rear			0.000		0.000		0.000		0.000	
	Safety equipment—used	−0.826	0.000	−0.733	0.000	0.836	0.000	−1.070	0.000	−0.660	0.000
	Safety equipment—not or bad used	0.000		0.000		0.000		0.000		0.000	

Table 5 continued

Categories	Variable	MBL fatal		MBL minor		MOL		MML—fatality versus NI		MML—minor versus NI	
		Coeff.	Sig.	Coeff.	Sig.	Coeff.	Sig.	Coeff.	Sig.	Coeff.	Sig.
Weather	Wind speed (km/h)			−0.006	0.003	0.005	0.012			−0.006	0.003
	Visibility (km)			−0.004	0.046	0.006	0.003			−0.004	0.046
	Hourly precipitation (cm/h)					0.169	0.039				
Traffic	ln(Traffic)	−0.418	0.000	−0.166	0.000	0.208	0.000	−0.474	0.000	−0.169	0.000
	μ_1					−1.566	0.000				
	μ_2					0.702	0.001				
	−2 × log likelihood (null)	−24,094.8		37,806.7		13,896.6		13,896.6			
	−2 × log likelihood (full)	−44,556.4		33,338.5		−4,166.57		−11,450.9			
	Variance at collision level	8.179	0.000	0.782	0.000	7.561	0	7.832		0	
	Variance at vehicle level	0	0	0.1	0	0	0	0		0	
	Variance at occupant level	3.29	0	3.29	0	3.29	0	3.29		0	
	Collision level observations	13,775		13,375		13,775		13,775			
	Vehicle level observations	19,635		19,046		19,635		19,635			
	Occupant level observations	39,564		37,979		39,564		39,564			

occupant level, whereas the collision level accounts for 19 % of the variation ($\rho = 0.19$) and vehicle level for 2 % ($\rho = 0.02$). Similarly, vehicle-based results (Table 4) show that around 94 % of the variation ($\rho = 0.94$) is accounted for at the vehicle level, whereas the collision level accounts for 0.06 % of the variation ($\rho = 0.06$). This flexibility offered by multilevel modeling improves the reliability of the modeling results obtained with such models as compared to single-level models [57–59].

Data used in a collision level severity analysis are, however, aggregated to the level of a collision. This takes care of the correlation within the data but can result in some immediate problems: (i) loss of information by reducing the number of observations, (ii) miss-specification of collision attributes resulting in erroneous share of high severity levels (Table 2), and (iii) the incapability to analyze different variables related to individual persons or vehicles at aggregate level such as seat belt use, position in the vehicle, vehicle age and type, etc. These could result in biased parameter estimates (see e.g., Mensah and Hauer [60] for some of these issues in collision frequency modeling). In this research, we utilized the multilevel framework to account for the correlation between occupants in the same vehicle and vehicles in the same collision. Treating occupant-based data results as the base case we compare modeling results from MML models for the three datasets.

The percent change in parameter estimates for fatality and major injury collisions show a difference ranging from −131 % to 214 % (average = 13 %) between occupant-based (as the base case) and vehicle-based and −9 % to 310 % (average = 62 %) between occupant-based and collision-based data. The difference between vehicle-based data (as the base case) and collision-based data is −52 % to 191 % (average reduction in size of the parameter estimate = 28 %). For minor injuries the difference is from −49 % to 139 % (average = 20 %) between occupant-based and vehicle-based data and from −29 % to 134 % (average = 54 %) between occupant-based and collision-based data, whereas for vehicle-based data (as the base case) and collision-based data this difference is from −3 % to 186 % (average = 64 %). This shows that aggregating the data results in underestimation of the parameters estimates. This could be of grave consequences if the purpose of the analysis is to evaluate the effects of some policies through some variables, in which case precise estimation of the magnitude of the parameter for the variable is of great importance. Besides data aggregation, another reason for this is the model setting (Table 1) where it can be seen that not all the variables used in the occupant-based data model are used for the other two level of aggregation. This will also result in parameter estimates for the remaining variables to be rescaled. This is evident from the results as well

Table 6 Elasticities for the three datasets

Variable	MBL fatal	MBL minor	MOL	MML—fatality versus NI	MML—minor versus NI
Occupant-based model					
Freeways	0.053	0.1	0.012	0.047	0.122
Multilane kings	0.288	0.098	0.089	0.257	0.09
Light/dawn	−0.172		−0.055	−0.181	
Accident location—intersections	−0.163		−0.125	−0.208	
Accident location—bridges/underpasses	0.561		0.28	0.582	
Road alignment—straight on hill		0.123	0.14		0.125
Road alignment—curve on level		0.165	0.213		0.152
Road alignment—curve on hill		0.18	0.183		0.173
Driver age (years)	0.261	0.122	0.122	0.297	0.121
Driver—male		−0.458	−0.28		−0.464
Driver condition—normal	−0.752		−0.137	−0.804	
Vehicle age (years)		0.072	0.064		0.068
Vehicle type—vans		−0.332	−0.231	−0.209	−0.334
Vehicle type—large trucks etc		−1.484	−0.546	−0.556	−1.425
Vehicle condition—non-defective	−0.508		−0.144	−0.52	
Position in vehicle—front		0.214	0.24	0.153	0.217
Safety equipment—used	−1.284	−1.081	−0.567	−1.915	−0.935
Speed limit	1.492	0.685	0.608	1.668	0.68
Number of lanes	−0.838	−0.443	−0.47	−0.963	−0.434
RSI		−0.193	−0.162		−0.198
Wind speed (km/h)		−0.079	−0.066		−0.079
Visibility (km)		−0.052	−0.078		−0.052
Hourly precipitation (cm/h)			−0.01		
ln(traffic)	−3.282	−1.064	−1.331	−3.699	−1.075
Vehicle-based model					
Freeways	0.029	0.152	0.026	0.105	0.164
Multilane kings	0.257	0.03	0.011	0.272	0.036
Light/dawn	−0.191			−0.202	
Weekdays	−0.172			−0.177	
Collision location—intersections	−0.16		−0.125	−0.255	
Collision location—bridges/underpasses	0.486		0.271	0.507	
Road alignment—straight on hill		0.183	0.213		0.18
Road alignment—curve on level		0.136	0.172		0.128
Road alignment—curve on hill		0.118	0.103		0.121
Driver age (years)	0.253	0.053	0.079	0.29	0.08
Driver—male	0.13	−0.273	-0.161		−0.274
Driver condition—normal	−0.831	−0.185	−0.248	−1.036	−0.194
Vehicle age (years)		0.054	0.05		0.054
Vehicle type—vans	0.175	0.11	0.138	0.195	0.106
Vehicle type—large trucks etc	0.041	−0.844	−0.39	−0.093	−0.813
Speed limit	1.27	0.527	0.529	1.362	0.533
Number of lanes	−0.722	−0.385	−0.411	−0.876	−0.389
RSI		−0.18	−0.149		−0.176
Wind speed (km/h)		−0.068	−0.068		−0.069
ln(Traffic)	−2.909	−0.765	−1.111	−3.388	−0.784
Collision-based model					
Freeways	−0.121	0.017	−0.147	−0.106	0.027

Table 6 continued

Variable	MBL fatal	MBL minor	MOL	MML—fatality versus NI	MML—minor versus NI
Multilane kings	0.249	−0.039	−0.043	0.23	−0.033
Light/dawn	−0.23			−0.198	
Weekdays	−0.218	−0.102	−0.128	−0.28	−0.101
Road alignment—straight on hill		0.17	0.197		0.158
Road alignment—curve on level		−0.004	0.014		−0.008
Road alignment—curve on hill		−0.021	−0.037		−0.014
Weather—freezing rain, snow			−0.081		
Speed limit	1.19	0.454	0.544	1.275	0.463
Number of lanes	−0.69	−0.345	−0.346	−0.847	−0.348
RSI	0.242			0.235	
Wind speed (km/h)		−0.036	−0.035		−0.037
ln(Traffic)	−2.412	−0.262	−0.604	−2.542	−0.257

Table 7 Prediction results from models versus observed results

Severity type	MBL (%)	MML (%)	MOL (%)	Observed (%)
Occupant-based model				
NI + minimal injury	78.3	77	80.0	76.4
Minor injury	19.9	20.5	17.5	19.5
Fatal + major injury	1.8	2.5	2.5	4.0
Vehicle-based model				
NI + minimal injury	63.5	64.3	69.1	64.7
Minor injury	31.2	30.4	27.1	30.1
Fatal + major injury	5.3	5.3	3.8	5.3
Collision-based model				
NI + minimal injury	55.1	55.9	56.9	53.4
Minor injury	40.7	39.4	37.9	40.2
Fatal + major injury	4.2	4.7	5.2	6.5

where the range is wider for the difference between occupant- and collision-based data than those from occupant- and vehicle-based data.

4.3 Comparison of significant factors

Despite different in quality of fitting and effect sizes of various safety factors from different models and data aggregation methods, there were consistent results in terms of the factors that were found to have statistically significant effect on collision severity. This section discusses the main findings on the contributing factors and the magnitude of their effects (Tables 3 through 6).

4.3.1 Driver characteristics and accident impact type

One percent change in driver age will cause an average increase of 0.297 in the probability of suffering a fatal/major injury and 0.121 increases in the probability of having minor injuries. For male drivers, the probability of suffering minor injuries are 0.46 less compared to female drivers. Alcohol can increase the probability of fatality/major injuries by 0.80. Collisions on bridges increase the probability of fatality/major injuries by 0.58, whereas those occurring at intersections reduce it by 0.21.

4.3.2 Road characteristics

Multilane-divided highways increase the probabilities of fatality/major injuries by 0.26 and minor injuries by 0.09, whereas for freeways these figures are 0.05 and 0.12 compared to undivided two-lane highways. Improvement in road surface condition causes the probability of minor injuries to reduce by 0.20. The presence of curves or hilly terrain increases the probability of minor injuries from 0.12 to 0.17. Increase in number of lanes decreases the probability of fatal/major injuries by 0.96 and minor injuries by 0.43. Increase in speed limit increases the probability of fatality/major injuries by 1.67 and minor injuries by 0.68.

4.3.3 Vehicle and individual

Heavy weight and non-defective vehicles decrease the probability of fatal/major injuries from 0.21 to 0.56 and minor injuries by 0.33–1.43. Increase in the age of a vehicle increases the chances of minor injuries by 0.07. Front position increases the chances of fatal/major injuries by 0.15 and minor injuries by 0.22, whereas the use of safety devices decreases the chances of fatal/major injuries by 1.92 and minor injuries by 0.94.

4.3.4 Weather and environment

Increase in wind speed and visibility decreases the probability of minor injuries by 0.08 and 0.05. The presence of lighting conditions reduces the chances of fatality/major injuries by 0.18.

4.3.5 Traffic volume

Traffic volume is the most influential factor of all and an increase in traffic volume decreases the probability of fatal/major injuries by 3.70 and minor injuries by 1.08. Intuitively, a higher traffic volume will lead to more congestion resulting in lower speeds.

5 Conclusions and future research

Three alternative logistic regression models, namely multinomial logit model, sequential binary logit model, and ordered logit model applied in a multilevel framework, were compared and evaluated for their performance for predicting the conditional probabilities of different severity levels of a given collision. These models were applied to collision data aggregated at three levels—occupant level, vehicle level, and collision level. These three levels were used to evaluate the effects of data aggregation and correlation on collision severity analysis. Collision data from six winter seasons (2,000–2,006) and 31 sites containing 13,775 collisions, involving 39,564 individuals and 19,635 vehicles was used for this analysis. Based on the modeling results, it was found that multilevel multinomial logit (MML) has the best overall fit to the data, and occupant-based data results are more reliable than vehicle- and collision-based data.

Moreover, it was found that data aggregation affects the parameter estimates, on the average, by as much as 13 % for vehicle-based aggregated data and 62 % collision-based aggregated data compared to occupant-based data. Similarly, from correlation perspective, around 79 % of the variation is accounted for when using occupant-based data compared to the 19 % variation accounted for by collision-based data. This could have significant implications for evaluating the effects of different safety-related policies and countermeasures when using, showing the importance of data analysis at a disaggregate level.

Our future efforts will be directed toward the comparison of data compiled from winter seasons and snow storm events using the results from this research. Moreover, other modeling types such as latent class models will also be evaluated and compared to the modeling results from this analysis.

Acknowledgments This research was supported by MTO in part through the Highway Infrastructure and Innovations Funding Program (HIIFP). The authors wish to acknowledge in particular the assistance of Max Perchanok from MTO.

References

1. Shankar V, Mannering F (1996) An exploratory multinomial logit analysis of single-vehicle motorcycle accident severity. J Saf Res 27(3):183–194
2. Lee, J. and F. Mannering (1999). Analysis of roadside accident frequency and severity and roadside safety management. Final Research Report. Research Project T9903, Task 97. Report Number WA-RD 475.1
3. Lee J, Mannering F (2002) Impact of roadside features on the frequency and severity of run-off-roadway accidents: an empirical analysis. Accid Anal Prev 34:149–161
4. Ulfarsson GF, Mannering FL (2004) Differences in male and female injury severities in sport-utility vehicle, minivan, pickup and passenger car accidents. Accid Anal Prev 36:135–147
5. Holdridge MJ, Shankar VN, Ulfarsson GF (2005) The crash severity impacts of fixed roadside objects. J Saf Res 36:139–147
6. Khorashadi A, Niemeier D, Shankar V, Mannering F (2005) Differences in rural and urban driver-injury severities in accidents involving large-trucks: an exploratory analysis. Accid Anal Prev 37:910–921
7. Ulfarsson GF, Kim S, Lentz ET (2006) Factors affecting common vehicle-to-vehicle collision types road safety priorities in an aging society. Transp Res Rec 1980:70–78
8. Malyshkina NV, Mannering F (2008) Effect of increases in speed limits on severities of injuries in accidents. Transp Res Rec 2083:122–127
9. Miranda-Moreno, Luis F., Liping Fu, Satish Ukkusuri, and Dominique Lord (2009). How to incorporate accident severity and vehicle occupancy into the hotspot identification process? 88th Annual Meeting of the Transportation Research Board. Paper No. 09 -2824
10. Mergia WY (2010) Exploring factors contributing to injury severity at freeway merging and diverging areas. MSc Thesis, University of Dayton
11. Lee C, Abdel-Aty M (2008) Two-level nested logit model to identify traffic flow parameters affecting crash occurrence on freeway ramps. Transp Res Rec 2083:145–152
12. Savolainen PT, Mannering FL, Lord D, Quddus MA (2011) The statistical analysis of highway crash-injury severities: a review and assessment of methodological alternatives. Accid Anal Prev 43(5):1666–1676
13. Nassar SA, Saccomanno FF, Shortreed JH (1994) Disaggregate analysis of road accident severities. Int J Impact Eng 15(6):815–826
14. Saccomanno FF, Nassar SA, Shortreed JH (1996) Reliability of statistical road accident injury severity models. Transp Res Rec 1542:14–23
15. Shankar V, Mannering F, Barfield W (1996) Statistical analysis of accident severity on rural freeways. Accid Anal Prev 28(3):391–401
16. Carson J, Mannering F (2001) The effect of ice warning signs on ice-accident frequencies and Severities. Accid Anal Prev 33:99–109

17. Dissanayake S, Lu J (2002). Analysis of severity of young driver crashes sequential binary logistic regression modeling. Transportation Research Record 1784. 108–114. Paper No. 02-2302
18. Jones AP, Jørgensen SH (2003) The use of multilevel models for the prediction of road accident outcomes. Accid Anal Prev 35:59–69
19. Donnell ET, Mason JM Jr (2004) Predicting the severity of median-related crashes in pennsylvania by using logistic regression. Transp Res Rec 1897:55–63
20. Lenguerrand E, Martin JL, Laumon B (2006) Modelling the hierarchical structure of road crash data—Application to severity analysis. Accid Anal Prev 38:43–53
21. Milton JC, Shankar VN, Mannering FL (2008) Highway accident severities and the mixed Logit model: an exploratory empirical analysis. Accid Anal Prev 40:260–266
22. Gibbons RD, Hedeker D (1997) Random effects probit and logistic regression models for three-level data. Biometrics 53:1527–1537
23. Usman T, Fu L, Miranda-Moreno L (2011) Accident prediction models for winter road safety: does temporal aggregation of data matters? Transp Res Rec 2237:144–151
24. Ronald H. H., Thomas and S. Loring (2000). "An Introduction to multilevel modeling techniques quantitative methodology series"
25. Newsom, J.T., and Nishishiba, M., (2002). Hierarchical linear modeling of dyadic data. nonconvergence and sample bias in hierarchical linear modeling of dyadic data, 2004, http://www.upa.pdx.edu/IOA/newsom/mlrdyad4.doc Accessed 29 Mar 2010
26. Schreiber JB, Griffin BW (2004) Review of multilevel modeling and multilevel studies in The Journal of Educational Research (1992–2002). J Educ Res 98:24–33
27. Gelman A, Hill J (2007) Data analysis using regression and multilevel/hierarchical models. Cambridge University Press, New York
28. Usman T, Fu L, Miranda-Moreno Luis F (2010) Quantifying safety benefit of winter road maintenance: accident frequency modeling. Accid Anal Prev 42(6):1878–1887
29. Usman T, Fu L, Miranda-Moreno LF (2012) A disaggregate model for quantifying the safety effects of winter road maintenance activities at an operational level. J Accid Anal Prev 48:368–378
30. Hutchings C, Knight S, Reading JC (2003) The use of generalized estimating equations in the analysis of motor vehicle crash data. Accident Anal. Prev. 35(1):3–8
31. Bijleveld FD (2005) The covariance between the number of accidents and the number of victims in multivariate analysis of accident related outcomes. Accid Anal Prev 37(4):591–600
32. Ma J, Kockelman KM (2006) Bayesian multivariate Poisson regression for models of injury count, by severity. Transp Res Rec 1950:24–34
33. Ma J, Kockelman KM, Damien P (2008) A multivariate Poisson-lognormal regression model for prediction of crash counts by severity, using Bayesian methods. Accid Anal Prev 40(3):964–975
34. Park ES, Lord D (2007) Multivariate Poisson-lognormal models for jointly modeling crash frequency by severity. Transp Res Rec 2019:1–6
35. Wong J, Chung Y (2008) Comparison of methodology approach to identify causal factors of accident severity. Transp Res Rec 2083:190–198
36. Yau KKW (2004) Risk factors affecting the severity of single vehicle traffic accidents in Hong Kong. Accid Anal Prev 36(2004):333–340
37. Rasbash J, Steele F, Browne WJ, Goldstein H (2009) A User's Guide to MLwiN, version 2.10 Centre for Multilevel Modelling. University of Bristol, London
38. Savolainen P, Mannering F (2007) Probabilistic models of motorcyclists' injury severities in single- and multi-vehicle crashes. Accid Anal Prev 39:955–963
39. O'Donnell CJ, Connor DH (1996) Predicting the Severity of Motor Vehicle Accident Injuries Using Models of Ordered Multiple Choice. Accid Anal Prev 28(6):739–753
40. Khattak A, Kantor P, Council FM (1998) Role of adverse weather in key crash types on limited access: roadways implications for advanced weather systems. Transp Res Rec 1621:10–19
41. Quddus MA, Wang C, Ison SG (2010) Road traffic congestion and crash severity: econometric analysis using ordered response models. J Transp Eng 136(5):424–435
42. Srinivasan, K. K. (2002). Injury severity analysis with variable and correlated thresholds ordered mixed logit formulation. Transportation Research Record 1784 Paper No. 02-3805
43. Wang X, Kockelman KM (2005) Occupant injury severity using a heteroscedastic ordered logit model: distinguishing the effects of vehicle weight and type. Transp Res Rec 1908:195–204
44. Zhang H (2010). Identifying and quantifying factors affecting traffic crash severity in louisiana. Ph.D. dissertation, Louisiana State University
45. Train KE (2009) Discrete choice methods with simulation, 2nd edn. Cambridge University Press, New York
46. Kosmelj K and Vadnal K (2003). Comparison of two generalized logistic regression models; a case study. In: 25th international conference on information technology interfaces IT1 2003, June 16-1 9, 2003, Cavtat, Croatia
47. Kamarudin MNBC, Ahmad I, Zaharim A, Abdullah S, Kamarudin H (2007). A comparison on two generalized logistic regression models: a case study on failure mode for multiple reflow effect on ball grid array (BGA) application. In: regional conference on engineering mathematics, mechanics, manufacturing & architecture
48. Dissanayake S (2004) Comparison of severity affecting factors between young and older drivers involved in single vehicle crashes. IATSS Res 28(2):48–54
49. Jung S, Qin X, Noyce DA (2010) Rainfall effect on single-vehicle crash severities using polychotomous response models. Accid Anal Prev 42(2010):213–224
50. McGraw KO, Wong SP (1996) Forming inferences about some intraclass correlation coefficients. Psychol Methods 1(1):30–46
51. Newsom JT, Nishishiba M 2002. Hierarchical linear modeling of dyadic data. Non-convergence and sample bias in hierarchical linear modeling of dyadic data, http://www.upa.pdx.edu/IOA/newsom/mlrdyad4.doc Accessed 29 Mar 2010
52. Miaou Shaw-pin, Song Joon Jin, Mallick Bani K (2003) Roadway traffic crash mapping: a space-time modeling approach. J Transp Stat 6(1):33–57
53. Andrew V, Bared J (1998). Accident models for two-lane rural segments and intersections. TRR 1635, Paper No. 98-0294
54. Pickery J, Loosveldt G (2002) A multilevel multinomial analysis of interviewer effects on various components of unit non-response. Qual Quant 36:427–437
55. Akaike H (1974) A new look at the statistical model of identification. IEEE Trans Autom Control 19:716–723
56. Schwarz GE (1978) Estimating the dimension of a model. Ann Stat 6(2):461–464
57. Washington SP, Karlaftis MG, Mannering FL (2010) Statistical and econometric methods for transportation data analysis, 2nd edn. Chapman Hall/CRC, Boca Raton
58. Aguero-Valverde J, Jovanis PP (2008) Analysis of road crash frequency with spatial models. Transp Res Rec 2061:55–63
59. Lord D, Mannering F (2010) the statistical analysis of crash-frequency data: a review and assessment of methodological alternatives. Transp Res Part A 44(2010):291–305
60. Mensah A, Hauer E (1998) Two problems of averaging arising in the estimation of the relationship between accidents and traffic flow. Transp Res Rec 1635:37–43

4

Derailment risk and dynamics of railway vehicles in curved tracks: Analysis of the effect of failed fasteners

Silvia Morales-Ivorra[1] · Julia Irene Real[1] · César Hernández[1] · Laura Montalbán[1]

Abstract The effect of the fastener's failure in a railway track on the dynamic forces produced in the wheel-rail contact is studied using the simulation software VAMPIRE to assess the derailment risk of two different vehicles in two curves with distinct characteristics. First, a 3D-FEM model of a real track is constructed, paying special attention to fasteners, and calibrated with displacement data obtained experimentally during a train passage. This numerical model is subsequently used to determine the track vertical and lateral stiffness. This study evidences that although the track can practically lose its lateral stiffness as a consequence of the failure of 7 consecutive fasteners, the vehicle stability would not be necessarily compromised in the flawed zone. Moreover, the results reveal that the uncompensated acceleration and the distance along which the fasteners are failed play an important role in the dynamic behavior of the vehicle-track system, influencing strongly the risk of derailment.

Keywords Railway dynamics · Fasteners · Derailment · Curved track

1 Introduction

Any railway manager must fulfill minimum requirements of comfort and safety. In this sense, curved layouts require special attention since they have a great influence on the dynamic behavior of the vehicle-track system. Correct designs as well as an effective maintenance plan are key aspects to prevent and correct the deterioration of the vehicle and the track.

There are many complex track-vehicle interaction models designed for curved tracks as stated in [1]. For instance, in [2] a half-car on a ballasted track is modeled in detail; in this model, the different masses of the vehicle-track system are interconnected via springs and dampers. The versatility of this model is so high that it has been adapted to predict rail wear in curves [3], to calculate the plastic deformation in welds located in curved stretches [4], and to analyze the development and propagation of corrugation in curves [5]. Other approaches for the study of track-vehicle interaction can be found in [6] for the study of dynamic non-linearities produced in curves and in [7] for the analysis of the lateral stability of a freight train in a curve.

Apart from the above-mentioned multi-body methods, there are commercial programs that permit the implementation and resolution of complex vehicle-track dynamic systems with many degrees of freedom. These software products are suitable to study the complex dynamic phenomena arising in curved tracks. For example, SIMPACK is used in [8] and [9] to study, respectively, the vibrations produced in a vehicle running on a curve and the influence of the curve parameters on the wear of the wheel-rail contact. Furthermore, the software of dynamic simulation VAMPIRE is used in [10] to study the wear produced in the elements of a turnout by the vehicle passage.

Vehicle-track dynamics in curves are closely related to derailments [11]. For this reason, in [12] track and vehicle conditions are related with the proclivity to derailment in curves of small radii. This study revealed that the derailment coefficient decreases as the curve radius increases. Another similar research was conducted in [13] using the

✉ Silvia Morales-Ivorra
silmoiv@cam.upv.es

1 University Institute for Multidisciplinary Mathematics, Polytechnic University of Valencia, 46022 Valencia, Spain

simulation software SIMPACK. A deeper analysis about the influence of the curve parameters and vehicle speed on derailment was presented in [14], employing the commercial software ADAMS/Rail. Derailment in a high-radius curve (7,000 m) including the effect of the fastener failure is analyzed in [15]; this study demonstrated that in this high-radius concrete curve, the stability of a high speed train is not compromised if the number of failed fasteners is lower than 15.

The relationship between flawed fasteners and derailment risk has been also studied in [16] in a tangent track using a coupled vehicle/track model. The results show the strong influence of the disabled fasteners on the derailment coefficient and on the track widening, which also increases the derailment risk. The research [17], conducted in slab tracks, concludes that failed fasteners cause increased displacements on the rail and the slab which become greater with increasing vehicle speed. In [18], the derailment risk in a curved track is assessed considering failures in the track supports, demonstrating their great influence on the running safety; the vehicle and the track infrastructure are modeled as a multi-body system, while the rails are modeled as Timoshenko beams resting on discrete supports. The importance of the lateral dynamics on vehicle stability has prompted the development of innovative elements, such as frictional sleepers [19] to increase the lateral resistance of railway tracks, and of new monitoring techniques to maintain the fasteners in perfect condition [20].

The present study aims to continue in this research stem to evaluate the dynamic behavior of the vehicle in curved tracks where fasteners have failed. However, two different vehicles will be considered in this case (passengers and freight), with their respective speeds, and two curves of different radii. The vehicle-track model will be implemented in VAMPIRE, while the track vertical and lateral stiffness are calculated using an auxiliary FEM model calibrated from experimental data. The objective of this research is to present a method to reproduce the track vertical and lateral stiffness in a FEM model and study how a fastener failure influences the dynamic stability in two different curves on which two different vehicles circulate with different speeds and static loads.

2 Experimental campaign

A ballasted track stretch with Iberian gauge (1,668 mm), UIC-60 rails, and monoblock prestressed concrete sleepers AI-04-EA was monitored during the experimental campaign. The railpad, with a thickness of 5 mm, has a vertical stiffness of 300 kN/mm, and the rail is attached to the sleeper with Vossloh VM fasteners, generally used in the Spanish ballasted tracks. Each fastener consists of a clip SKL-1, a sheath V2, an angled mounting plate A2, a T2 screw with a locking washer, and an elastic plate PAE-2. The fastener sketch is depicted in Fig. 1.

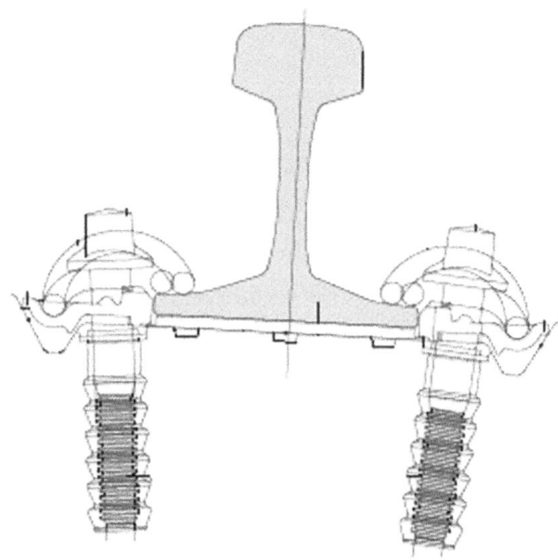

Fig. 1 Vossloh VM fastener

The track stretch was instrumented with displacement transducers in the rail and with rod extensometers located between the granular layers (ballast and subballast). Therefore, the total deflection of the infrastructure caused by the train passages can be calculated as the sum of these relative displacements. In Table 1, the experimental results corresponding to the passage of a locomotive *Renfe S-252* at 160 km/h are shown.

The interface between the subballast and the platform was also equipped with pressure cells. These cells registered the stresses induced in this zone, allowing the calculation of the track vertical stiffness, which resulted to be 90 kN/mm. Moreover, during the data acquisition campaign, the tightening torque of the fastener screws was revised, obtaining an average value of 220 Nm. Experimental data will be the fundamental support to construct a numerical model able to represent accurately the real track behavior.

3 Models

3.1 Model of the studied track

The objective of the FEM model is to calculate the values of track vertical and lateral stiffness in different scenarios

Table 1 Deflections produced in the infrastructure by the locomotive passage

Relative strains (mm)		Total deflection (mm)
Rail	Granular layers	
0.479	0.765	1.244

Fig. 2 FEM model of the track superstructure with springs supporting the sleeper

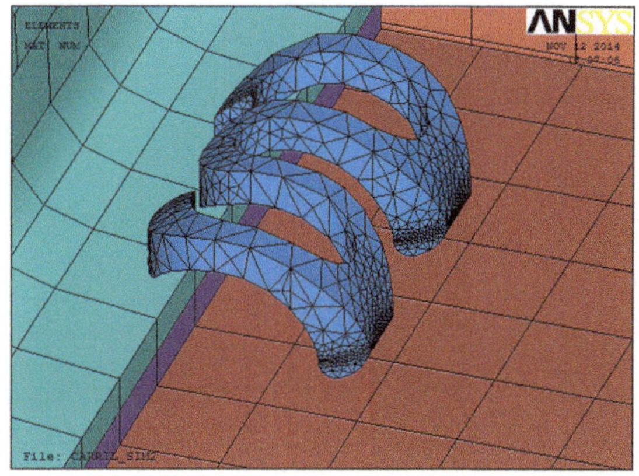

Fig. 3 Detail of the FEM model of the clip

in which fasteners fail. For this reason, a simplified model is constructed to represent in detail the upper part of the track superstructure, including the rails, the fasteners, and the sleepers which rest on elastic supports that simulate the granular layers (see Fig. 2). These elastic supports consist of spring elements attached to the sleepers and oriented in the vertical and lateral direction.

The cross sections of the UIC-60 rail, the AI-04-EA sleeper, and the railpad sections have been drawn in detail using a CAD software and then imported to the FEM software ANSYS LS-DYNA V.14. These sections are extruded, generating the real elements, which are divided in hexahedral elements SOLID45 with a maximum side of 0.04 m. The mechanical properties (Young modulus, E; Poisson coefficient, v; and density, ρ) of the track elements are summarized in Table 2. Regarding the boundary conditions, the length of the model is 9 m, long enough to avoid the influence of the extreme boundaries on the static results, which are always calculated in the center of the model. Rotations and longitudinal movements of the rail extremes are constrained in the model boundaries, while the sleeper displacements are conditioned by the spring elements that account for the elastic support.

As observed in Fig. 3, in order to model the fastener, the clip has been represented in detail and the screw effect has been represented by compressive forces acting on each clip. The lateral and plan view of the fastener is imported in

Table 2 Mechanical properties of the track superstructure elements

Element	E (MPa)	v	ρ (kg/m^3)
Rail	2.10×10^4	0.3	7,830
Railpad	119	0.49	900
Sleeper	2.70×10^4	0.35	2,400
Fastener	2.10×10^4	0.3	7,830

a CAD file to construct the model in ANSYS. In this case, the element used is also the SOLID45 but the mesh is self-adapted by the program in order to adapt the mesh to the curved contours of the fastener in an efficient manner that provides numerical convergence. This force has been calculated from the tightening torque of the screws measured in the experimental campaign. Furthermore, the condition of strain compatibility has been imposed in the nodes located in the clip-sleeper and clip-rail contact zones.

The crucial step to calibrate the model is to determine the stiffness of the springs attached to the sleepers. From the experimental campaign, it is known that the track vertical stiffness is 90 kN/mm. However, assigning directly this stiffness to the springs in the model is incorrect, since it includes the contribution of elements such as the sleeper, the fastener, and the rail which are already represented in the FEM model. Instead, a linear behavior is assumed to estimate the force F transmitted by the wheel of the locomotive *Renfe S-252* to the rail head in the monitored section from Eq. (1):

$$F = k_v \delta, \tag{1}$$

where k_v is the vertical track stiffness (90 kN/mm) and δ is the deflection produced according to Table 1 (1.244 mm). Therefore, the transmitted force is $F = 112$ kN.

On the one hand, knowing the vertical force acting on the wheel-rail contact, it is possible to calculate the vertical stiffness of the springs that support the sleepers in the FEM model. To do so, a force $F = 112$ kN is introduced in the model and, subsequently, the deflections in the rail head are calculated by varying the stiffness of the springs. When a deflection result equal to the experimentally measured 1.244 mm is reached, it is considered that the model is calibrated and that the vertical stiffness of the springs at that moment is adequate to reproduce the real behavior of the system. In this case, the stiffness of the vertical springs when all the fasteners are in perfect conditions is $k_{v\_spring} = 2.3962\times10^{-2}$ kN/mm.

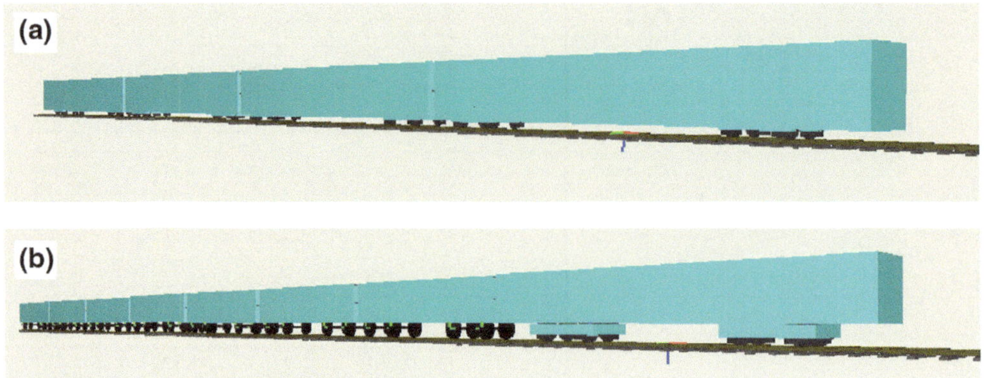

Fig. 4 Model of the vehicles in VAMPIRE. **a** Passenger train. **b** Freight train

On the other hand, obtaining the track lateral stiffness experimentally requires very complex and costly tests [21] that were not done during the experimental campaign. Consequently, the authors have searched in the existing literature experimental values of this parameter already used to validate other models. The Ref. [22] states that in a ballasted track with UIC-60 rails, middle-stiffness railpads and monoblock prestressed concrete sleepers, as in the studied case, the lateral stiffness of the ballast layer is $k_{h\_ballast}$ = 110 kN/mm per linear meter. It should be highlighted that this value for lateral stiffness is given per unit of length. This figure can be assimilated to the equivalent stiffness of all the springs located in the lateral direction in the FEM model. Considering that there are 1,695 parallel springs in the sleepers' lateral sides and that the model length is $L = 9$ m, the lateral stiffness of each spring can be calculated by Eq. (2):

$$k_{h\_spring} = k_{h\_ballast}\frac{L}{N} = 0.584071 \text{ kN/mm}, \qquad (2)$$

where N is the number springs.

Apart from the lateral stability provided by the ballast layer and the fastener, there are friction forces at the rail-railpad and sleeper-railpad surfaces that oppose the lateral displacements. These friction forces depend mainly on the mechanical properties of the railpads' material [23] and therefore are implicitly considered in the studied model.

The result of the calibration is a finite element model able to reproduce the real behavior of the track. This model is used in Sect. 4 to calculate the vertical and lateral track stiffness in different scenarios with failed fasteners.

3.2 Track-vehicle interaction model

Track-vehicle interaction has been analyzed using the dynamic simulation package VAMPIRE. In this program, two of the vehicles that normally run in this line have been represented: the passenger vehicle *Renfe S-120* and a convoy of well cars carrying containers and hauled by a locomotive *Renfe S-253* (see Fig. 4). Both vehicles have been represented as dynamic systems of three masses (unsprung, semi-sprung, and sprung) connected by spring-damper elements that simulate the contact with the rail and the vehicle suspensions. The estimated values of the vehicle dynamic systems studied are summarized in Table 3.

4 Calculation of the track stiffness in different scenarios of fastener failure

In the model described in Sect. 3, fasteners will be steadily removed to calculate the variations produced on the vertical stiffness k_v and on the lateral stiffness k_h when the

Table 3 Inputs of the vehicle-track dynamic interaction model in VAMPIRE

Vehicle type	Car type	Wheelset mass (kg)	Bogie mass (kg)	Carbody mass (kg)	Primary stiffness (N/m)	Primary damping (Ns/m)	Secondary stiffness (N/m)	Secondary damping (Ns/m)
Renfe S-120	Locomotive	1,600	4,000	48,000	1.69×10^6	5.20×10^4	2.38×10^6	2.00×10^5
	Carriage	1,600	4,000	44,500	1.69×10^6	5.20×10^4	2.38×10^6	2.00×10^5
Renfe S-253	Locomotive	1,800	5,000	40,000	1.20×10^6	1.50×10^4	3.00×10^5	1.00×10^4
	Freight wagon	1,800	2,100	32,000	2.60×10^6	–	–	–

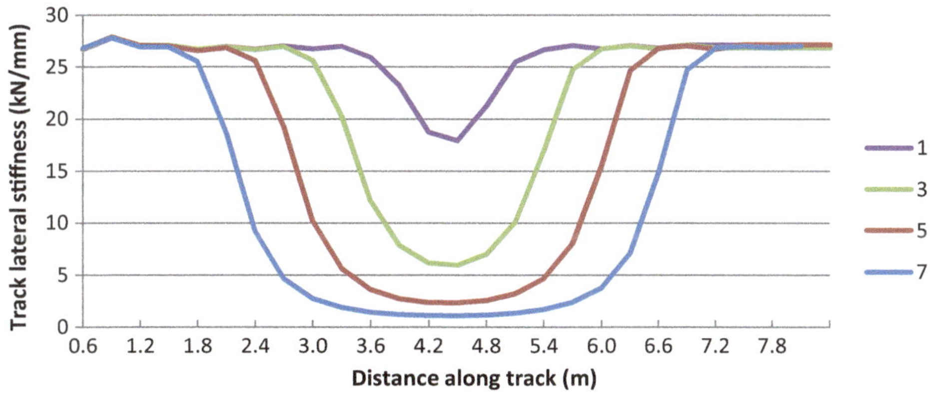

Fig. 5 Evolution of the track lateral stiffness along the track for different numbers of consecutive failed fasteners

fasteners fail. To do so, a unit load is applied on the centerline of the rail head in the vertical direction and another in the lateral direction, for calculating the displacements caused in the corresponding directions with the FEM model. From this displacements and knowing that loads are unit, the track vertical and lateral stiffness can be calculated applying Eq. (1). In Fig. 5, the variation of the track lateral stiffness along the track is represented for the different numbers of consecutive failed fasteners studied. As previously mentioned, failed fasteners are supposed to be in the center of the model in order to avoid the effect of the boundaries on the results and thus, the distance along track represented in Fig. 5 does not include the areas closer to the model extremes. The lateral stiffness approaches asymptotically the 0 value, since there is always a minimum of resistance caused by the rail lateral bending stiffness and the friction forces in the pad surfaces, the track lateral stiffness is not totally lost. Nevertheless, for the purposes of this study, it is assumed that the lateral stiffness is zero if more than 7 consecutive fasteners fail, since the remaining resistance is negligible if compared to the track with fasteners in perfect conditions.

The minimum values obtained at the point where fasteners fail are shown in Table 4 for different numbers of consecutive failed fasteners.

As the number of consecutive failed fasteners increases, the track lateral stiffness decreases and the vertical remains constant according to Fig. 5 and Table 4. When a fastener fails, vibrations and rail movements may cause that closer fasteners have higher stresses than in normal conditions. This might lead to a progressive fastener failure until the track losses completely its lateral resistance. As evidenced by the FEM model, for the studied track, this situation occurs if more than 7 consecutive fasteners fail; in this case, it can be assumed that the track does not oppose any resistance to lateral forces and the stability of the vehicle might be compromised.

Table 4 Minimum track vertical and lateral stiffness as a function of the number of consecutive failed fasteners

Number of failed fasteners	k_v (kN/mm)	k_h (kN/mm)
0	90	27.7
1	90	17.5
3	90	5.8
5	90	2.1
7	90	1.1

5 Dynamic analysis of the vehicles in curved tracks

The parameters of two studied curves, which also can be found in the real line, as well as the running speeds of the passenger and freight trains are shown in Table 5. Speeds have been specifically defined to induce that the passenger train negotiates the curves with cant deficit (exerting a greater force on the high rail as a consequence of the centrifuge acceleration), while the freight train circulates in the curve with cant excess. The publication [14] revealed, implicitly, that both the cant excess and deficiency cause high forces in the low and high rails, respectively, modifying the derailment coefficient. In order to assess the less favorable scenario, fasteners are supposed to fail in the high rail for the analysis of the passenger train and in the low rail for the analysis of the freight train.

Cant excess or deficiency can be quantified by means of uncompensated acceleration (γ). This parameter represents the share of the total centrifuge acceleration which is not compensated by the track geometry in each case. Its value can be calculated by Eq. (3), in which V is the vehicle speed, R the curve radius, h the cant, g the acceleration of the gravity (9.81 m/s^2), and d the track gage (1,668 mm):

$$\gamma = \frac{V^2}{R} - \frac{h}{d}g. \quad (3)$$

Table 5 Parameters and maximum train speeds of the studied curves

Curve No.	Radius (m)	Cant (mm)	Passenger train speed (km/h)	Freight train speed (km/h)
Curve 1	2,500	160	200	80
Curve 2	1,000	100	125	80

Table 6 Uncompensated acceleration (γ) for the different studied cases (units: m/s^2)

Curve no.	Passenger train	Freight train
Curve 1	0.294	−0.743
Curve 2	0.618	−0.09

The results for the uncompensated acceleration in the different cases proposed in this study are presented in Table 6. The higher uncompensated acceleration is, the higher centrifugal force results, pushing the train out of the track and causing greater forces in the wheel-rail contact.

5.1 Analysis of the dynamic forces

Firstly, the evolution of the vertical and lateral forces when the fasteners fail will be analyzed. For this purpose, the details obtained from the dynamic simulation in VAMPIRE for the case in which 7 fasteners fail are thoroughly studied since it is at this point when the track losses all its lateral resistance and the increase of the forces is more clearly appreciated. Moreover, it will be distinguished which part of these forces is exclusively due to the fasteners failure and not to the geometric conditions of the track and the vehicle. The development of the dynamic forces along the curve is shown in Fig. 6 for the passenger train and in Fig. 7 for the freight train. In the case of the passenger train, the results correspond to the fourth wheelset of the first vehicle, while in the freight train the results are obtained for the second wheelset of the third vehicle. These wheelsets have been found to be the closest to derailment in both curves studied. Results are shown along 200 m in the center of the curve, considering that failed fasteners begin at the point $x = 2{,}000$ m.

From Figs. 6 and 7, it can be deduced that, in all cases, the peak in lateral forces is not produced exactly at the point where fasteners start to fail ($x = 2{,}000$ m) but 25 m later for the passenger train and up to 40 m for the freight train. Before reaching this point, in which an abrupt change in dynamic forces occur, forces remain constant around a value which is proportional to the uncompensated acceleration caused by the cant excess or deficit of the vehicle in the curve (see Table 6). This effect can be explained by the fact that, as the train circulates on the failed-fastener stretch, no force opposes the lateral movement of the wheel-rail system; however, when the train leaves the failed-fastener stretch, the track exerts a force which tries to reestablish the train direction and oppose the yaw movements that the vehicle is experiencing in the flawed stretch. This point is where great lateral forces appear in the graphs of Figs. 6 and 7.

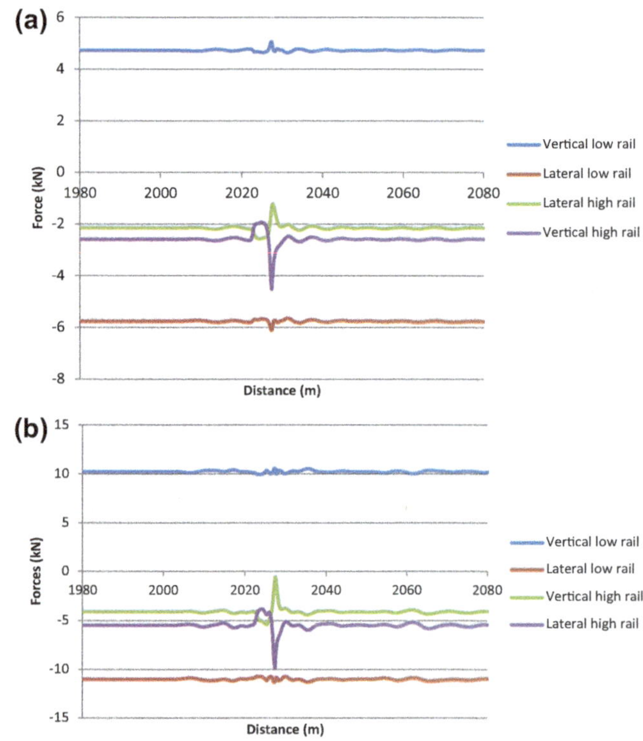

Fig. 6 Forces in Curve 1 (**a**) and in Curve 2 (**b**) for the passenger train when 7 consecutive fasteners fail

Peaks of vertical forces appear at the same time that peaks of lateral forces do but are significantly lower. Furthermore, it can be also affirmed that vertical forces are proportional to the uncompensated acceleration and remain almost unaltered by the fasteners' failure.

Once studied in detail, the dynamic effects in the less favorable case, the evolution of lateral dynamic forces as a function of the number of failed fasteners up to the point in which the track losses its lateral resistance (i.e., with 7 consecutive failed fasteners) is presented in Fig. 8. Vertical dynamic forces are not included in this discussion because they remain almost unaffected by the lateral stiffness as seen above.

Results in Fig. 8 are coherent with the uncompensated accelerations calculated in Table 6. The passenger train

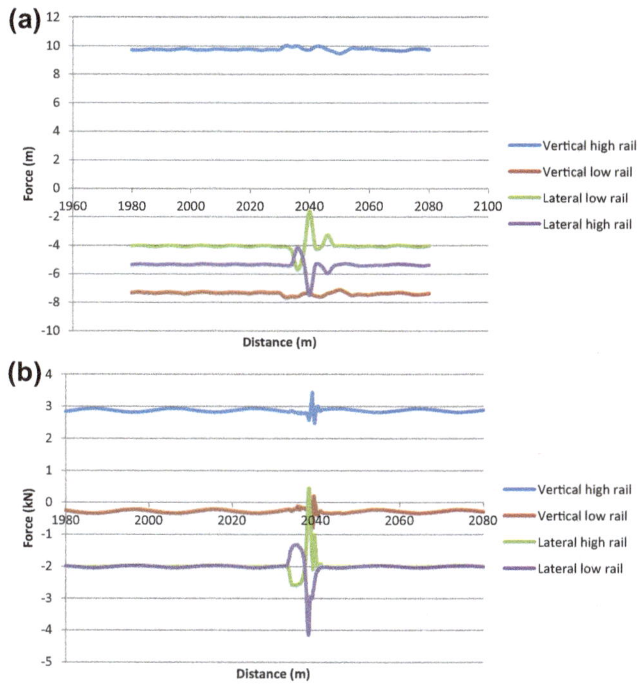

Fig. 7 Forces in Curve 1 (**a**) and in Curve 2 (**b**) for the freight train when 7 consecutive fasteners fail

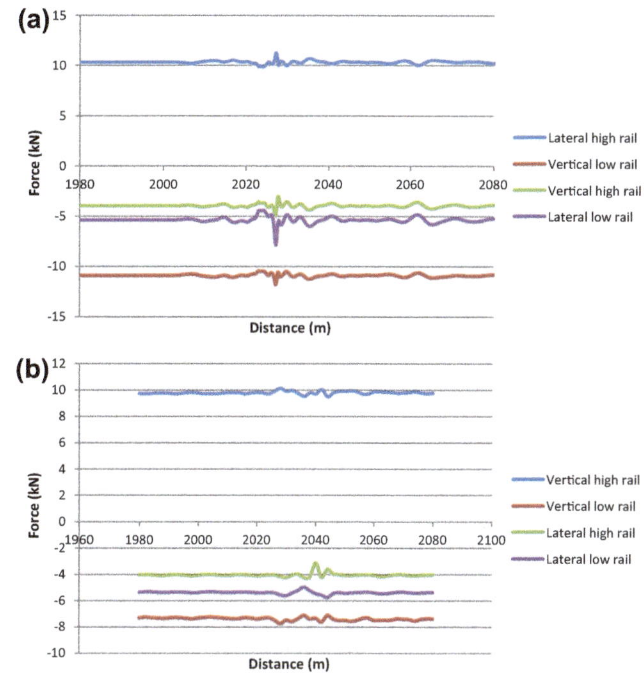

Fig. 9 Forces in Curve 2 for the passenger train (**a**) and in Curve 1 for the freight train (**b**) when 7 consecutive fasteners fail simultaneously in each rail

transmits higher lateral forces in Curve 2 as a consequence of the greater uncompensated acceleration in this curve, while in the case of the freight train, the greater uncompensated acceleration appears in Curve 1, resulting in higher forces associated to the fasteners' failure. In any case, lateral forces are under the maximum of 62 kN fixed by the UIC (International Union of Railways) [24], which means that the momentary lack of track lateral stiffness caused by the fasteners' failure does not represent a serious danger for the stability of the track-vehicle system.

Additionally, the simultaneous failure of seven fasteners in the high and the low rail has been studied in the scenarios that resulted more prone to derailment: Curve 2 for the passenger train and Curve 1 for the freight train. Results in Fig. 9 reveal that the maximum dynamic forces caused by the simultaneous failure of the fasteners in both rails are not significantly different from those obtained in Figs. 6 and 7, which were calculated supposing flawed fasteners in a single rail.

5.2 Analysis of the derailment coefficient

Derailment coefficient D is defined as the ratio of the lateral force L to the vertical force V. This coefficient is used to evaluate the risk of derailment of a railway vehicle by comparing its value with the so-called limit derailment coefficients, which depend on many factors according to the derailment mechanism considered in each case. For instance, the Nadal's criterion considers the flange climb and, therefore, the limit derailment coefficient depends on

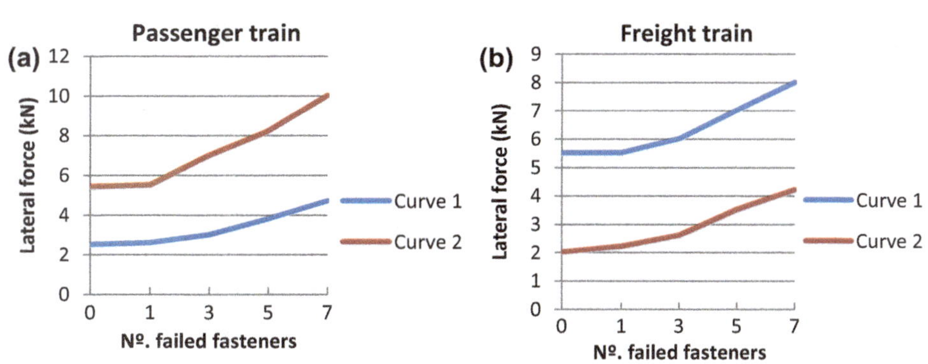

Fig. 8 Evolution of track lateral stiffness with the number of failed fasteners for passenger train (**a**) and freight train (**b**)

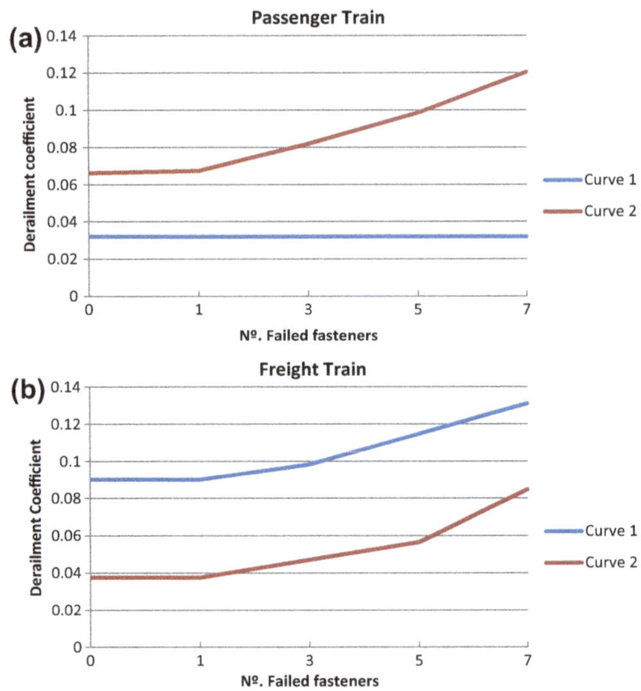

Fig. 10 Evolution of the derailment coefficient with the number of failed fasteners for passenger train (**a**) and freight train (**b**) in the studied scenarios

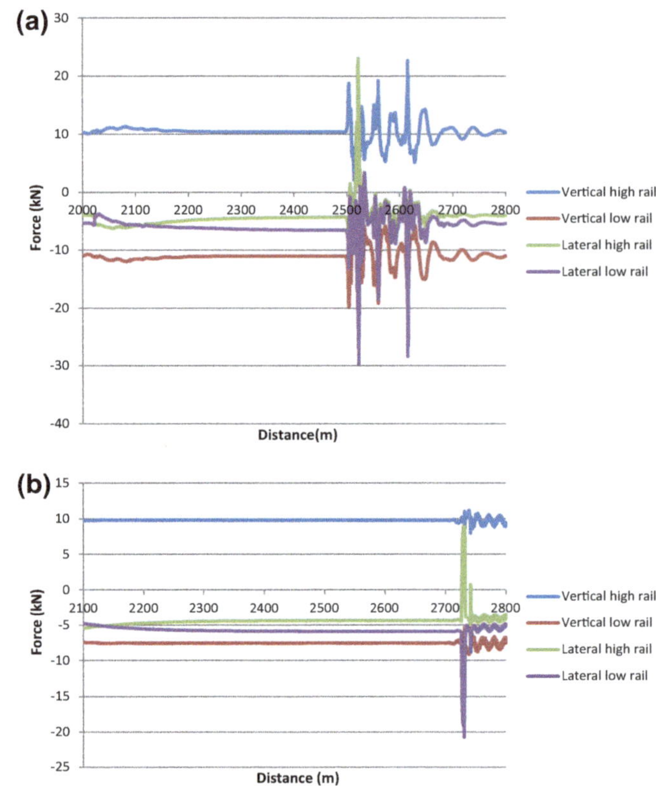

Fig. 11 Dynamic forces for the passenger train in Curve 2 (**a**) and for the freight train in Curve 1 (**b**) with failed fasteners along 500 and 750 m, respectively

the friction of the wheel and the rail (See [25]). However, some studies consider the derailment by wheel lifting and consequently include in the calculation of the limit derailment coefficient lateral acceleration and axle forces [26].

The UIC leaflet 518 [27] sets the limit value for the limit derailment coefficient to 0.8 and specifies that this limit must not be exceeded along a length of 2 m; from that moment, the vehicle stability might be compromised. Besides, the standard UNE-EN 14363 [28] establishes that the wheels with worn flanges may derail when the derailment coefficient is over 0.4. Hence, the value of 0.4 will be set as the limit derailment coefficient in this study.

Figure 10 shows the value of the derailment coefficient in the studied scenarios, taking into account that the wheel derailment would be produced in the high rail for the passenger train and in the low rail for the freight train since both vehicles negotiate the curve with cant deficit and excess, respectively.

As depicted in Fig. 10, derailment coefficient remains far below the limit of 0.4. For this reason, it can be concluded that there is not risk of derailment in the studied cases when 7 consecutive fasteners fail, although the track had lost the lateral resistance in this point.

However, if the number of failed fasteners is high enough and the track losses its lateral resistance along a certain distance, the derailment coefficient may exceed the limit value, affecting the train safety. It has been proven that the limit value of 0.4 is surpassed for distances of failed fasteners of 500 m in the case of the passenger train negotiating Curve 2 and of 700 m in the case of the freight train negotiating Curve 1. These cases are illustrated in Fig. 11, in which the reader can see how the maximum peak is produced again after the track recovers its lateral resistance and how the instability persists during the following meters. The evolution of the derailment coefficient along the curves is illustrated in Fig. 12 for these limit cases. As in the previous cases, the derailment risk appears in the low rail for the freight train and in the high rail for the passenger trains.

6 Conclusions

In this study, track vertical stiffness has been estimated from experimental results of track settlements using a FEM model. A new strategy to model the fastener as a clip and a compressive force calculated from the tightening torque has been explained, before studying the vehicle-track system dynamics in curves and the derailment risk. From the results obtained, it can be concluded that

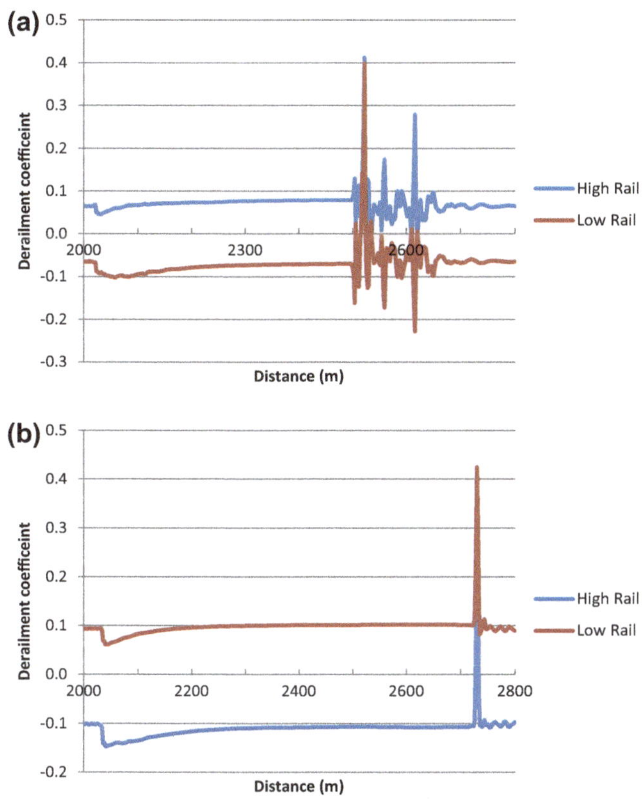

Fig. 12 Evolution of the derailment coefficient in the limit scenarios for passenger train (**a**) and freight train (**b**)

- The lateral forces are caused by the fasteners' failure, and their value is proportional to the uncompensated acceleration regardless of whether the vehicle negotiates the curve with cant excess or deficiency, according to [12], [13] and [14]. Vertical forces are not significantly influenced by the fasteners' failure.
- Although the studied track losses its lateral resistance when 7 consecutive fasteners fail, the vehicle stability is not compromised in this situation. However, if this situation of null lateral stiffness prevails along a certain distance, the vehicle may be in serious risk of derailment. In the studied examples, this critical distance was higher than 500 m for the two studied vehicles.
- The peak force due to the failed fasteners is not produced at the point where flawed fasteners are located but several meters after this point. An explanation to this phenomenon can be found in the fact that until the lateral stiffness is not totally recovered, there is no force opposing the yaw and lateral movements that the vehicle suffers on the stretch with flawed fasteners. It means that a potential derailment would not occur within the flawed zone but some meters beyond.
- The influence of fastener failure on the vehicle dynamics is similar when the flawed fasteners are located only in the rail which receives the extra load of the uncompensated acceleration and when the flawed fasteners are located in both rails simultaneously.
- The derailment coefficient increased with the distance along which fasteners fail. Even though lateral resistance is lost after 7 consecutive flawed fasteners, a higher number of failed fasteners must be considered in order to asses accurately the risk of derailment.

References

1. Iwnicki S (2006) Handbook of railway vehicle dynamics.CRC Press, Boca Ratón
2. Jin XS, Wen ZF, Wang KY, Zhou ZR, Liu QY, Li CH (2006) Three-dimensional train-track model for study of rail corrugation. J Sound Vib 293:830–855
3. Jin X, Wen Z, Xiao X (2007) A numerical method for prediction of curved rail wear. Multibody Sys Dyn 18:531–557
4. Li W, Xiao G, Wen Z, Xiao X, Jin X (2011) Plastic deformation of curved rail at rail weld caused by train-track dynamic interaction. Wear 271:311–318
5. Jin X, Wen Z (2008) Effect of discrete track support by sleepers on rail corrugation at a curved track. J Sound Vib 315:279–300
6. Di Gialleonardo E, Bruni S, True H (2014) Analysis of the nonlinear dynamics of a 2-axle freight wagon in curves. Veh Syst Dyn 52(1):125–141
7. Wang K, Liu P (2012) Lateral stability analysis of heavy-haul vehicle on curved track based on wheel/rail coupled dynamics. J Transp Technol 2:150–157
8. Kurzeck B, Hecht M (2010) Dynamic simulation of friction-induced vibrations in a light railway bogie while curving compared with measurement results. Veh Syst Dyn 48:121–138
9. Chen P, Gao L, Hao J (2007) Simulation study on parameters influencing wheel/rail wear in railway curve. China Railway Science 5
10. Sun Y, Cole C, Boyd P (2011) A numerical method using VAMPIRE modelling for prediction of turnout curve wheel-rail wear. Wear 271(1–2):482–491
11. Brabie D (2007) On derailment-worthiness in railway vehicle design. Analysis of vehicle features influencing derailment processes and consequences, PhD Thesis, Royal Institute of Technology, Stockholm
12. Matsumoto A, Sato Y, Ohno H, Shimizu M, Kurihara J, Saitou T, Michitsuji Y, Matsui R, Tanimoto M, Mizuno M (2014) Actual states of wheel/rail contact forces and friction on sharp curves—continuous monitoring from in-service trains and numerical simulations. Wear 314:189–197
13. Wang L, Huang A, Liu G (2013) Analysis on curve negotiation ability of the rail vehicle based on SIMPACK. Adv Mater Res 721:551–555
14. Eom BG, Lee H (2010) Assessment of running safety of railway vehicles using multibody dynamics. Int J Precis Eng Manuf 11(2):315–320

15. Zhou L, Shen Z (2013) Dynamic analysis of a high-speed train operating on a curved track with failed fasteners. J Zhejiang Univ Sci A 14(6):447–458
16. Xiao X, Jin X, Wen Z (2007) Effect of disabled fastening systems and ballast on vehicle derailment. J Vib Acoust 129:217–229
17. Shi W, Cai C (2011) Influence of slab track fastener failure on track dynamic performance. In: Advances in environmental vibration-proceedings of the 5th international symposium on environmental vibration, Chengdu, pp 686–692
18. Xiao X, Jin X, Deng Y, Zhou Z (2008) Effect of curved track support failure on vehicle derailment. Veh Syst Dyn 46(11):1029–1059
19. Zakeri J, Fakhari M, Mirfattahi B (2012) Lateral resistance of railway track with frictional sleepers. Proc Inst Civ Eng Transp 165(2):151–155
20. Gibert X, Patel V, Chellappa R (2015) Robust fastener detection for autonomous visual railway track inspection. In: IEEE Winter Conference on Applications of Computer Vision, Waikoloa Beach Hawai, pp 694–701
21. Thompson D, Hemsworth B, Vincent N (1996) Experimental validation of the TWINS prediction program for rolling noise, part 1: description of the model and method. J Sound Vib 193(1):123–135
22. Zakeri JA (2012) Lateral resistance of railway track. In: Perpinya DX (ed) Reliability and safety in railway. InTech Europe, Rijeka, pp 357–374
23. Kernes RG, Edwards JR, Dersh MS, Lange DA, Barkan CP (2011) Investigation of the dynamic frictional properties of a concrete crosstie rail seat and pad and its effect on rail seat deterioration (RSD). Transportation Research Board 91st annual meeting. Railtec, Chicago, Illinois
24. Carrascal-Vaquero I (2006) Optimization and analysis of the behavior of fastener systems for Spanish high-speed railways. PhD Thesis, University of Cantabria, Santander (In Spanish)
25. Sany J (1996) Another look at the single wheel derailment criteria. In: Proceedings of the IEEE/ASME Joint Railroad Conference, pp 17–22
26. Koo J, Oh H (2014) A new derailment coefficient considering dynamic and geometrical effects of a single wheelset. J Mech Sci Technol 28(9):3483–3498
27. UIC-Leaflet-518 (2009) Testing and approval of railway vehicles from the point of view of their dynamic behaviour-safety-track fatigue-ride quality. Railway Technical Publications, Paris
28. UNE-EN14363:2007 (2007) Railway applications—testing for the acceptance of running characteristics of railway vehicles—testing of running behaviour and stationary tests. AENOR, Madrid

5

Bluetooth as a traffic sensor for stream travel time estimation under Bogazici Bosporus conditions in Turkey

İlker Erkan[1] · Hasan Hastemoglu[1]

Abstract Travel time estimation is an integral part of Intelligent Transportation Systems, and has been an important component in traffic management and operations for many years. Travel time, being spatial in nature, requires spatial sensors to measure it accurately. Bluetooth is emerging as a promising technology for the direct measurement of travel time, and is reported in a few studies from homogenous traffic conditions. At the same time, there have been no studies on the applicability of Bluetooth for travel time estimation in heterogeneous traffic seen in Istanbul and even that Turkey. Bluetooth data collected from a busy urban road in Istanbul city have been analyzed and the penetration rate was found to be about 5 %. Two wheelers and light motor vehicles have been detected using the Bluetooth sensor and the data have been extrapolated to estimate travel times of other classes of vehicles. The study developed linear relationships between speeds of different classes of vehicles through weighted linear regression methods and were used for the estimation of stream travel time. The results obtained were promising and show that Bluetooth is a cost-effective technology for estimation of travel time for heterogeneous traffic conditions.

Keywords Travel time · Public transport · Traffic management · Istanbul traffic · Traffic time estimation

✉ İlker Erkan
ilkererkan@sdu.edu.tr

Hasan Hastemoglu
hasanhastemoglu@sdu.edu.tr

1 Faculty of Architecture, Suleyman Demirel University, 32260 Cunur, İsparta, Turkey

1 Introduction

The time taken to commute between two points in a traffic stream—the "travel time"—is useful information that can help the traveler plan their trip better. Travel time information is also an important parameter in Intelligent Transportation Systems, traffic demand modeling and prediction, traffic simulation, traffic signal timing control, incident detection, congestion management, and dynamic route assignment [1].

Travel time is a dynamic, spatial parameter that is difficult to measure directly for an entire travel stream. Sensors affixed at predetermined spots en-route a stream or within vehicles moving along the stream, can be used for direct travel time measurements. Automatic vehicle identification (AVI), License Plate Recognition Systems, Signature Matching Systems, Platoon Identification Systems, Global Positioning Systems (GPS), mobile phones, and Bluetooth are some common examples of devices that can be used as spatial sensors for travel stream analyses. Despite the apparent applicability of these sensors in measurement of travel time, many of them suffer from one or more of the following drawbacks: they provide no information on count, occupancy and flow, require user participation, have privacy issues and capture only a small fraction of the traffic [2, 3].

Bluetooth is increasingly being explored for travel time measurement applications. Preliminary studies in the West have proven cost-efficiency benefits of Bluetooth in spatial sensing in homogenous traffic conditions [4]. Agencies in the USA such as The Illinois Department of Transportation [5] and Houston Transtar already use Bluetooth sensors to collect travel time information.

However, there are no reports yet, on the use or study of Bluetooth sensors in the heterogeneous Istanbul Bosporus traffic scenario. The present study attempts to overcome this stalemate by evaluating the applicability of the Bluetooth sensor as a traffic data source under The Bogazici Bosporus traffic conditions.

Bluetooth can be used to measure travel times of those vehicles that have Bluetooth devices. This may only be a small fraction of all vehicles plying any particular route. Hence, it is important to understand the percentage of vehicles from which data can be collected using this method, namely penetration rate analysis. Also, it is important to know the class-wise distribution of vehicles from which data can be collected. This is particularly important for a country like Turkey, whose roads are characterized by heterogeneous traffic conditions—two (bicycles, motorbikes), three (auto rickshaws), and four wheelers (cars) share the road with light and heavy motor vehicles and a heavy pedestrian population.

This study performs penetration analysis and determines the class-wise distribution of Bluetooth-based data. A method for the estimation of stream travel time from the sampled data is also developed.

2 Background and scope

The state of the traffic is estimated using data from various sensors that range from traditional inductive loops to advanced Bluetooth MAC Scanners. Models have been proposed to estimate travel time (or speeds) [6, 7, 8–11] and density [12–14] from loop detector data.

Bluetooth uses low-power radio waves to wirelessly link various electronic devices over short distances (1–100 m). A Bluetooth device uses a 12-digit electronic identifier, called a Media Access Control address, or MAC address that serves as an electronic nickname for the device. An "inquiry" establishes continuous connection between MAC ID's. The anonymity of Mac ID's ensures privacy [5] and can be used as a handle to obtain traffic information.

In the present study, MAC ID's, which are captured by a Bluetooth sensor device, are replaced by an automatically generated random number, thus further obviating possible privacy intrusions. Travel time can be computed from Bluetooth sensor data by matching the IDs at two locations and calculating the difference between the arrival time stamps of the MAC ID at these locations.

Most existing studies on Bluetooth for traffic applications have focused on quality control of Bluetooth data and estimation of average speed or origin-destination [4, 5]. In many of these studies, only a percentage of traffic stream—the sampling rate or penetration rate—is captured and used as the traffic data source. For example, a Bluetooth hourly sampling rate as low as 2 %–4 % was observed in a study at the University of Maryland [4]. In this study, it was found that the travel times obtained using Bluetooth sensors were comparable to those obtained using GPS. The study also proposed a two-step travel time filtering mechanism to estimate the upper and lower bounds from the distribution of travel time. The accuracy of measured travel time was reported to be better with increased distance between the two Bluetooth detectors and decreased vehicle speed.

Stevanovic and Martin [15] compared the travel times measured by Bluetooth MAC readers with that obtained using floating cars equipped with GPS. They reported that the travel times from the Bluetooth readers did not vary significantly (95 % level) from the GPS floating car travel time for 83 % of the cases. Wang et al. [16] demonstrated that the travel time obtained from Bluetooth sensors with sampled data and that estimated from loop detectors are roughly similar.

An analysis of the impact of vehicle speed and effective range of the device on the number of in-range scan intervals for improvements in device detection was reported in [17].

Welsh et al. [18] worked on the improvements of the Bluetooth technology connection times between moving devices and proposed the creation of a mesh network of Bluetooth-connected devices. Ahmed et al. [19] further investigated the concept, envisioning the utilization of Bluetooth technology, in order to create a static mesh network for ITS data collection. The Bluetooth detection technology has been shown to be effective in several research studies [19, 20]. Analyses for freeways and arterials demonstrated that methodologies based on data recorded by Bluetooth detectors are capable of capturing traffic conditions as accurately as other data sources, including intrusive methods, such as loop detectors as well as non-intrusive, such as Floating Car Data [4, 21–23].

Sadabadi et al. [24] estimated the upper bounds for the errors by using the relationship among the segment length, average speed, and travel time and showed that the travel time measurement error is negligible when the average speed is around 45 kmph and the distance between two detectors is 2–3 miles. Quayle et al. [21, 22] identified the upper and lower limits of travel times measured with a Bluetooth sensor, using a moving standard deviation [25].

Jaume et al. [26] studied the quality of the data produced by the Bluetooth and Wi-Fi detection of mobile devices to estimate time-dependent Origin-Destination matrices based on Kalman Filtering. Sawant et al. [27] used the concepts of wireless sensor networks and Bluetooth protocols to develop a novel approach that increased safety of road travel. Bullock et al. [28] studied the applicability of

Bluetooth in measuring time spent waiting in security screening lines, transiting the security screening checkpoint, and walking to concourse in airports. A study by Blogg et al. [29] reported on the passive observations of Bluetooth protocol devices embedded in vehicles and motorists' mobile devices to collect the OD data.

Li et al. [30] compared the performance of various traffic data collection methods for urban road network environments, showing similarities, advantages, and disadvantages. This reference studied the quality of decomposed link travel times based on empirical data in the Chinese city of Changsha. They found that the accuracy typically is within a range of just 1 s.

As with any travel time measurement method, sources of error exist and outlying data points must be addressed to ensure that accurate travel times are reported. Errors may appear in Bluetooth travel time measurements on arterials as a result of signal delay and non-uniform traffic flow [31]. The 10.24 s required to complete the Bluetooth inquiry process may introduce a major source of error and may result in inaccurate travel times, though this error decreases as the distance between Bluetooth stations increases [32, 33]. In recent years, applications of Bluetooth and cell phones have attracted many researchers in estimating travel time. Consequently, various studies have been presented in detecting and addressing outliers [31, 34]. Li et al. [35] examined traffic data gathered from taxicabs over 24 days by applying the Temporal Outlier Discovery (TOD) concept to detect temporal outliers. It has been widely reported that point-to-point data sources like the Bluetooth are self-sufficient, as far as the prediction of travel times is concerned [36–38]. However, the assumption that is often made is that the data samples are large enough for computing the statistics of interest [39].

On the other hand, in travel time estimation method, there are several data collected methods. In Liu et al. [40], the data collected by TRANSMIT readers, Bluetooth sensors, and INRIX were assessed by comparing each to the 'ground truth' travel times collected by probe vehicles carrying GPS-based navigation devices.

All studies reported above are from the Western traffic condition which is largely homogeneous in nature. While the basic principles of using Bluetooth for travel time prediction remain the same for heterogeneous traffic conditions, significant adaptations are required to suit the Turkey road. This study is a pioneer in the investigation of Bluetooth as traffic sensor under Turkey traffic conditions and presents penetration and class-wise distribution analyses, in addition to proposing a method to estimate stream travel time from sampled Bluetooth data.

3 Data collection

3.1 Field data

The Bogazici Bosporus is a gravity-anchored suspension bridge with steel towers and inclined hangers [41]. It is 1,560 m but the study stretch selected was the initial section of the Bosporus in Istanbul, Turkey. The two locations were selected (at a distance of 850 m) (Fig. 1).

Bluetooth sensors were connected to laptops installed with 'Bluetooth View' software and placed at the two locations. When a vehicle fitted with a Bluetooth device crossed the sensor, the MAC ID and time stamp were automatically recorded by the software. Matching pairs were identified using data collected from the two locations. The difference between the arrival time stamps at the two locations for the corresponding matched MAC ID was considered as the travel time of the vehicle.

In all data collection activities, the traffic was simultaneously recorded on video in order to manually extract actual values of the total vehicle volume, classification, and travel time. Such information was useful for penetration analysis, class identification, and validation of estimated stream travel times.

Data were collected during two peak traffic periods (9.00–1.30 a.m. and 1.30–4.00 p.m.) for penetration analysis and also, someday, more data were collected (9.30 a.m.–9.30 p.m.) for class identification of vehicles.

3.2 Simulated data

The estimated stream travel time obtained from simulations studies was used for corroboration, because collection of data from field was strenuous and time-consuming. VISSIM was used to generate the data. Simulations were

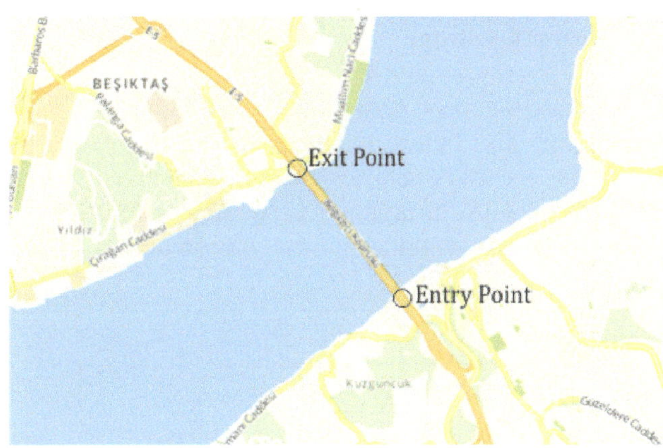

Fig. 1 Study site details

performed using actual field flow information obtained on September 20th and 21st 2014 from an automated sensor. Travel time, classified speed, and flow data were generated for 24 h periods on both days and used to estimate stream travel time and develop and validate models.

4 Results and discussion

4.1 Penetration rate analysis

The percentage of the actual traffic captured by the Bluetooth sensor is generally called as the sampling rate or penetration rate [5] and is an important parameter to ascertain the adequacy of sample size for analysis. The penetration rate analysis was performed separately at entry and exit locations. From the raw data obtained using the Bluetooth sensor, the number of vehicles captured by Bluetooth at each location was obtained. The total number of vehicles passing each of these locations was obtained by manually counting the vehicles in the video. Such counts were made at 5-min intervals over the entire duration of data collection. The number of samples captured by Bluetooth divided by total vehicle flow during a particular 5-min period was taken as the penetration rate for that period. The maximum and minimum penetration rates at the two locations during the morning and evening hours are shown in Fig. 2.

It can be seen that the average penetration rate of Bluetooth was about 5 %. This value is comparable to the reported penetration rate in previous studies [4]. The sample size obtained is more than the GPS penetration rate in Turkey cities, which is available only from public transport vehicles. The Bluetooth sensor is thus a promising source for traffic data collection for Turkey conditions.

4.2 Class identification of vehicles captured by Bluetooth sensor

After knowing the penetration rate, it is important to understand the distribution of these Bluetooth devices among various classes of vehicles to ensure applicability of Bluetooth technology in accurate travel time estimation. Four major classes of vehicles were considered in this study:

- Two wheelers (2W),
- Three wheelers (3W),
- Light motor vehicles (LMV), and
- Heavy motor vehicles (HMV).

To conduct this analysis, Bluetooth data and video data were collected simultaneously at the entry and exit locations. Matching pairs of MAC ID's and their corresponding

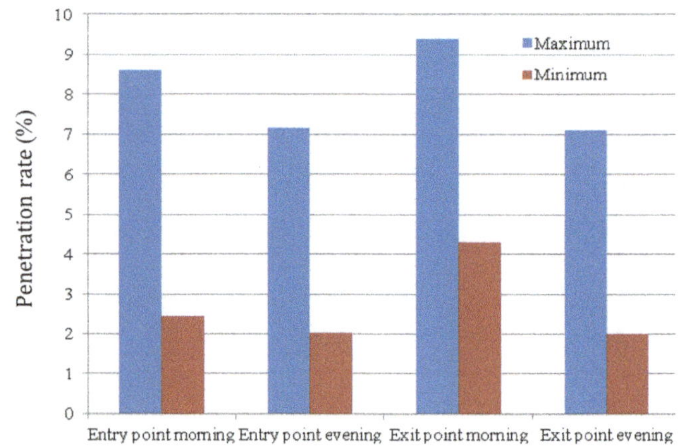

Fig. 2 Minimum and maximum penetration rates (%)

arrival time stamps obtained from Bluetooth were first identified for the two locations. The arrival time stamp of a vehicle detected at the first location was matched with the time in the corresponding video to identify the possible class of the detected vehicle. This was repeated at the second location for the same Bluetooth ID and the common vehicle matching the time stamp in both the videos was identified as the vehicle corresponding to that Bluetooth ID. The videos at the two locations were visually compared in parallel for the corresponding arrival time stamps in order to identify the matching vehicle and class. The per-cent-wise distribution of different classes captured by the Bluetooth on a sample day is shown in Fig. 3. Similar observations were obtained on other days too.

It can be seen that more than 91 % of vehicles captured using Bluetooth sensors were either LMV or 2W, indicating a possible bias of the Bluetooth to fast moving vehicles comprising mainly passenger cars and motorbikes. It is therefore necessary to project the data of these fast moving vehicles to the other two classes namely 3W and HMV,

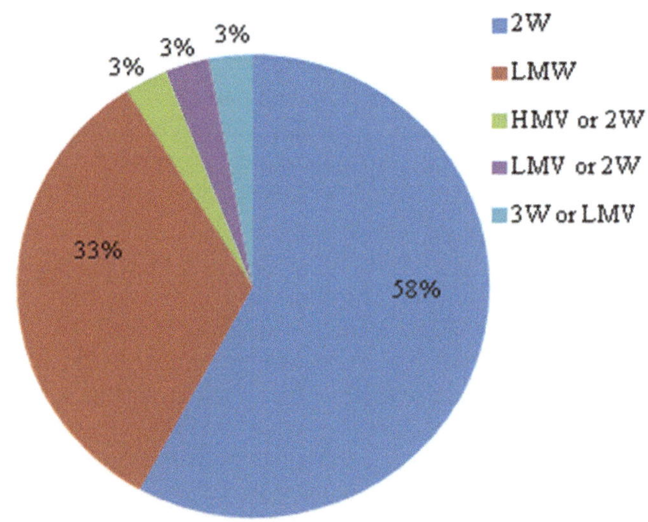

Fig. 3 Classes of vehicle identified by Bluetooth

which move relatively slow, before using the data for stream travel time estimation described in the following section.

4.3 Estimation of stream travel time

It can be seen from penetration rate analysis and class identification studies that the Bluetooth has an average penetration rate of 5 % in Turkey conditions, and captures mainly 2W and LMV, which are grouped as High Speed Vehicles (HSV) in this study. The travel time captured by Bluetooth is thus the average travel time of HSV. The travel time of slow moving vehicles must be estimated from the measured travel time in order to compute the complete stream travel time.

The average speeds of HSV were calculated from the measured HSV average travel times using the known distance between the data collection points as

$$V_{HSV} = \frac{D_{12}}{T_{Bluetooth}}, \quad (1)$$

where $T_{Bluetooth}$ is travel time measured by Bluetooth sensor, V_{HSV} speed of high speed vehicles and D_{12} distance between the two detector stations.

The speed of the entire stream must be estimated from this speed of the HSV. The traffic composition in the study corridor was observed to be 45 % 2W, 6 % 3W, 47 % LMV, and 2 % HMV. Thus, a weighted average speed of the entire stream can be written as

$$V_{stream} = 0.45V_{2W} + 0.06V_{3W} + 0.02V_{HMV} + 0.47V_{LMV}, \quad (2)$$

where V_{2W} is speed of 2-wheelers, V_{3W} speed of 3-wheelers, V_{HMV} speed of heavy motor vehicles, V_{LMV} speed of light motor vehicles and V_{stream} average stream speed.

The corresponding stream travel time can be expressed as follows:

$$T_{stream} = \frac{D_{12}}{V_{stream}}, \quad (3)$$

where T_{stream} is travel time of the stream.

To estimate the speed of slow moving 3W and HMV from the travel time data of HSV, weighted linear regression was adopted. The speed of the 3W/HMV was taken as the dependent variable and the speed of the HSV as the independent variable as

$$V_{3W} = \alpha_1 V_{HSV}, \quad (4)$$

$$V_{HMV} = \alpha_2 V_{HSV}, \quad (5)$$

where α_1 is coefficient obtained from regression between speed of 3W and HSV; α_2 coefficient obtained from regression between speed of HMV and HSV.

However, in linear regression, it is assumed that each data point provides precisely equal information, which may not always be the case. In such cases, to reduce error, weighted linear regression can be used, where each point is assigned a weight which regulates its influence in the estimation process. The scatter plot for the speed of 3W and HSV is given in Fig. 4, where the scatter plot is funnel shaped indicating that the points have varying influence in the estimation process.

A similar shape was also observed in the case of HMV vs HSV. Hence, the weighted linear regression was used for establishing the relationship between the speeds of 3W and HSV and speeds of HMV and HSV. Also, traffic conditions are dependent on the time of the day and there will be variations between peak periods and off-peak periods and nights. To identify these regimes, full day data obtained from VISSIM are plotted. A sample plot of travel time is given in Fig. 5. A travel time of 150 s was selected as a threshold between peak and off-peak flow regimes.

Based on this, the congested and normal flow regimes were identified and the relationship between the speeds of 3W and HMV with HSV were formulated separately for each of the flow regimes. The above scheme was implemented using simulated data in order to obtain sufficient

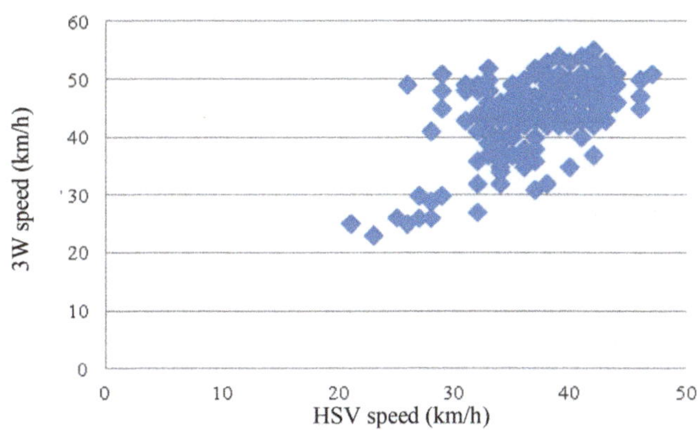

Fig. 4 Scatter plot of 3W and HSV Speed

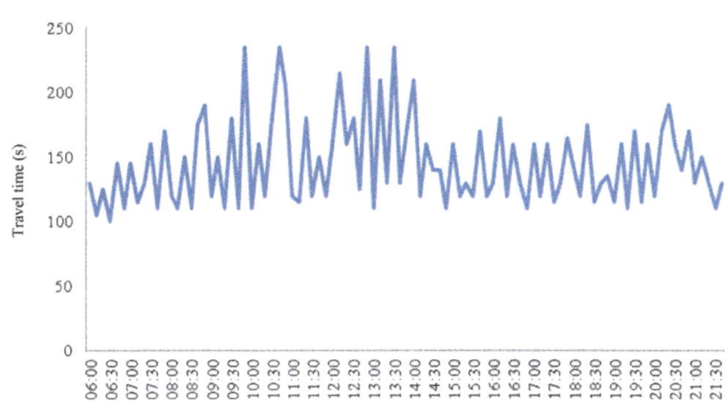

Fig. 5 Different flow regimes during the day

sample size. The speed of HSV was assumed to be the average speed of LMV and 2W. The weights were assigned as the square inverse of the speeds and weighted linear regression was carried out using R-software.

An estimation of the linear relation between the speeds of different classes of vehicles was made from the linear regression. The average stream speed was estimated using Eq. (2) using the known speed of HSV and estimated speeds of 3W and HMV. The different coefficients used to estimate average stream speed from speed of HSV for different times of the day are tabulated in Table 1. Tables 2 and 3 list the parameters obtained from the weighted linear regression for the 1:30–4:00 p.m. time period.

Validation was carried out using simulated data collected on a different day. Classified travel time and speed data were generated for the entire day and speed of HSV calculated as distance divided by travel time of HSV (average travel time of LMV and 2W). The speed of 3W and HMV was estimated using the coefficients α_1 and α_2 obtained as above, and the total stream speed was calculated using Eq. (2). The estimated and actual stream speed for sample duration of 9:00 a.m.–1:30 p.m. are shown in Fig. 6.

The estimated speeds agree well with the actual speed values, as shown in Fig. 6. The corresponding stream travel time was calculated using Eq. (3) above and is shown in Fig. 7, along with the actual stream travel. Mean Absolute Percentage Error (MAPE), given by the following equation was used to quantify errors.

$$\mathrm{MAPE} = \frac{1}{n}\sum_{i=1}^{n}\left|\frac{T_e - T_o}{T_o}\right| \times 100, \tag{6}$$

Table 1 Coefficients for 3W (α_1) and HMV (α_2) for different flow regimes

Class	9:00 a.m.–1:30 p.m.	1:30–4:00 p.m.	7:00–9:30 p.m.
Linear coefficients			
α_1	0.84	0.783	0.81
α_2	0.83	0.793	0.83

Table 2 R-output for linear regression for HMV

Residuals					
Min		1Q	Median	3Q	Max
−0.79380		−0.07010	0.02565	0.08200	0.26518
Coefficients					
	Estimate		SE	t value	Pr(>|t|)
HSV1	0.7938		0.0194	40.91	<2e−16***

Residual standard error: 0.08983 on 144 degree of freedom (DF) (87 observations deleted due to missingness)

Multiple R-squared: 0.9871, Adjusted R-squared: 0.987

F-statistic: 1.105e+04 on 1 and 144 DF, p value: <2.2e−16

Signif. codes: 0 '***,' 0.001 '**' 0.01 '*' 0.05 '.' 0.1 '·' 1

Table 3 R-output for linear regression for 3W

Residuals					
Min		1Q	Median	3Q	Max
−0.27594		−0.05098	0.00462	0.04104	0.33671
Coefficients					
	Estimate		SE	t value	Pr(>|t|)
HSV2	0.78406		0.00746	105.1	<2e−16***

Residual standard error: 0.08983 on 144 DF

Multiple R-squared: 0.9871, Adjusted R-squared: 0.987

F-statistic: 1.105e+04 on 1 and 144 DF, p value: <2.2e−16

Signif. codes: 0 '***' 0.001 '**' 0.01 '*' 0.05 '.' 0.1 '·' 1

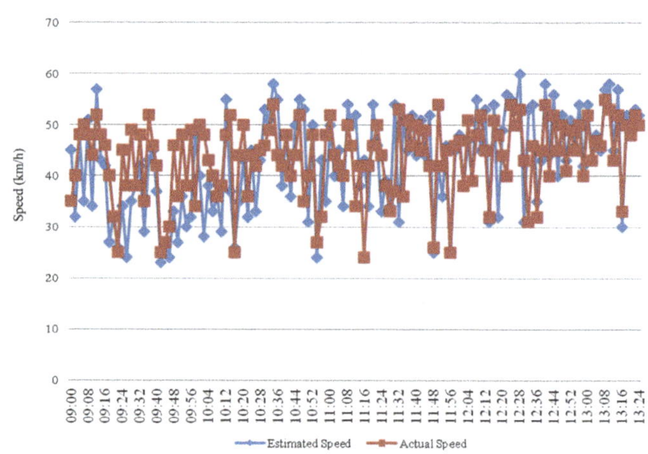

Fig. 6 Estimated and actual stream speed during 9:00 a.m.–1:30 p.m.

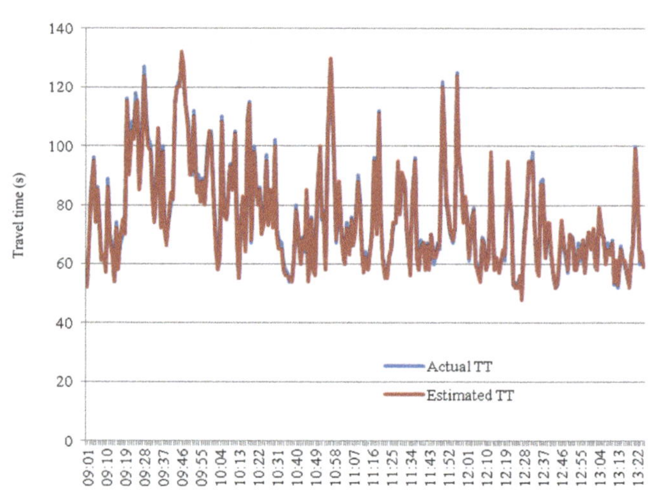

Fig. 7 Actual and estimated travel time for 9:00 a.m.–1:30 p.m.

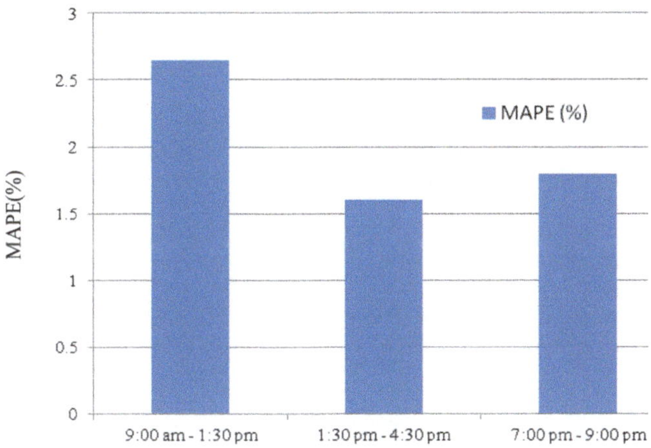

Fig. 8 MAPE values for different time zones

where T_e and T_o are the estimated and observed travel times.

Similar results were obtained for other time zones and the MAPE obtained for different time zones are shown in Fig. 8. It can be seen that the average error is around 2 %, indicating excellent performance of the scheme.

5 Conclusions

In this study, Bluetooth data collected from a busy road in Istanbul Bogazici Bosporus were analyzed for possible use as a traffic sensor. Penetration rate analysis, class identification, and estimation of stream travel time using Bluetooth have been demonstrated for the first time in Turkey. The penetration rate was found to be about 5 %. Majority of the detected vehicles were found to be two wheelers and light motor vehicles, indicating a bias in the data.

Estimating the stream travel time for an entire stream from limited Bluetooth data is a challenge and not many studies have addressed this problem until now. In this study, the data obtained from 2W and LMV were extrapolated to estimate travel times of other classes of vehicles such as three wheelers and heavy motor vehicles to get a true estimate of stream travel time. The analysis was carried out separately for peak and off-peak traffic flow conditions. The study established linear relationships between speeds of different classes of vehicles through weighted linear regression analysis and estimated stream travel time as distance divided by stream speed. This technique was successful in estimating stream travel time accurately with an average MAPE of 2 %.

This study has shown that the Bluetooth is a cost-effective technology for estimation of travel time for heterogeneous traffic conditions. This study is the first of its kind to demonstrate the applicability of Bluetooth for travel time estimation under heterogeneous traffic conditions.

6 Future study

Much (more) transportation departments use inductive loops, traffic cameras, or stationary sensors to measure travel time for estimate travel time. Although these traditional systems are proven techniques of collecting traffic data, they each also have some shortcomings. Recently and alternatively, Floating Car Data method, Bluetooth, and GPS technologies have introduced an effortless method for traffic data collection and hereby traffic performance measurement. In the future study is going to try to investigate the use of GPS technology for stream travel time estimation under Turkey traffic conditions. In this case, some comparable or differences can be found between two technologies such as Bluetooth and GPS. After all, companies like Waze, TomTom, and Google are already able to provide city and country level traffic condition using GPS and other sources of data. For this reason, comparison is made among all technologies for travel time estimation under different conditions.

References

1. Lin HE, Zito R, Taylor MAP (2005) A review of travel time prediction in transport and logistics. Proc East Asia Soc Transp Stud 5:1433–1448
2. Soriguera Martí F, Rosas Díaz DM, Abeijón Monjas D, Thorson Bofarull L, Robusté Antón F (2007) Travel time estimation from multiple data sources. In: 11th world conference on transport research. University of California, Berkeley
3. Stiener A (2008) An improved method for travel time estimation on long freeway sections. In: conference paper STRC 2008, 8th Swiss transport research conference, Monte Verità, Ascona
4. Haghani A, Hamedi M, Sadabadi KF, Young S, Tarnoff P (2010) Data collection of freeway travel time ground truth with Bluetooth sensors, no. 2160, transportation research record, transportation research board of the national academies, Washington, D.C, pp 60–68
5. Young S (2008) Bluetooth traffic monitoring technology: concept of operation and deployment guidelines. University of Maryland—Centre for Advanced Transportation Technology, Maryland
6. Coifman B, Kim S (2009) Speed estimation and length based vehicle classification from freeway single-loop detectors. Transp Res Part C Emerg Technol 17:349–364
7. Coifman B, Krishnamurthy S (2007) Vehicle reidentification and travel time measurement across freeway junctions using the

existing detector infrastructure. Transp Res Part C Emerg Technol 15(3):135–153
8. Bhaskar A, Qu M, Nantes A, Miska M, Chung E (2015) Is bus overrepresented in Bluetooth MAC scanner data? Is MAC-ID really unique? Int J Intell Transp Syst Res 13(2):119–130
9. Cortes C, Lavanya R, Oh JS, Jayakrishnan R (2002) General-purpose methodology for estimating link travel time with multiple-point detection of traffic. Transp Res Rec J Transp Res Board 1802:181–189
10. Dailey DJ (1999) A statistical algorithm for estimating speed from single loop volume and occupancy measurements. Transp Res Part B Methodol 33(5):313–322
11. Wang Y, Nihan NL (2003) Can single-loop detectors do the work of dual-loop detectors? J Transp Eng 129(2):169–176
12. Qian QQ, Lin S, He ZY, Li XP (2012) Travelling wave time-frequency characteristic-based fault location method for transmission lines. Gener Transm Distrib IET 6(8):764–772
13. Sharma A, Bullock D, Bonneson J (2007) Input-output and hybrid techniques for real-time prediction of delay and maximum queue length at signalized intersections. Transp Res Rec J Transp Res Board 2035:69–80
14. Vigos G, Papageorgiou M, Wang Y (2008) Real-time estimation of vehicle-count within signalized links. Transp Res Part C Emerg Technol 16(1):18–35
15. Stevanovic AZ, Martin PT (2008) Assessment of the suitability of microsimulation as a tool for the evaluation of macroscopically optimized traffic signal timings. J Transp Eng 134(2):59–67
16. Wang Y, Malinovskiy Y, Lee UK, Wu YJ (2011) Investigation of bluetooth-based travel time estimation error on a short corridor. In: Transportation research board 90th annual meeting (No. 11–3056)
17. Bakula C, Schneider IV WH, Roth J (2011) Probabilistic model based on the effective range and vehicle speed to determine Bluetooth MAC address matches from roadside traffic monitoring. J Transp Eng 138(1):43–49
18. Welsh E, Murphy P, Frantz JP (2002) Improving connection times for Bluetooth devices in mobile environment. In: Proceedings of the 2002 international conference, on fundamentals of electronics communications and computer science of IEICE (ICFS 2002), 27–28 Mar 2002, Tokyo, Japan, pp 1–5
19. Ahmed H, El-Darieby M, Abdulhai B, Morgan Y (2008) Bluetooth- and Wi-Fi-based mesh network platform for traffic monitoring. In: TRB 87th annual meeting compendium of papers DVD, 13–17 Jan 2008, Washington, DC, pp 1–11
20. Sharifi E, Hamedi M, Haghani A, Sadrsadat H (2011) Analysis of vehicle detection rate for Bluetooth traffic sensors: a case study in Maryland and Delaware. In: Proceedings of the 18th world congress on intelligent transport systems, 16–20 Oct 2011, Orlando, FL, pp 1–12
21. Quayle SM, Koonce P, DePencier D, Bullock DM (2010) Arterial performance measures with media access control readers: Portland, Oregon, pilot study. Transp Res Rec J Transp Res Board 2192:185–193. doi:10.3141/2192-18
22. Quayle SM, Koonce P, DePencier D, Bullock D (2010) Arterial performance measures using MAC readers: Portland pilot study. In: Transportation research board annual meeting proceedings CD-ROM
23. Wasson JS, Sturdevant JR, Bullock DM (2008) Real-time travel time estimates using media access control address matching. ITE J 78(6):20–23
24. Sadabadi KG, Hamedi M, Haghani A (2010) Evaluating moving average techniques in short-term travel time prediction using an AVI dataset. In: Transportation research board, annual meeting proceedings CD-ROM
25. Huston TranStar and Bluetooth Traffic Monitoring (2012). http://traffic.houstontranstar.org/bluetooth/transtar_bluetooth.html. Accessed 22 July 2013
26. Jaume B, Lídia M, Laura M, Carlos C(2010) A Kalman-filter approach for dynamic OD estimation in corridors based on Bluetooth and WiFi data collection. 12th WCTR, Lisbon, Portugal
27. Sawant H, Tan J, Yang Q, Wang Q (2004) Using Bluetooth and sensor networks for intelligent transportation systems. In: IEEE intelligent transportation systems conference, Washington, D.C., USA
28. Bullock D, Haseman R, Wasson J, Spitler R (2010) Automated measurement of wait times at airport security: deployment at Indianapolis international airport, Indiana. Transp Res Rec J Transp Res Board 2177:60–68
29. Blogg M, Semler C, Hingorani M, Troutbeck R (2010) Travel time and origin-destination data collection using Bluetooth MAC address readers. In: Australasian transport research forum, pp 1–15
30. Li J, Van Zuylen H, Liu C, Lu S (2011) Monitoring travel times in an urban network using video, GPS and Bluetooth. Procedia Soc Behav Sci 20:630–637. doi:10.1016/j.sbspro.2011.08.070
31. Van Boxel D, Schneider W IV, Bakula C (2011) Innovative real-time methodology for detecting travel time outliers on interstate highways and urban arterials. Transp Res Rec J Transp Res Board 2256:60–67
32. Malinovskiy Y, Lee UK, Wu YJ, Wang Y (2011) Investigation of Bluetooth-based travel time estimation error on a short corridor. In: Transportation research board 90th annual meeting (No. 11-3056)
33. Puckett DD, Vickich MJ (2010) Bluetooth®-based travel time/speed measuring systems development (No. UTCM 09-00-17)
34. Horn C, Klampfl S, Cik M, Reiter T (2014) Detecting outliers in cell phone data: correcting trajectories to improve traffic modeling. Transp Res Rec J Transp Res Board 2405:49–56
35. Li X, Li Z, Han J, Lee JG (2009) Temporal outlier detection in vehicle traffic data. In: IEEE 25th international conference on data engineering, 2009. ICDE'09, IEEE, pp 1319–1322
36. Khoei AM, Bhaskar A, Chung E (2013) Travel time prediction on signalised urban arterials by applying SARIMA modelling on Bluetooth data. In: 36th Australasian Transport Research Forum (ATRF) 2013
37. Khosravi A, Mazloumi E, Nahavandi S, Creighton D, Van Lint JWC (2011) A genetic algorithm-based method for improving quality of travel time prediction intervals. Transp Res Part C Emerg Technol 19(6):1364–1376
38. Qiao W, Haghani A, Hamedi M (2013) A nonparametric model for short-term travel time prediction using bluetooth data. J Intell Transp Syst 17(2):165–175
39. Nantes A, Ngoduy D, Miska M, Chung E (2015) Probabilistic travel time progression and its application to automatic vehicle identification data. Transp Res Part B Methodol 81:131–145
40. Liu X, Chien S, Kim K (2012) Evaluation of floating car technologies for travel time estimation. J Mod Transp 20(1):49–56
41. GDH (2014) General directorate of highways in Turkey: project information about the Bosphorus bridge report

6

Evolution of the electrical contact of dynamic pantograph–catenary system

Guangning Wu[1] · Wenfu Wei[1] · Guoqiang Gao[1] · Jie Wu[1] · Yue Zhou[1]

Abstract A good contact between the pantograph and catenary is critically important for the working reliability of electric trains, while the basic understanding on the electrical contact evolution during the pantograph–catenary system working is still ambiguous so far. In this paper, the evolution of electric contact was studied in respects of the contact resistance, temperature rise, and microstructure variation, based on a home-made pantograph–catenary simulation system. Pure carbon strips and copper alloy contact wires were used, and the experimental electrical current, sliding speed, and normal force were set as 80 A, 30 km/h, and 80 N, respectively. The contact resistance presented a fluctuation without obvious regularity, concentrating in the region of 25 and 50 mΩ. Temperature rise of the contact point experienced a fast increase at the first several minutes and finally reached a steady state. The surface damage of carbon trips in microstructure analysis revealed a complicated interaction of the sliding friction, joule heating, and arc erosion.

Keywords Pantograph–catenary system · Electrical contact · Electric trains · Contact resistance · Temperature rise · Microstructure

1 Introduction

Pantograph–catenary system bears the task of energy transfer from the electrical power grid to electric trains; therefore, the electrical contact between the pantograph and catenary is directly related with the working reliability and safety of the trains. In the working situation, the overhead contact wire and sliding strip suffer from the multiple impacts of zigzag sliding friction, joule heating, and offline arcing erosion, which make the evolution of electrical contact rather complex. Especially the random arcing become more predominant at higher speed, heavier load, and cold weather conditions, which has already turned into a great obstacle for the stable current collection [1].

Among the great amounts of researches related with the pantograph–catenary system, large efforts were made on the mechanical vibration coupling analysis, aerodynamic force calculation, and active control design [2–5]. Relatively, researches on the electrical contact itself were paid less attention to. Besides the comparison of friction wear properties under different current and speed conditions [6, 7], there were also constant interests in the electrical temperature characteristics [8], conducted and radiated electromagnetic emission, etc. [9, 10]. All the above studies were closely related to the electrical contact of the pantograph–catenary system and therefore, possess significant importance to prompt the development of high-speed railway technology.

However, most previous researches were focused on only one or two aspects of the electrical contact and could reveal limited information on the dynamic evolution process. Simultaneous investigations into the different aspects of the electrical contact, such as the conductivity variation, temperature rise, as well as the microstructure characteristics, would be absolutely helpful for a better understanding of the pantograph–catenary system.

In this work, a comprehensive study on the sliding electrical contact of a pantograph–catenary system was carried out. Simultaneous voltage and current were

✉ Wenfu Wei
wfwei@home.swjtu.edu.cn

[1] School of Electrical Engineering, Southwest Jiaotong University, Chengdu 610031, China

recorded to obtain the conductivity property, while infrared thermometry was utilized to get the temperature rise. In the end, the surface damages and material transfer were studied through microstructure analysis. Experimental results of this paper may help reveal the physical mechanisms for the sliding electrical contact of pantograph–catenary system and also provide a benchmark for the modeling of pantograph–catenary system.

2 Experimental setup

2.1 The pantograph–catenary simulation test system

Considering the real working conditions of a pantograph–catenary system, a set of pantograph–catenary contact test apparatus was set up. As shown in Fig. 1, this experimental system consisted of an AC power source, adjustable resistance, inductance loads, a rotating wheel, and a carbon strip; a partial photo of the experimental apparatus is given in Fig. 2. The AC power source adopted the electrical–electronic technology and effectively realized a stable output current from 0 to 200 A in a compact structure. The adjustable resistance and inductance were used to simulate the different train loads. The copper alloy wire with a realistic geometry was fixed in the groove around the outer edge of the rotating wheel. The pantograph strip, made of carbon materials, was housed on a carriage driven by two different motors. The rotating movement of contact wire and the horizontal translation of pantograph strip successfully realized the zigzag motion in the real working condition. Furthermore, the pantograph strip could move in the vertical direction to simulate the switch status. During the experiments, the electrical current came from the constant current power source, passing through the contact wire, the pantograph strip, and the adjustable resistive–inductance load, and eventually went back to the power source. The

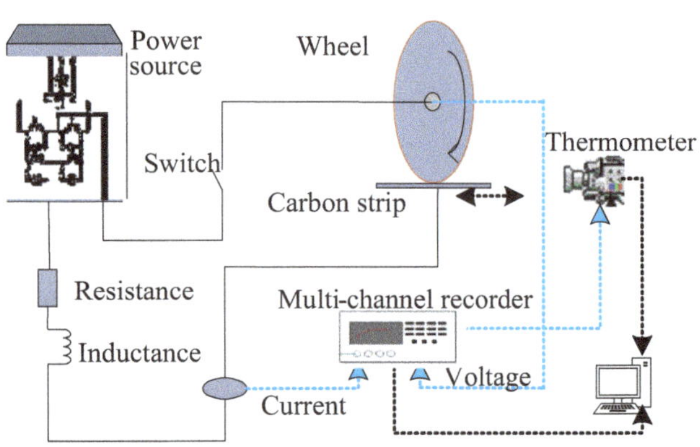

Fig. 1 Layout of the experimental setup

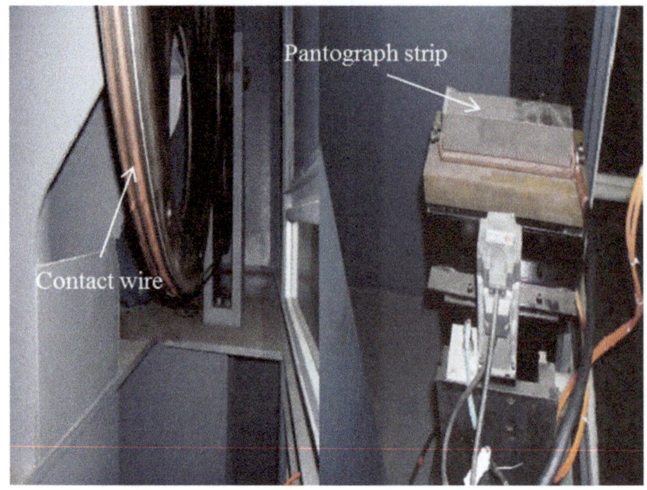

Fig. 2 Partial physical photo of the experimental apparatus

above electrical layout was similar to that of a real working pantograph–catenary system.

In this work, experimental parameters of the electrical current, sliding speed, and normal force was fixed as 80 A, 30 km/h, and 80 N, respectively. The running time of each experiment was lasted for 10 min.

2.2 Measurement methods

The contact voltage across the wheel and strip was measured by a 1,000:1 high voltage probe (TPP0250, Tektronix), while the current was measured by a Rogowski coil (i400s, Fluke). The two-dimensional infrared images containing temperature information were recorded by an infrared thermometer (NS9500, NEC InfraRed Camera) with a record speed of five frames per second. It is worth noting that since the most previous study on the temperature characteristics utilized a thermocouple probe buried in the strip, our infrared measurement provided a non-intrusive as well as spatial-resolution detection.

The electrical signals were both stored into a multichannel recorder, with a constant sampling frequency of 1 kHz. Thermometer could be triggered either manually or automatically. After the experiments, the carbon strip was disassembled for microstructure examination by a scanning electron microscope.

3 Results and discussions

3.1 Characteristics of electrical parameters

Parts of the contact voltage and current waveforms are shown in Fig. 3. Both of the waveforms hold a sinusoidal shape on the whole, except for the several voltage

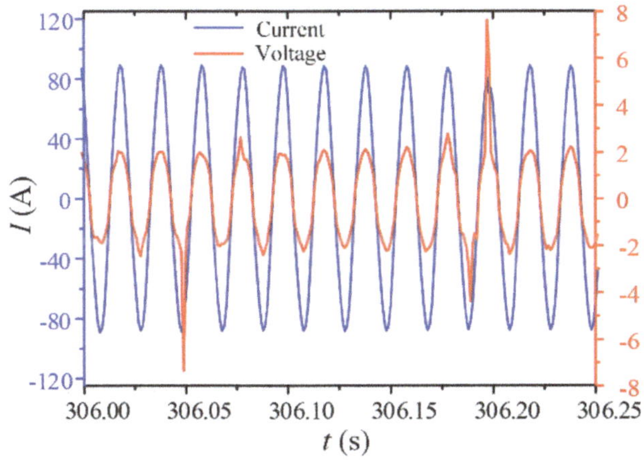

Fig. 3 Contact voltage and electrical current with time

distortion regions which were probably accompanied by pantograph arcing or sparking. The phase of voltage waveform stays the same as that of current waveform, due to the resistive properties of the contact area. Amplitude of the contact voltage reaches around 2 V. The sharp changes of contact voltage, for example at 306.05 and 306.19 s, were directly resulted from the steep drop of contact force, which was expected to induce the occurrence of pantograph arcing. According to our previous study on the pantograph arcing, the resistance of arc plasma was generally one order of magnitude higher than the normal contact resistance.

The occurrence of pantograph arcing not only distorted the electrical power quality, but also resulted in the severe temperature rise on the electrodes surfaces, ablated material, and worsen the current collection system damages. The primary cause for the pantograph arcing should be attributed to the unexpected vibration between contact wire and strip due to the geometry singularity. Therefore, the occurrence of pantograph arcing was quite random, which

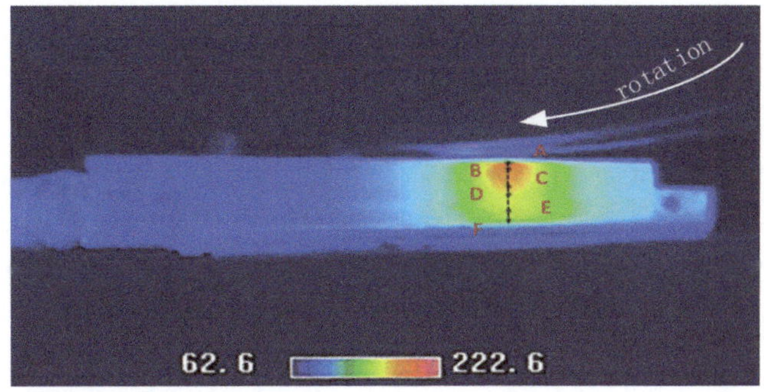

Fig. 5 Typical infrared image of the pantograph strip

brought in great difficulties in the study of electrical contact properties. Previous researches have shown that the occurrence probability increased as the running speed got higher, while the arc erosion became much more serious [11]. Although the comprehensive impacts of arcing on the surface damages were still not clarified well, the electrical parameter characteristics had been well analyzed based on the classic Cassie-mayr's equation [12, 13], while the evolution of spatial temperature distribution could be further surveyed by the magnetic-hydrodynamic theory.

3.2 Contact resistance with time

From the above discussion, the electrical contact exhibited an ideal resistive property. Therefore, the contact resistance R could be directly obtained using the Ohm's law, $R = U/I$, where U and I were the experimentally recorded contact voltage and current, respectively.

The deduced contact resistance with time is shown in Fig. 4, with an inset for the variation of the contact resistance in a shorter time. Results indicate that the contact resistance was concentrated in the region of 25 and 50 mΩ

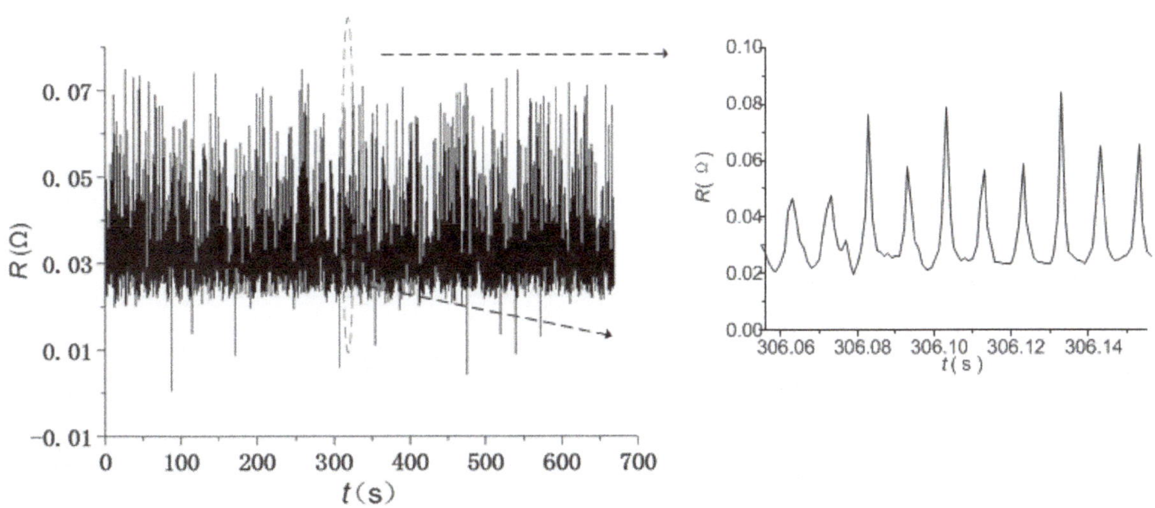

Fig. 4 Variation of the contact resistance with time

during the entire period and experienced a relatively small fluctuation. The fluctuation period was 10 ms, just a half cycle of the AC power frequency. We regarded that the regular crossing zero of AC current contributed to the regular rise of contact resistance. The fluctuation irregularity in resistance values could be explained from the variation of effective contact area. The effective contact area was the actual contact parts from the microscopic aspect, which suffered from the friction work, joule heat, and/or arcing erosion. During the constant sliding process with current, the effective contact area was always going through a dynamic variation of being destroyed and reformation from one place to another. Therefore, the contact resistance depended on the effective contact area and experienced an obvious fluctuation. On the other hand, the average contact resistance was highly depended on the contact force, material resistivity, running speed, and current. The former two parameters determined primarily the static contact resistance, while the latter two parameters influenced the dynamic variation process significantly.

3.3 Temperature distribution of pure carbon strip

Material properties, such as resistivity, hardness, and friction coefficient, are temperature dependent, and will influence the final wear loss level of the material. Hence an investigation on temperature characteristic during the pantograph–catenary working is quite necessary. Typical infrared image of the pantograph strip is shown in Fig. 5. It can be seen that the highest temperature was just located at the contact point of the pantograph strip and contact wire. With the distance away from the contact point increasing, the temperature decreased gradually. On the whole, a Gaussian-like temperature distribution in the pantograph strip could be found. The temperature of the contact wire remained at quite low level due to the constant rotation.

The temperature variation of the pantograph strip with time is shown in Fig. 6, where typical infrared images at different stages are also given. It should be noticed that the time at zero was shortly after the previous experiment, and local regions of the pantograph strip were still in warm state. Nevertheless, the overall variation in Fig. 6 still properly represents the temperature rise process. It can be found that after a fast temperature rise stage during the first several minutes, the pantograph strip came to a relative steady state of around 175 °C finally. During this process, pantograph arcing could take place, resulting in the surface temperature to reach 200 °C for a short period. For the final steady state, spatial temperatures at six typical locations as shown in Fig. 5 are given in Table 1.

Table 1 Temperature distribution of pure carbon strip

	A	B	C	D	E	F
T (°C)	157.5	178.2	147.9	138.5	129.4	91.7

It is worth mentioning that the contact resistance was significantly influenced by both the material resistivity and the condition of contact spots according to the classical electrical contact theory. On the one hand, the resistivity of copper changed little with the temperature rising from 100 to 200 °C, considering that the temperature coefficient of copper resistance was only 0.00393 /°C. On the other hand, the number and radius of contact spots could change obviously due to the high temperature at the contact area, accompanied with material softening. To make clear the accurate effects of temperature on the whole contact resistance, it is necessary to conduct experiments at constant temperatures of different levels in the future work.

3.4 SEM observation of the worn surfaces of pure carbon strips

Surface conditions of the carbon strip before and after the experiments are given in Fig. 7. As shown in Fig. 7a, the surface of the original pure carbon strip is very smooth without wear scar. After several minutes of sliding between

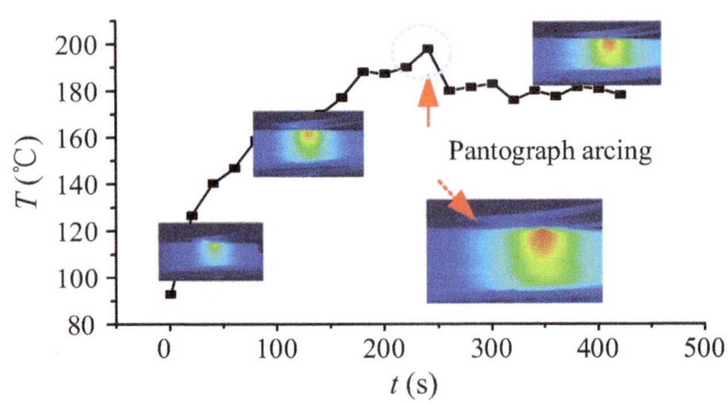

Fig. 6 Maximum temperature of pantograph strip, with typical infrared images inset

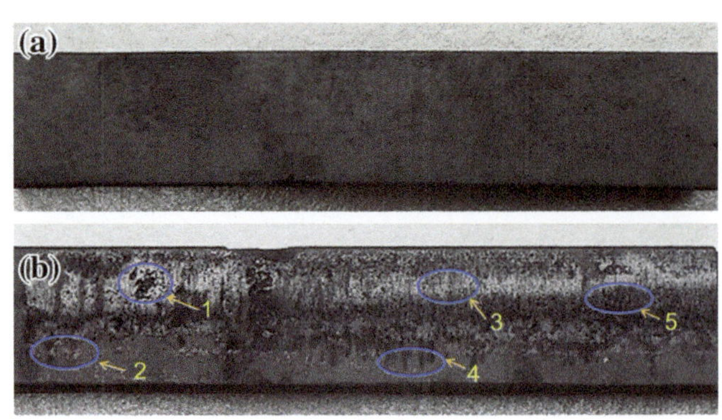

Fig. 7 Surface morphologies of pure carbon strips before (**a**) and after (**b**) test

pantograph strip and catenary wire, a new surface morphology with pits and craters appeared, as shown in Fig. 7b. The new morphologies were generally classified as the following five types (corresponding to regions 1–5 in Fig. 7, respectively):

- *Craters* Diameters of typical craters could reach as high as several millimeters. Serious arc discharge accompanying the high-speed friction would be major contributions to the formation of craters.
- *Dull-red area* There were irregular dull-red areas distributed on the strip surface, indicating that the material of copper alloy wire was gradually transferred to the strip. This area would be referenced as material transfer area hereinafter.
- *Bright area* White bright area was much brighter than the basic carbon materials, which could possibly result from the deposition of oxidative materials.
- *Dark stream line* Dark stream line was in the same direction of the contact wire rotation.
- *Pits* There were great amounts of pits, with much smaller diameter than the crater, widely distributed on the strip surface.

Both the dark stream line and the pits could be the direct evidence of arc discharge occurrence.

The typical microstructures in different areas of the pantograph strip are shown in Fig. 8. Figure 8a presents the original strip surface used for comparison, and Fig. 8b–d is the enlarged pictures of material transfer area, bright area, and arc ablation area, respectively. By comparing the surface pictures before and after test, we can find that the surface of strip became bumpy, and a large number of wear debris, arc ablation craters, and spalling blocks appeared. Meanwhile, some flake-like fragments can also be found from the worn surface.

When the pantograph strip rubbed against the catenary wire with the current, arc discharges occurred frequently because of the air gap of pantograph–catenary system breakdown. A lot of heat was generated on the contact surface, which included arc heat, Joule heat, and friction heat, resulting in the plastic deformation of material. The high temperature generated by the pantograph–catenary arc would ablate pantograph strip and catenary wire, simultaneously melting metal material and gasifying carbon

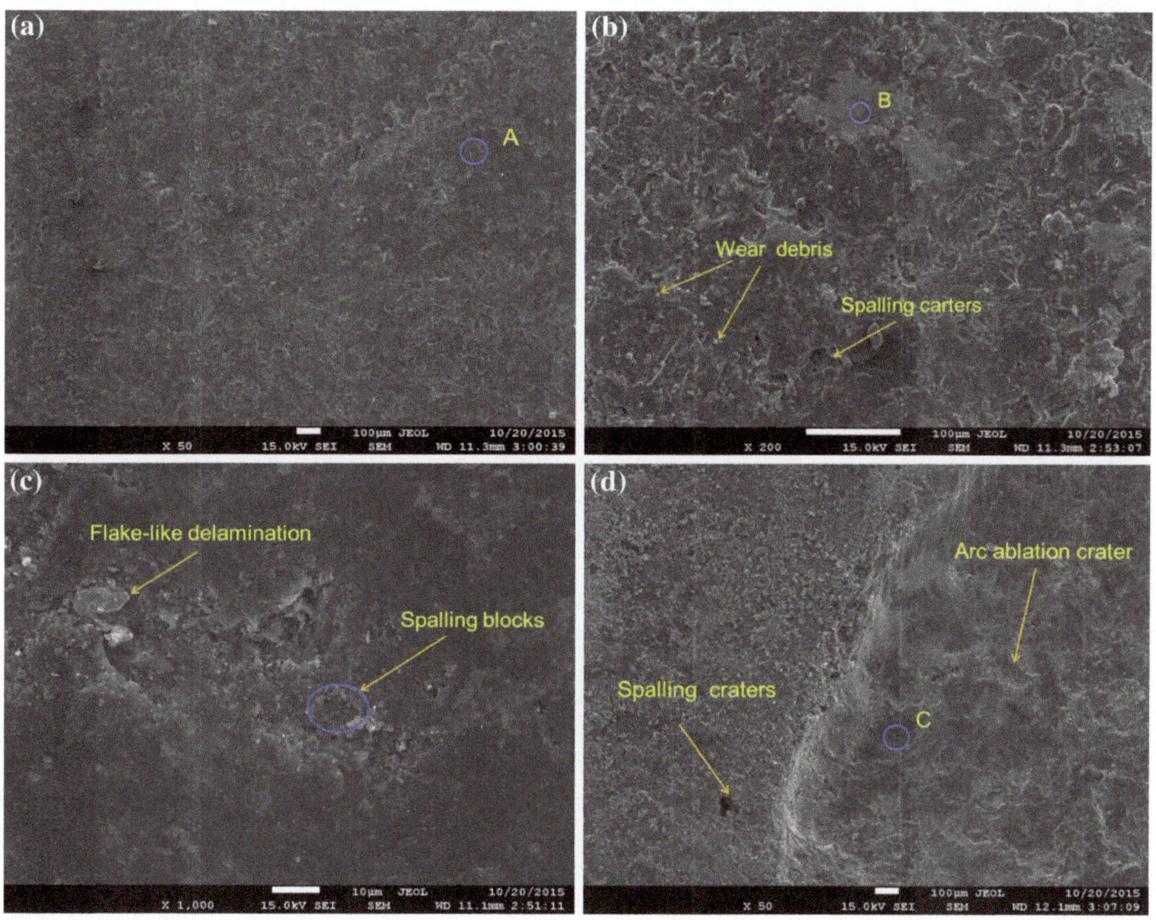

Fig. 8 SEM images of pure carbon strips in four different cases. **a** Original surface before test. **b** Material transfer area. **c** Bright area. **d** Arc ablation area

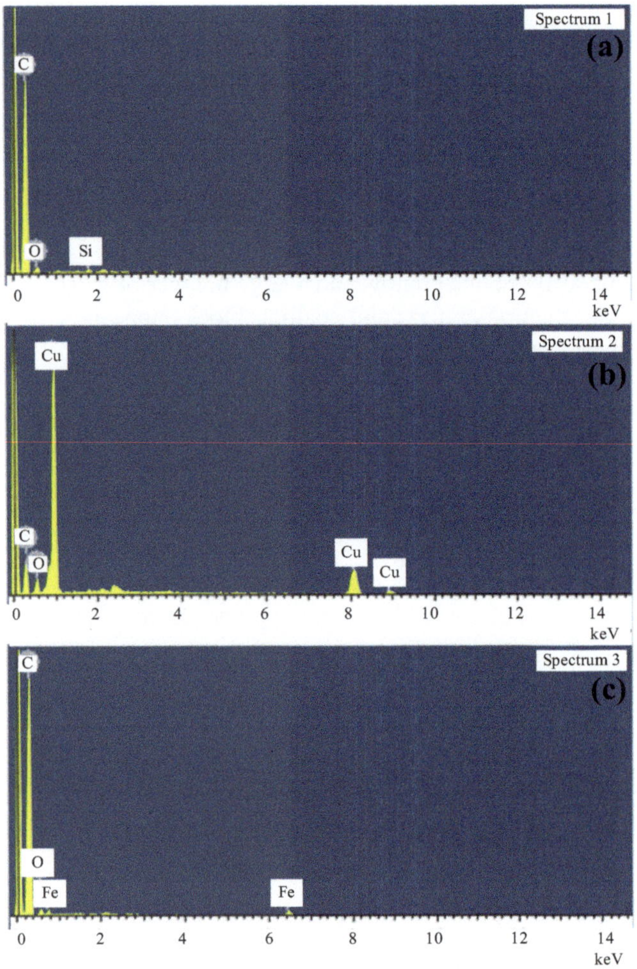

Fig. 9 EDX spectrums of pure carbon strips in three regions. **a** Original area. **b** Material transfer area. **c** Arc ablation area

Table 2 Elementary composition of the original area

Element	Mass percentage (%)	Atomic percentage (%)
C	39.69	48.24
O	52.00	47.44
Si	8.31	4.32

material from the pantograph strip and contact wire [14]. As a result, the number of real contact points on the contact surface decreased, and the electrical contact condition of the pantograph–catenary system tended to deteriorate. The friction force, friction heat, and shear force all increased, producing a few of wear cracks. With the sliding continually going on, flake-like delaminations (Fig. 8c) were formed on the contact surface. That was attributed to the serious material adhesion caused by high temperature, and the fracturing delamination caused by shearing force and normal stress. The flake-like delaminations would further evolve into spalling blocks by shearing force, as shown in

Table 3 Elementary composition of the material transfer area

Element	Mass percentage (%)	Atomic percentage (%)
C	1.33	5.56
O	7.15	22.37
Cu	91.51	72.07

Table 4 Elementary composition of the arc ablation area

Element	Mass percentage (%)	Atomic percentage (%)
C	22.81	44.55
O	23.04	33.79
Fe	32.79	13.77
Cu	21.36	7.89

Fig. 8b and c. In addition, we can find a large number of debris on the surface of the strip in Fig. 8b. It could be concluded that abrasive wear had played an important part in material transfer process. There are several craters distributed on the strip in Fig. 8d, which could be the evidence of intense arc discharges. In summary, in the whole process of rubbing, the temperature rise of the contact couple was an important factor which had a significant influence on the wear rate of the pantograph–catenary system.

Furthermore, the original area, arc ablation area, and material transfer area were measured by energy dispersive X-ray spectroscopy (EDX), and the results are shown in Fig. 9. The analyzed point in Fig. 9a corresponsds to the point A in Fig. 8a, where dominant carbon elements and slight oxygen and silicon elements could be seen. Specific element mass percentages are given in Table 2. As for the material transfer area corresponding to the point B in Fig. 8b, the EDX result revealed abundant copper material. The material transfer could be a result of a series of processes. During the process of rubbing, the hardness of materials could be effectively decreased by the temperature rise from several heat sources, and the interface between strip and wire would become more adhesive. The adhesive interface combined with frictional force could result in the material transformation from catenary wire to pantograph strip. The element percentage of material transfer area is shown in Table 3. In the arc ablation area, the spalling crater could be found on the surface of strip obviously. At the same time, the arc ablation pits and dark stream-lines were also the products of arc discharge. From the EDX measurement for point C in Fig. 8d, abundant carbon elements as well copper and iron elements could be discovered at the crater bottom area (Fig. 9c). Specific element mass percentages are given in Table 4. It suggests that material transfer also occurred from the copper contact wire to the strip during the crater formation, besides the previously described material transfer area. Therefore, the arc discharge could be expected as a significant factor resulting in the material transfer of pantograph–catenary system. However, the damage mechanism of the pantograph–catenary system included both mechanical and electrical wear. The accurate wear process would possibly be well understood by a multiple physics coupling simulation in the future work.

4 Conclusion

In this paper, the evolution of the electrical contact between pantograph and catenary was studied in respects of the electrical conductivity, temperature rise, as well as the microstructure variation. The following conclusions can be drawn.

- The contact resistance was majorly concentrated in the region of 25 and 50 mΩ, and a relative fluctuation without obvious regularity could be found.
- The temperature of the contact point experienced a fast rise int the first several minutes but finally reached a steady state of 175 °C. Spatial temperature exhibited a Gaussian-like distribution, and the random arc discharge could induce a sharp increase of local temperature.
- The evolution mechanisms were discussed from the microstructure analysis of five typical morphologies, and arc discharges was expected to be responsible for the formations of dark stream line, pits, and craters, while the material transfer process was dependent on adhesion wear and arc discharge.

Acknowledgments This work was supported by the National Natural Science Foundation of China (Nos. U1234202 and 51577158), the National Science Foundation for Distinguished Young Scholars of China (No. 51325704), and the Fundamental Research Funds for the Central Universities (No. A0920502051505-19).

References

1. Midya S, Bormann D, Larsson A, Schutte T, Thottappillil R (2008) Understanding pantograph arcing in electrified railways—influence of various parameters. In: 2008 IEEE international symposium on electromagnetic compatibility, vols 1–3, pp 592–597
2. Zhang W, Zhou N, Li R, Mei G, Song D (2011) Pantograph and catenary system with double pantographs for high-speed trains at 350 km/h or higher. J Mod Transp 19(1):7–11
3. Collina A, Bruni S (2002) Numerical simulation of pantograph-overhead equipment interaction. Veh Syst Dyn 38(4):261–291
4. Chater E, Ghani D, Giri F, Haloua M (2015) Output feedback control of pantograph–catenary system with adaptive estimation of catenary parameters. J Mod Transp 160(4):257–266
5. Zhai WM, Cai CB (1998) Effect of locomotive vibrations on pantograph–catenary system dynamics. Veh Syst Dyn 29:47–58
6. Yang HJ, Chen GX, Gao GQ, Wu GN, Zhang WH (2015) Experimental research on the friction and wear properties of a contact strip of a pantograph–catenary system at the sliding speed of 350 km/h with electric current. Wear 332:949–955
7. Yang HJ, Hu Y, Chen GX, Zhang WH, Wu GN (2014) Correlation between the wear and vibration of the contact strip in a contact wire rubbing against a contact strip with electrical current. Tribol Trans 57(1):86–93
8. Ge X, Liu WZ, Yang ZP, Wang YF (2014) The study on electrical temperature characteristics of high speed pantograph. In: 2014 IEEE transportation electrification conference and Expo (Itec) Asia-Pacific 2014
9. Tellini B, Macucci M, Giannetti R, Antonacci GA (2001) Conducted and radiated interference measurements in the line-pantograph system. IEEE Trans Instrum Meas 50(6):1661–1664
10. Midya S, Thottappillil R (2008) An overview of electromagnetic compatibility challenges in European rail traffic management system. Transp Res Part C Emerg Technol 16(5):515–534
11. Hu Y, Chen GX, Gao GQ, Wu GN, Zhang WH, Zhou ZR (2015) Study on material transfer in the process of contact strips rubbing against a contact wire with electric current. Proc Inst Mech Eng Part J J Eng Tribol 230:202–211
12. Liu YJ, Chang GW, Huang HM (2010) Mayr's equation-based model for pantograph arc of high-speed railway traction system. IEEE Trans Power Deliv 25(3):2025–2027
13. Li TZ, Wu GN, Zhou LJ, Gao GQ, Wang WG, Wang B, Liu DL, Li DJ (2011) Pantograph arcing's impact on locomotive equipments. In: 2011 IEEE 57th Holm conference on electrical contacts
14. Ding T, Chen GX, Bu J, Zhang WH (2011) Effect of temperature and arc discharge on friction and wear behaviours of carbon strip/copper contact wire in pantograph–catenary systems. Wear 271(9–10):1629–1636

7

Driving skill classification in curve driving scenes using machine learning

Naiwala P. Chandrasiri[1] · **Kazunari Nawa**[2] · **Akira Ishii**[3]

Abstract Driver support and infotainment systems can be adapted to the specific needs of individual drivers by assessing driver skill and state. In this paper, we present a machine learning approach to classifying the skill at maneuvering by drivers using both longitudinal and lateral controls in a vehicle. Conceptually, a model of drivers is constructed on the basis of sensor data related to the driving environment, the drivers' behaviors, and the vehicles' responses to the environment and behavior together. Once the model is built, the driving skills of an unknown driver can be classified automatically from the driving data. In this paper, we demonstrate the feasibility of using the proposed method to assess driving skill from the results of a driving simulator. We experiment with curve driving scenes, using both full curve and segmented curve scenarios. Six curves with different radii and angular changes were set up for the experiment. In the full curve driving scene, principal component analysis and a support vector machine-based method accurately classified drivers in 95.7 % of cases when using driving data about high- and low/average-skilled driver groups. In the cases with segmented curves, classification accuracy was 89 %.

Keywords Driving behavior · Driving skill · Driving simulator

✉ Naiwala P. Chandrasiri
chandrasiri@cc.kogakuin.ac.jp

1 The Kogakuin University, Tokyo, Japan

2 Toyota InfoTechnology Center, Co., Ltd., Tokyo, Japan

3 AISoftware Inc., Kobe, Japan

1 Introduction

Driving support and infotainment systems inside vehicles are expected to improve the safety of driving and the comfort of those in the vehicle. As the data communication infrastructure matures, high-bandwidth communication is increasingly available between vehicles and the outside world. The functionality of some in-vehicle systems can be realized at data centers, with data exchanged between the vehicles and the data center at high speed. In Japan, some services of this type are already being offered, such as Toyota Motor Corporation's G-Book [1], Honda Motor Corporations' Internavi [2], and Nissan Motor Corporation's CARWINGS [3]. Not only conventional vehicle data buses but also various sensors, such as Global Positioning System (GPS) sensors, radar, and cameras are now equipped in vehicles, mainly as safety equipment. New services can be provided by uploading data from these sensors over a data communication network. As examples, G-Book can locate a stolen vehicle by providing its position and can detect unlocked doors or windows in a vehicle and then send an alert to a user.

Along with more recent advances in the field of active vehicle control, many other driver assist systems exist, such as anti-lock braking systems, vehicle stability control, adaptive cruise control, and lane keeping assist. These improve the safety of driving. It is expected that novel services to provide driving support and infotainment will be implemented in the future.

Our overall aim is to create a framework for providing drivers with the right information, at the right time, in the right way to make driving safer, more comfortable, and more enjoyable. Toward realizing this goal, a conceptual schematic drawing of a data center-based cyber-physical information processing system is shown in Fig. 1 [4]. Vehicle- and driver-related sensor data are uploaded to the data center

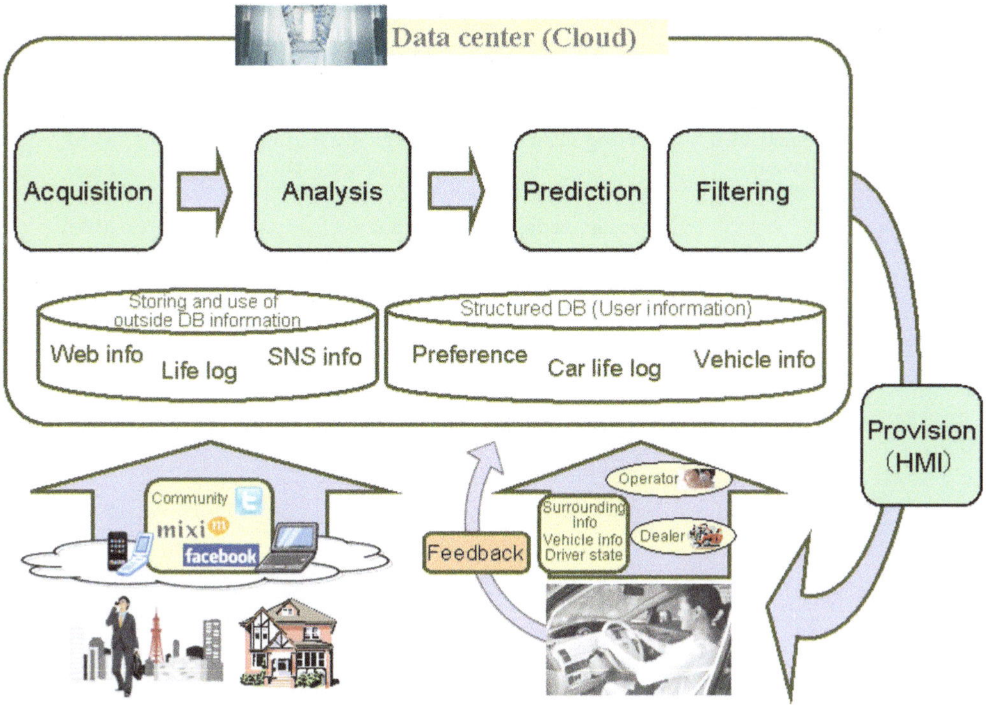

Fig. 1 Cyber-physical system for vehicle application

and stored in structured databases. Data related to the outside world, such as data about the social networks of the driver and Web data, are also uploaded to the data center. Inside the data center, the main data-processing steps can be summarized as acquisition and accumulating the data, analyzing the data, making predictions, and then filtering the output according to driver needs and preferences. Optimal human–machine interfaces are expected to provide the right information to each driver at the right time, and the drivers' feedback data are to be uploaded to the data center for further processing.

Classification of driving skills and driver state is a basic issue in building systems for driver support and infotainment that can be adapted to the individual needs of specific drivers. In this paper, we present a machine learning approach to this problem. Conceptually, in this approach, a driver model is trained from sensor data related to the driving environment, vehicle response, and driving behavior for driver. Once the model is trained, the driving skill of drivers can be classified automatically for novel situations.

The results of this system can be used in building user interfaces and interactive vehicular applications.

2 Related work

To understand the driving skill of a given driver, her continuous process, which is characterized as the Perception–Decision–Action cycle, has to be taken into account. A wide variety of research on driver modeling has been reported [5–18]. However, very few studies have been explicitly characterized driving skill in terms of driving action [14–20] and applied this using machine learning methods.

Table 1 shows a detailed comparison of machine learning methods used in prior studies of classification of driving maneuver skill. Tang et al. [17] and Zhang et al. [16] have dealt with analyzing driving data in the lateral direction (steering angle only) by setting the speed of the driving simulator to a constant. In contrast, Chandrasiri et al. [15, 18] have dealt with a more natural case: motions in both longitudinal and lateral directions of the driving simulator/real vehicle are considered with multiple attributes (specifically, speed is not fixed, and the driver can control the vehicle in both longitudinal and lateral directions). In making this possible, a novel approach has been used, converting time-indexed data into distance-indexed data. The details of this are discussed in Sect. 3.1.

Most of the target driving scenes involve curves or double lane changes, although one includes a narrow straight lane [15]. Each scene is intended to elicit a display of maneuvering skill. Another novel aspect of our research is that it presents scenes with curves of different radiuses and angular changes. Analysis of results from these scenes can determine which require more maneuvering skill.

In using machine learning to characterize driving behavior, there are two main steps.

Table 1 Comparison of machine learning methods for classifying driving maneuver skill

Research	Data source	Driver's vehicle control	Target driving scene	Data	Features extraction	Classifier	Classification accuracy (%)[a]
This work	Driving simulator	Both longitudinal and lateral directions	Curving section	16 drivers, Total of 160 runs for each curve	PCA	SVM	96
						k-NN	88
			Curving section with segmentation	16 drivers, Total of 160 runs for each segment		SVM	89
						k-NN	82
Chandrasiri et al. [18]	Driving simulator	Both longitudinal and lateral directions	Curving section	16 drivers, Total of 160 runs for each curve	PCA	SVM	94
						k-NN	86
			Curving section with segmentation	16 drivers, Total of 160 runs for each segment		SVM	87
						k-NN	82
Chandrasiri et al. [15]	Real vehicle with sensor equipment	Both longitudinal and lateral directions	Narrow straight lane	4 drivers, Total of 72 runs	DFT	PNN	57
					(None)	PNN	75
						SVM	53
			Double lane change	2 drivers, Total 72 runs	DFT	PNN	88
						SVM	94
					(None)	PNN	50
						SVM	94
Zhang et al. [16]	Driving simulator	Lateral direction only	Double lane change	12 drivers, Total 551 runs	DFT	FFNN	88
						Decision tree	87
			Lane change on the curve	12 drivers, Total 514 runs		SVM	88
Tang [17]	Driving simulator	Lateral direction only	Double lane change	12 drivers, Total of 403 runs	WT	FFNN	85
						RBFN	90
						SVM	91
					WT + DFT	FFNN	83
						RBFN	88
						SVM	88

PCA principal components analysis, *DFT* discrete Fourier transform, *WT* wavelet transform, *k-NN* *k*-nearest neighbor, *SVM* support vector machine, *PNN* Probabilistic neural network, *FFNN* feed forward neural network, *RBFN* radial basis function networks

[a] The Classification accuracy is shown based on the best case in the original paper

1. Identify attributes and features that can elicit displays of driving skill.
2. Design a classifier that outputs driving skill level when provided the features extracted during the prior step as input.

Various features and classifiers have been used. Tang et al. used wavelet transforms (WTs) and WTs with a discrete Fourier transform (WT + DFT), applying this to the steering angle and using a neural network and support vector machine (SVM) for learning. Zhang et al. have applied DFT to the steering angle and used a neural network, SVM, and decision trees for learning [14, 16]. Chandrasiri et al. have used DFT with probabilistic neural networks and SVM in their work [15].

In our work, we have experimented with a combination of multiple features that cover both lateral and longitudinal controls as discriminative features. For dealing with multivariate time-series data, we first apply the well-known principal component analysis (PCA) to reduce the dimensionality of the data. In doing so, we also aim to overcome the problem of data sparseness.

In this work, we use the *k*-nearest neighbor (*k*-NN) classifier, which is simple to implement, and SVM, which is known to be powerful. The driving scenarios and data segmentation methods that are used in our experiment distinguish this study from Refs. [16, 17]. In those studies, a full set of data from single runs in driving simulator scenes involving a double lane change and a lane change in a curve (a single curve of common radius and angular change is used for all trials) is used for learning and prediction. In our work, we simulate typical curves to elicit typical driving maneuvers. Specifically, we use full curve data from six different curves of distinct radii and angular changes. By choosing driving scenes in this way, we can analyze differences in driving skill that are demonstrated in response to the difficulty of the driving scene, which can vary according to the radius and angular change of the curve. We also analyzed driving data after dividing the curves into small segments. This is a challenging task because the data from a small segment of a curve must be applied to analyzing the skill level of the driver. Despite the difficulty in finding it, knowing the driving skill displayed from a small driving segment would benefit driver support and infotainment applications.

As an example application, information overload could be prevented by taking instantaneous driver skill into account. Even for a highly skilled driver, real-time fluctuation in the classification of driving skill can be expected for reasons such as fatigue and driving conditions.

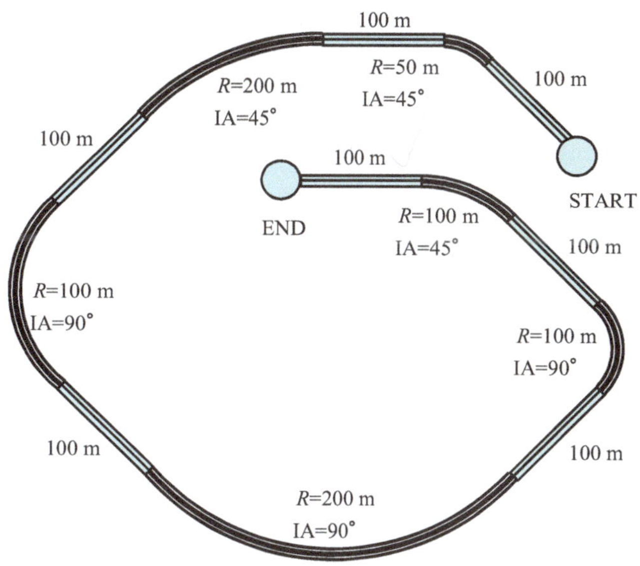

Fig. 3 Driving course

3 Driving simulator experiment

We collected data on driving in curve-containing scenes using the driving simulator shown in Fig. 2. The collected data include steering angle, speed, longitudinal acceleration, lateral acceleration, yaw rate, accelerator control, brake control, lateral displacement, longitudinal displacement, accelerator control speed, and brake control speed. These data were sampled at 60 Hz. Sixteen adult participants (men and women 20–40 years of age) took part in the experiment. As shown in Fig. 3, the driving course consisted of 6 curves (radius = 50, 100, or 200 m; angular change = 45° or 90°). In the simulator, each driver ran up to 12 trials in an urban scene with 2 trials per scene. In total, 960 (16 × 10 × 6) runs of driving data from curves were used for the analysis. Traffic signs indicating a speed limit of 60 km/h are set at 100 m and 50 m before the starting point of each curve (Fig. 4). This controls the speed at which drivers enter the curves. We asked drivers to drive in the left lane and not to cross the yellow center line.

Fig. 4 Experimental scenario

Fig. 2 Driving simulator

3.1 Data conversion

The data were coded by time when acquired from the driving simulator. However, we want to compare these data across different test runs and different drivers for a given location of the driving course. To allow this, we converted the time-indexed data into distance-indexed[1] data. Figure 5

[1] In real vehicles, it will soon be possible to extract centimeter-level accuracy of the driving location using high-precision GPS sensors.

shows an example of data before and after the conversion. From the time-indexed sensor data, we convert to distance-indexed sensor data by linear interpolation.

3.2 Driving skill tagging

This section explains how the simulation runs of drivers are categorized into different skill classes. The results of analysis of a single run for each driver by an expert on driving skill were used to divide the group of sixteen drivers who took part in the driving simulator experiment into two skill classes. Five drivers were classed as high-skilled drivers, and the other 11 as low/average-skilled drivers. Fluctuations in driving skill between different runs of a driver are not considered in this experiment. Tagging the driving skill on that basis of an individual run instead of all runs together of a driver may cause the final output results to be more accurate. However, there is a tradeoff between the tagging cost and the final outcome.

After tagging, there were 300 runs of curve driving data for high-skilled drivers and 660 for low/average-skilled drivers. With curve segmentation, this was increased to 1500 and 3300 data, respectively.

4 Data analysis methods

The flow of data analysis in this paper is depicted in Fig. 6. We used driving data acquired by a driving simulator as explained in Sect. 3. Ten driving data attributes analyzed here are those previously mentioned (steering angle, speed, longitudinal acceleration, lateral acceleration, yaw rate, accelerator control, brake control, lateral displacement, accelerator control speed, and brake control speed), which covers both longitudinal and lateral control of the vehicle. In the pre-processing step, the collected driving data are converted from time-index data into distance-indexed data. This enables comparison among different runs of drivers at the same point on a curve (see Sect. 3.1 for details). The segmentation method is depicted in Fig. 7. As shown in figure, segments 1 and 5 are included in the data analysis along with the curve Sects. 2–4. In those end segments, the vehicle enters and leaves the curve. Therefore, in the full curve case, analysis of driving skill via machine learning considers transition effects from straight segments to curves and vice versa. In our future experiments, the design of the road could be smoother, such as using clothoids.

To merge the data from different sensors into a single feature vector, some normalization method should be employed. For our analysis, we use the standardized score (z-scores) normalization at the points in the distance-indexed data. Feature extraction using PCA and driving skill classification using *k*-NN and SVM are explained in Sects. 4.1 and 4.2, respectively (Fig. 8).

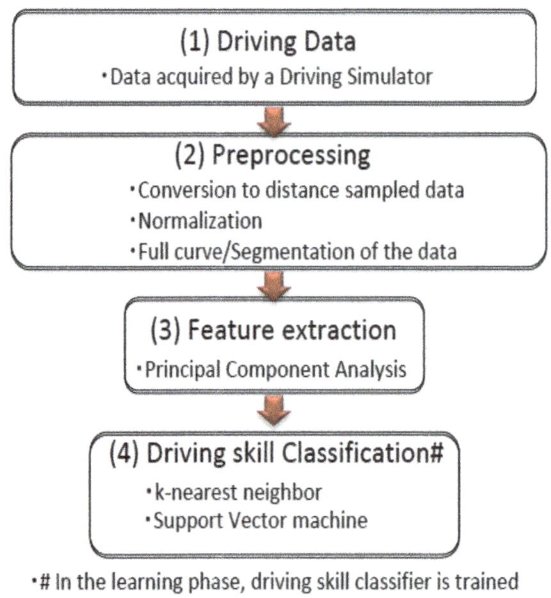

Fig. 6 Data analysis flow

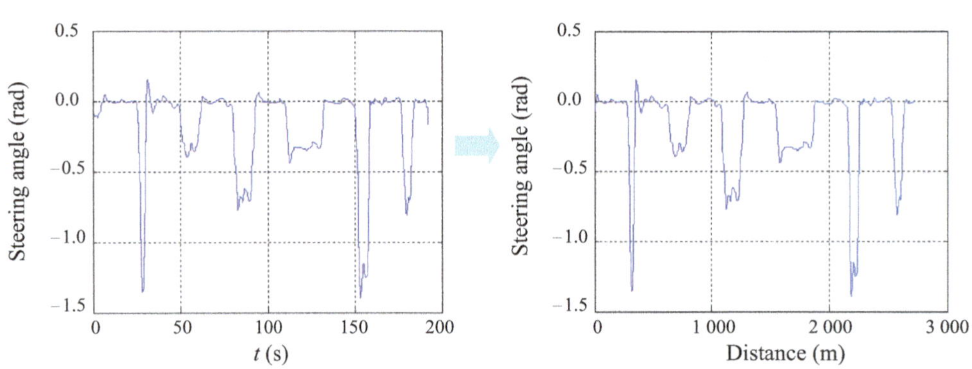

Fig. 5 Data before (left) and after (right) conversion

4.1 Feature extraction

For extracting features to use in analyzing driving skill with respect to both longitudinal and lateral maneuvers in a curve, we use multivariate time-series data from the sensors. Because the dataset is high dimensional and the data points are sparsely distributed, we use PCA to reduce the dimensionality, as is conventional to do for data with these characteristics.

We combine these normalized multiple sensor data into a one-dimensional feature vector, characterizing the ith run by a vector $\boldsymbol{x}_i$.

The covariance matrix of n data vectors, $(x_1, x_2, \ldots, x_n)$, can be written as

$$\boldsymbol{C} = \frac{1}{n}\sum_{i=1}^{n}(x_i - \mu)(x_i - \mu)^{\mathrm{T}}, \tag{1}$$

where T is the transpose operator and μ is the mean of the data.

By eigenvector decomposition of $\boldsymbol{C}$, we calculate m eigenvectors and then sort them in order of decreasing

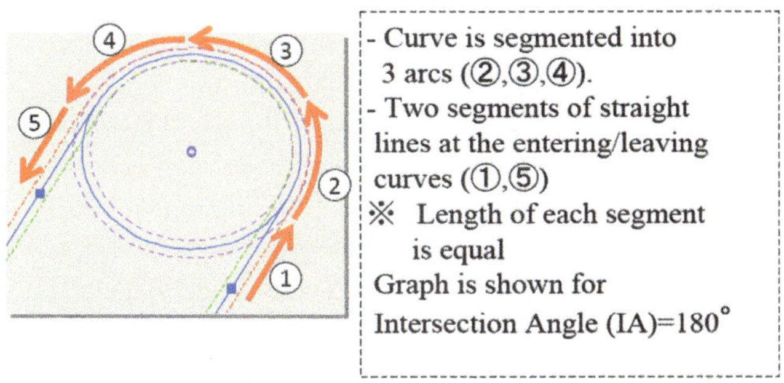

Fig. 7 Data segmentation at a curve

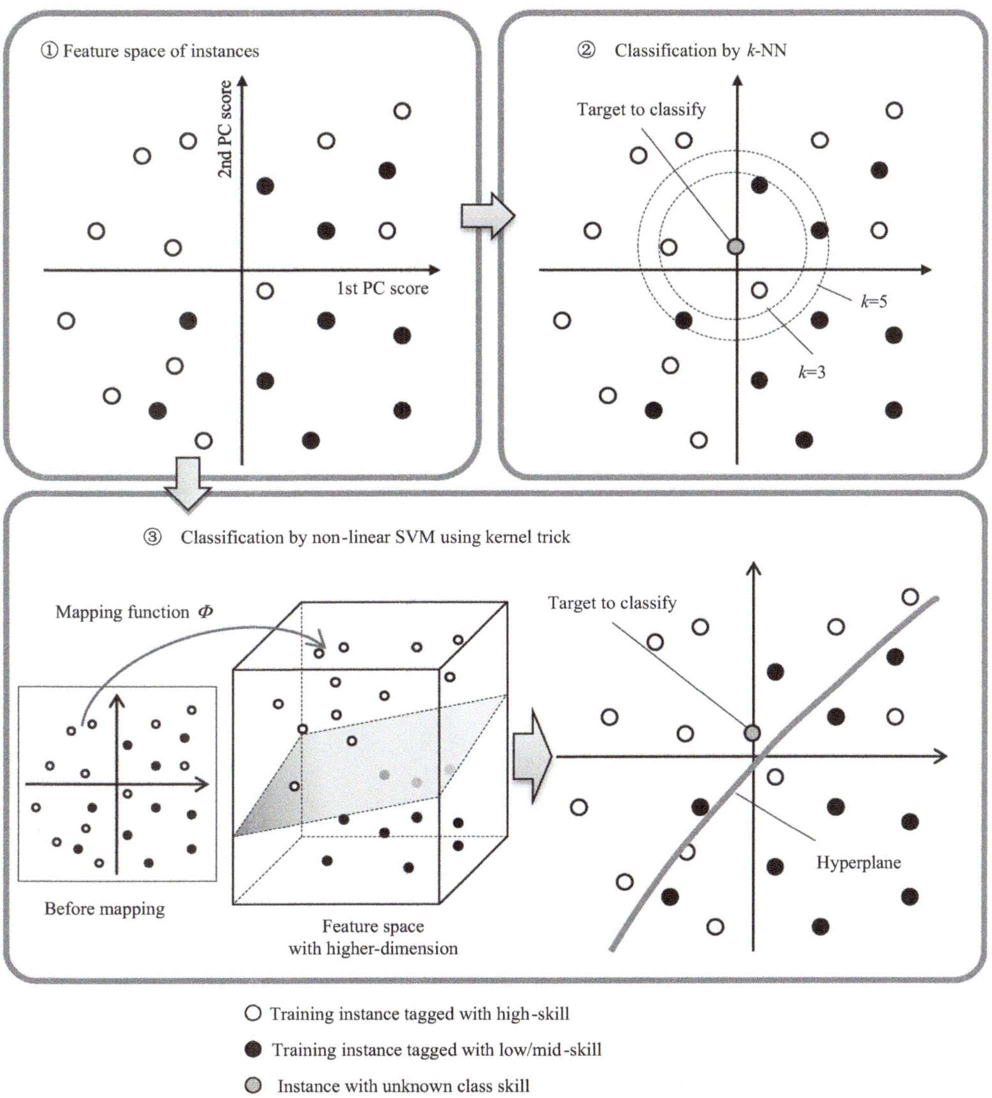

Fig. 8 k-NN and SVM classifiers

magnitude of eigenvalue. By projecting the original observation data onto the eigenspace spanned by the top *d* eigenvectors, we can represent them in a lower-dimensional space effectively. In this paper, we selected the number of principal components *d* to optimize learning (choosing 10 %–15 % of the components) in presenting the original data. Enough principal components to account for over 90 % of the variance in the data were used for the analysis in our previous work [18]. Overall accuracy could be increased by optimizing the number of principal components that are used (see Fig. 9).

Depending on how we define $\boldsymbol{x}_i$, there are two main cases studied in this paper.

1. Full curve analysis: $\boldsymbol{x}_i$ is defined for the sensor data across the entire curve for each curve.
2. Segmented curve analysis: $\boldsymbol{x}_i$ is defined for each segment of the curve and PCA analysis is performed separately for each segment of each curve.

4.2 Driving skill classifier

One method of classifying skill is to manually set up rules for classifying driving skills on the basis of extracted features of the driving sensor data. However, it is widely known that machine learning/data mining algorithms can effectively formulate these rules automatically using some training dataset. Tang et al. have surveyed the ten most influential data mining algorithms in the research community [25]. In this work, we use *k*-NN and SVM, which are both in the top ten. The *k*-NN algorithm is easy to implement, and the model can be built easily. In contrast, SVM is more difficult to implement but is one of the most robust and accurate methods known so far.

In the *k*-NN algorithm, to classify unlabeled data, the Euclidean distances between the unknown data and labeled data are calculated in the feature space, the *k*-nearest neighbors are identified, and the class labels of these nearest neighbors are then used to determine the class label of the unknown data.

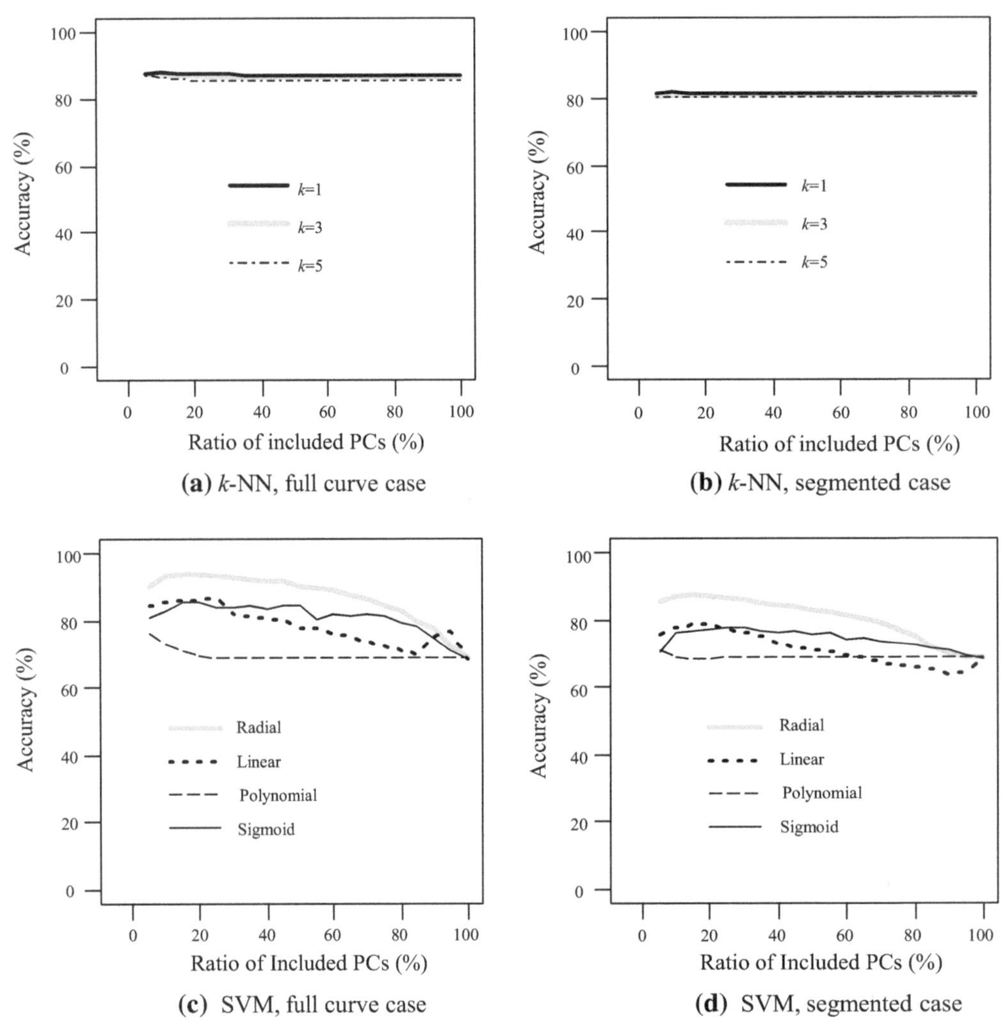

Fig. 9 Grid-search result for optimization of threshold for including principal components in the analysis, *k* for *k*-NNs, and the kernel for SVMs

In a two-class learning task, SVM finds the best hyperplane to distinguish between the two classes by maximizing the margin between the two classes from the training data. SVM also has one of the best generalization abilities for correctly classifying future data. We compared different kernels (radial, linear, polynomial, and sigmoid) in terms of their classification accuracy and used the radial kernel, which had the best performance, in this work (Fig. 9). Details of these two algorithms can be found in [21–25].

In this paper, we use the leave-one-out method for evaluating the accuracy of the driving skill classification algorithms. We build classifiers (k-NN and SVM) by leaving out the data to be classified and using the rest of the data to build the classifiers for training. In the testing phase, the left-out data are used to test the classification ability of the classifier. This process is repeated for all available data.

Driving skill classification accuracy is defined as follows:

$$\text{Accuracy} = \frac{N_{\text{HH}} + N_{\text{LL}}}{N}. \quad (2)$$

Here, the following are used:

N_{HH}: Number of correctly classified high-skilled driving runs.

N_{LL}: Number of correctly classified low/average-skilled driving runs.

N: Total number of runs.

5 Results and discussions

In this section, we show and discuss some of the results from the analysis of driving skill that was performed using the experimental method discussed in Sect. 4. Driving course contained six curves with different radiuses and intersection angles as explained in Sect. 3. In this paper, we report classification accuracy for six full curves and curve-segmented cases.

Figure 10 shows how combinations of different attributes from feature extraction and two driving skill classifiers contribute to the accuracy of driving skill classification. In previous research, except for those using our proposed methods (this work and [18]), which are shown in Table 1, only steering angle data were used. We wanted to compare how combinations of multiple parameters work, and this can be seen in Fig. 10. We selected a combination of parameters that can be easily acquired in a real driving scenario using currently available sensor data, such as driver's control data (steering angle, acceleration pedal control, and brake pedal control) and basic parameters that account for both longitudinal and lateral movement of the vehicle (steering angle, speed, longitudinal acceleration, lateral acceleration, and yaw rate). Additionally, we also included lateral displacement, which

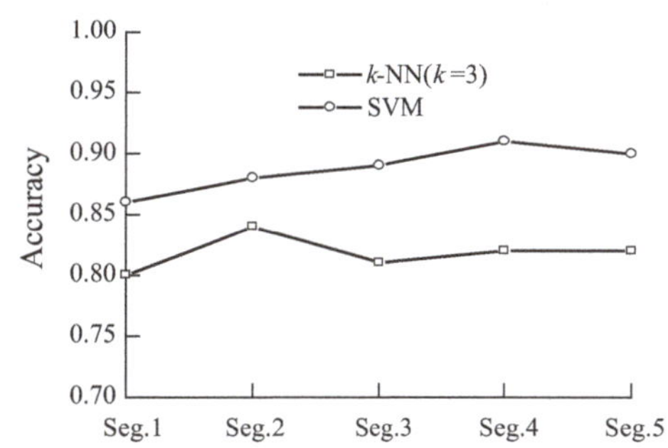

Fig. 11 Accuracy of driving skill classification for segmented curve cases

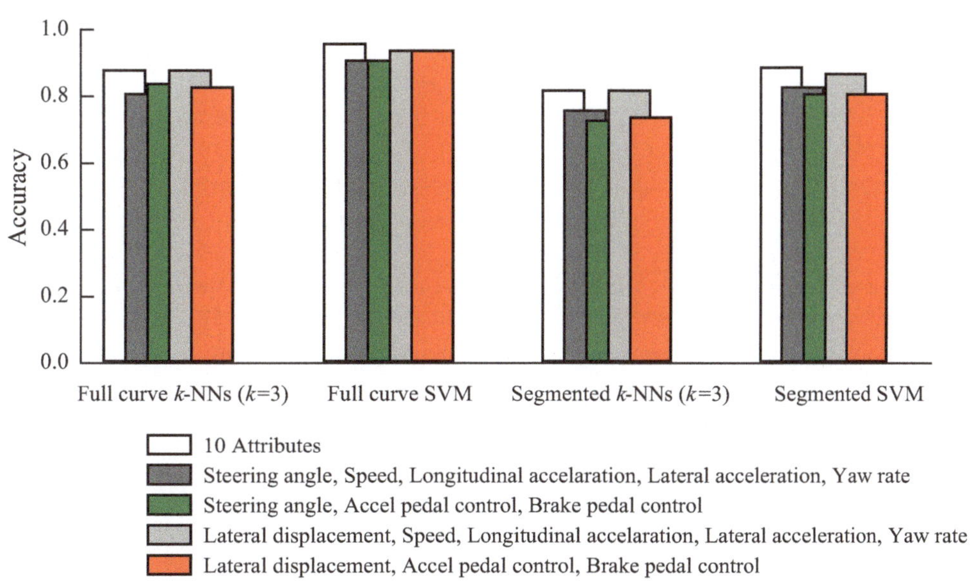

Fig. 10 Driving skill classification accuracy for full curve and curve-segmented cases

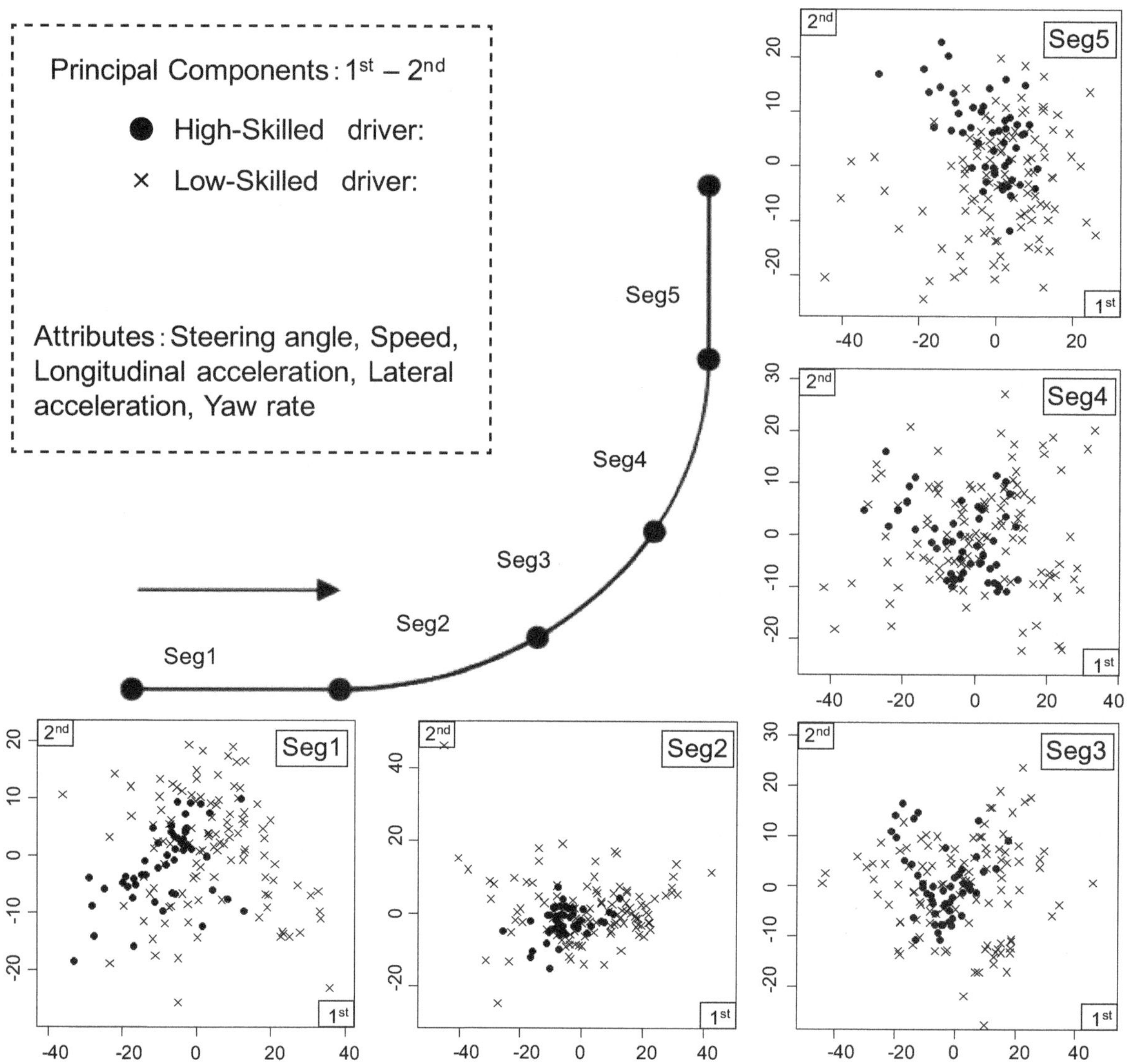

Fig. 12 Segmented curve with driving data analysis: PCA space for the sixth curve ($R = 100$ m, 45°), projected to 2D

accurately characterizes lateral movement of the vehicle in a driving simulator scene.

In terms of the classifier, SVM performs better than *k*-NN in all cases, which is not surprising. However, since we have extracted low-dimensional discriminative features via PCA, *k*-NN does not offer much worse performance.

In analysis of the full curve data, using all ten attributes gives the highest accuracy rates among tested cases. Overall, low/average-skilled and high-skilled drivers were classified with 95.7 % accuracy using SVM.

In analysis of the segmented curves, an average accuracy of 89 % was obtained, using the same classifier. These results show the feasibility of using only a segment of a curve for classification, in spite of the small data window. Figure 11 illustrates how driving skill recognition accuracy depends on segment number. Using SVM for classification, the accuracy varies from 85 % to 91 % in different segments of the curves, with the maximum accuracy of 91 % recorded at the fourth segment. This segment is near the curve exit, which may require higher use of skill, particularly if errors had accumulated in earlier segments.

An example of PCA-based projection of the driving data onto a low number of dimensions (here, 2) is depicted in Fig. 12 for the segmented curve case. Clustering of drivers by skill level is visible, even in two dimensions. However, we cannot directly compare the accuracies of driving skill through the qualitative analysis of data for each segment in Fig. 11 because we use several different dimensionalities, with the components to use determined from the cumulative contribution ratio of the principal components.

Table 1 shows a comparison of machine learning methods for classifying driving skill at maneuvers. In the analysis by the proposed method, both lateral and longitudinal control of a vehicle are considered. In contrast, only

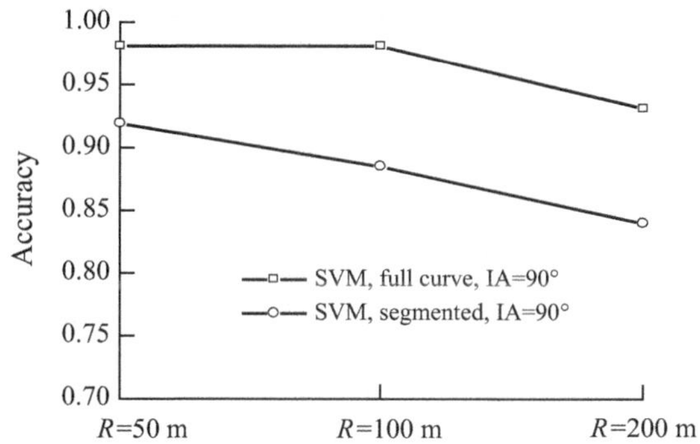

Fig. 13 Accuracy of driving skill classification of different radii of curves (angular change, IA = 90°)

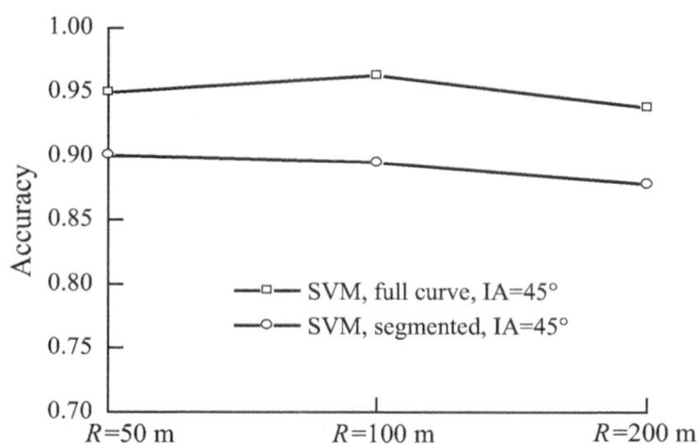

Fig. 14 Accuracy of driving skill classification with different radii of curves (angular change = 45°)

lateral control is taken into account in the works by Zang et al. [16] and Tang [17].

Our work deals with a simple scenario of driving on a curve, and a highest accuracy in classifying driving skill compares favorably with results from prior research.

Driving skill classification results for curves of different radii and angular change [intersection angle (IA)] are depicted in Figs. 13 and 14. There is a tendency toward higher accuracy when the radius is smaller. This could be because the difficulty of driving increases as the radius becomes smaller. For curves of the same radius but different angular change, a change by 90° has higher difficulty than a change by 45° due to the longer distance of the curve in the first case. The above tendency can be seen explicitly in the analysis of the segmented curves. With the full curves, differences between the 50-m and 200-m radii are marked. From these results, we can select curves with a smaller radius and a larger angular change to increase the accuracy of driving skill classification.

In our experiment, all curves were to the left for convenience in building the simulator scenario. This may affect both the potential learning effects and the anticipation of control input needed. We intend to address these issues in future research.

However, this work and [18] are attempts to analyze driving skill on multiple curves and multiple segments and to compare driving skill in curves of different radii and angular change (as shown in Figs. 13, 14) using machine learning methods.

6 Conclusions

In this paper, we analyzed the skills of drivers at longitudinal and lateral maneuvers, using sensor data from a driving simulator.

We demonstrated the feasibility of classifying driving skill in both full curve and segmented curve cases by comparing features composed of different attributes and classifiers.

As features, principal components of combination of attributes such as steering angle, speed, longitudinal acceleration, lateral acceleration, and yaw rate were used. As classifiers, *k*-NN and SVM were used. In classification into two classes of driving skill with full curve scenes, an overall accuracy of 95.7 % was obtained using PCA components of ten attributes and an SVM classifier. The average accuracy was 89 % for the cases with segmented curves.

Analysis of driving skill in complex traffic environments, such as scenarios that contain surrounding vehicles and different road gradients, is left to future work.

In the future, we aim to build driver support and infotainment systems that can be adapted to individual user needs.

Acknowledgments Special thanks go to Dr. Shuguang Li, Dr. Shigeyuki Yamabe, Prof. Yoshihiro Suda, and Prof. Yoichi Sato (all of the University of Tokyo at the time of this research) for providing us with the driving simulator and the experimental environment. We also thank them for fruitful discussions on various topics related to the research. We are very thankful to Takeshi Matsumura and Koji Taguchi of Toyota Motor Corporation for their design of the driving scenarios. We thank Prof. Takashi Washio of Osaka University for constructive discussions.

References

1. Toyota's G-Book. http://g-book.com
2. Honda's Internavi. http://www.honda.co.jp/internavi/
3. Nissan's carwings. http://drive.nissan-carwings.com/WEB/index.htm
4. Nawa K, Chandrasiri NP, Yanagihara T, Komori T, Oguchi K (2012) Cyber physical system for vehicle application. In: IEEE international conference on cyber technology in automation, control, and intelligent systems

5. Savkoor AR, Ausejo S (2000) Analysis of driver's steering and speed control strategies in curve negotiation. Veh Syst Dyn 33:94–109
6. Van Winsum W, Godthelp H (1996) Speed choice and steering behavior in curve driving. J Hum Factors Ergonom Soc 38(3):434–441
7. Salvucci Dario D, Gray Rob (2004) A two-point visual control model of steering. Perception 33(10):1233–1248
8. Erseus a (2010) Driver-vehicle interaction, PhD thesis, Royal Institute of Technology, Sweden
9. Miyajima C, Nishiwaki Y, Ozawa K, Wakita T, Itou K, Takeda K, Itakura F (2007) Driver modeling based on driving behavior and its evaluation in driver identification. Proc IEEE 95(2):427–437
10. McCall JC, Trivedi MM (2007) Driver behavior and situation aware brake assistance for intelligent vehicles. Proc IEEE 95(2):374–387
11. Miyaji M, Danno M, Kawanaka H, Oguri K (2008) Driver's cognitive distraction detection using adaboost on pattern recognition basis. In: Proceedings of the IEEE international conference on vehicular electronics and safety, pp 51–56
12. Wang Y, Peng P, Liang LJ, Zhang W, Wu S. (2007) Road hazard reaction testing using driving simulation: The novice versus the experienced drivers. In: Proceedings of the IEEE international conference on industrial engineering and engineering management, pp 477–481
13. Pradhan AK, Fisher DL, Pollatsek A (2006) Risk perception training for novice drivers: evaluating duration of effects of training on a driving simulator. Transp Res Rec 1969:58–64
14. Zhang Y, Lin WC, Chin Y-KS (2008) Driving skill characterization: a feasibility study. In: 2008 IEEE international conference on robotics and automation, pp 2571–2576
15. Chandrasiri NP, Nawa K, Kageyama I (2010) Classification of driving skills base on machine learning. In: FISITA world automotive congress
16. Zhang Y, Lin WC, Chin Y-KS (2010) A pattern-recognition approach for driving skill characterization. IEEE Trans Intell Transp Syst 11(4):905–916
17. Tang X (2009) Driving skill recognition: new approaches and their comparison. In: American control conference
18. Chandrasiri NP, Nawa K, Li AIS, Yamabe S, Hirasawa T, Sato Y, Suda Y, Matsumura T, Taguchi K (2012) Driving skill analysis using machine learning the full curve and curve segmented cases. In: International conference on ITS telecommunications (ITST), Taipe, pp 542–547
19. Support Vector Machine. http://www.svms.org/
20. Hsu CW, Chang CC, Lin CJ (2010) A practical guide to support vector classification. http://www.csie.ntu.edu.tw/~cjlin/libsvm/
21. Bishop CM (2007) Pattern recognition and machine learning. Springer, New York
22. Burges CJC (1998) A tutorial on support vector machines for pattern recognition. Data Min Knowl Discov 2:121–167
23. Murphey Y L et al. (2009) Driver's style classification using jerk analysis. In: IEEE workshop on computational intelligence in vehicles and vehicular systems, pp 23–28
24. Li S, Yamabe S, Sato Y, Hirasawa T, Suda Y, Chandrasiri NP, Nawa K, Matsumura T, Taguchi K (2012) Feature extraction focused on driver's steering behavior in curving. JSAE Annual Congress (Spring)
25. Wu X et al (2008) Top 10 algorithms in data mining. Knowl Inf Syste 14:1–37

8

Industrial frequency single-phase AC traction power supply system for urban rail transit and its key technologies

Qunzhan Li[1]

Abstract To avoid stray current and maintain the benefit of no phase-split in the DC traction power supply system, an AC traction power supply system was proposed for the urban public transport such as metro and light rail transit. The proposed system consists of a main substation (MSS) and cable traction network (CTN). The MSS includes a single-phase main traction transformer and a negative-sequence compensation device, while the CTN includes double-core cables, traction transformers, overhead catenary system, rails, etc. Several key techniques for the proposed system were put forward and discussed, which can be summarized as (1) the power supply principle, equivalent circuit and transmission ability of the CTN, the cable-catenary matching technique, and the selection of catenary voltage level; (2) the segmentation technology and status identification method for traction power supply network, distributed and centralized protection schemes, etc.; (3) a power supply scheme for single-line MSS and a power supply scheme of MSS shared by two or more lines. The proposed industrial frequency single-phase AC traction power supply system shows an excellent technical performance, good economy, and high reliability, hence provides a new alternative for metro and urban rail transit power supply systems.

Keywords Urban rail transit · 50 Hz industrial frequency single-phase AC traction power supply · Cable traction network · Power supply state monitoring and control · Main substation · Stray current

The Chinese version of this paper was published in Journal of Southwest Jiaotong University (2015) 50(2).

✉ Qunzhan Li
lqz3431@263.net

1 School of Electrical Engineering, Southwest Jiaotong University, Chengdu 610031, China

1 Introduction

From worldwide application point of view, DC system, low-frequency single-phase AC system, and industrial frequency single-phase AC system are three main power supply systems for trunk railways, metros, and light rails. Each of them has its own characteristics and some comparative advantages changing with the development of electrotechnics [1]. Today, except the low-frequency single-phase AC system still being used in trunk railways in Germany and some of its neighborhood countries, the DC system and the industrial frequency single-phase AC system have been mainly used around the world. The DC system is mainly used for metros and light railways, while the industrial frequency single-phase AC system is mostly for trunk railways [2–4], both with an increasing automation level [4–6].

Metros and light railways in urban transit systems have prominent features of large traffic density and frequent starts and stops. Their DC traction power supply systems have the advantage of no phase-split along the overhead catenary system (OCS), so that trains can run smoothly. On the other hand, DC traction systems also have the disadvantage of the long-term adverse effects caused by the stray current (earth fan stream) [3, 6]. Anti-stray current measures, however, are usually complex with high costs and limited effects because the stray current is hard to control. After 1960s, except that parts of the trunk railway lines still retain the DC power system in Europe and the former Soviet Union, almost all of newly built trunk railways, especially high-speed railways or heavy-haul railways, have adopted the industrial frequency single-phase AC system, because of its advantages of strong supply capacity, low costs, being able to meet the large transport capacity requirements, non-existence of stray current, and directly efficient utilization of regenerative braking power. However, its disadvantages, switched into the public power grid in order to reduce impacts of negative-sequence component, create the phase-split in traction power supply side, which leads to power supply breakpoints for running trains. Therefore, techniques for auto-passing phase-splits have to be developed due to the increasing speed of trains. Another drawback is that the harmonic current of traction load could produce electromagnetic interference in the vicinity of communication lines, even though with the use of optical cables it is not a problem for the protection to interfere.

Traction motors, as the power source of trains [7], can be categorized both as DC and AC types. Compared with a DC traction motor, an AC traction motor has properties of high-power density, low costs, simple and reliable structure, good anti-idling performance, etc. In addition, the absence of mechanical parts in DC traction motors, such as the commutator, makes AC traction motors more suitable for operation of high-power, high-speed trains. Therefore, the AC traction motor is replacing the DC traction motor and has become the main power source for trunk railways and urban rail transit.

The train, powered by AC traction motors and composed of an electric locomotive or electric multiple units (EMUs), is called an AC–DC–AC train, which is developed mainly for trunk railways and high-speed railways. It is easy for AC–DC–AC trains to store the brake energy back to the grid, it is not only simple, convenient, safe but also reliable to operate. In contrast, for a DC traction motor, to achieve these goals, very special and expensive equipment and complex controlling devices have to be extra utilized, and some of these goals are even unachievable ultimately.

Considering the characteristics of the traction power supply system for trunk railways, metros, and light rails in urban transit system, and comparing the merits and demerits between the DC system and the industrial frequency single-phase AC system, it is naturally necessary to provide a new industrial frequency single-phase AC traction power supply system without phase-split or stray current in urban transit system.

Compared to trunk railways, the power supply lines of metros and light rails are shorter, usually less than 50 km, and does not have line-crossing or line-joining. To achieve an AC traction power supply system for metros and light rails without phase-split, a direct approach would be used to co-phase power supply technique that was proposed for trunk railways [8–10] and the recently developed traction power supply system of new generation [11]. However, the power supply system for metros and light rails is significantly different from that for trunk railways, because it has shorter power supply line, lower voltage grade, and denser traction substations. In order to reduce the impact of negative-sequence current, negative-sequence compensation devices have to be installed at every traction substation. Additionally, bilateral supply mode needs to be carried out at section posts (between two adjacent substations) by connecting different power supply lines at the high-voltage side, which is known as loop closing [12, 13]. Feasibility study must be carried out after obtaining consent from power sectors. To avoid complications and difficulties mentioned above, a new industrial frequency-based single-phase AC traction power supply system using 35 kV cables is proposed in this paper for urban rail transit, due to the fact that 35 kV cables have widely been used in metro systems.

This study is focused on the proposed AC traction power supply system and its key technologies, which include cable traction network, segmented power supply, status identification and protection, and main substation power supply scheme. In addition, a brief discussion on system's economy and reliability, etc. is also given.

2 Cable traction network

The proposed industrial frequency single-phase AC traction power supply system for the urban rail transit consists of a main substation (MSS) and cable traction network (CTN). The latter, as shown in Fig. 1, comprises the cables, traction transformers (TT), overhead catenary system (OCS), rails (R), etc. The type of the cables can be a two-core cable (TCC), or a coaxial cable (CC). TCC is made of a supply core (SC), a return core (RC), and a protective layer (PL). The TT has single-phase connection. The TCC and the OCS are erected in parallel. The primary winding of the TT is connected in parallel between the SC and RC of TCC, while its secondary winding is connected in parallel between the OCS and the R. TTs are distributed in proper distances along TCC and OCS. The SC of TCC is connected with the OCS in parallel by adjacent TTs, so are the RC of TCC and the R. Electric trains (ET), either electric locomotives or EMUs, are powered between the OCS and the R. In Fig. 1, BB1 and BB2 represent main traction busbars; W and K represent circuit breakers; S represents OCS slicer; P represents section post; and ET represents an electric train, which is an electric locomotive or an EMU. MSS and its power supply scheme will be discussed in Sect. 3.

The primary side of MSS should be connected to three-phase lines in a utility grid of 110 or 220 kV, while the secondary side should use 35 kV TCC for traction power

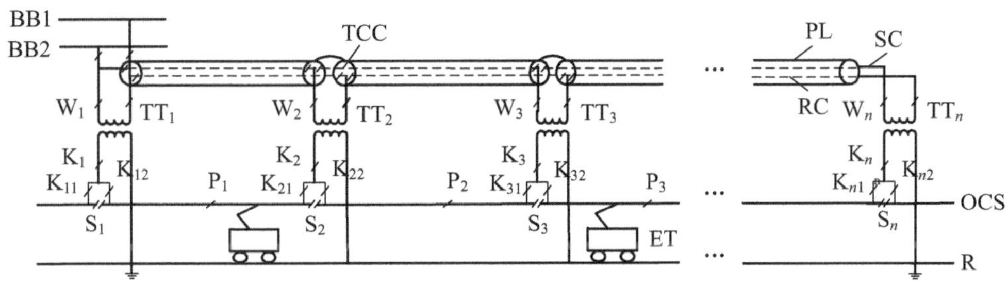

Fig. 1 Traction power supply network with TCCs

supply and use 35 kV three-phase cables for three-phase power and single-phase lighting supply. The single-phase traction power supply and the three-phase power and lighting supply should be kept independent to avoid the direct impact of negative sequence generated in traction power supply on the power and lighting system, also avoid the mutual interference between the two systems during operation and maintenance.

The transmission capacity refers to the maximum transmission power delivered through cables, overhead lines, catenaries, or other types of transmission lines [14]. The natural power is one of the most important characteristic parameters for measuring the transmission capacity. It is defined as the transmitted power when the load impedance at the transmission line terminal is equal to the wave impedance. Given the distributed resistance R_0, the distributed inductance L_0, the distributed capacitance C_0 between lines, the distributed conductance G_0, as to per unit length of cables, overhead lines or other transmission lines, the wave impedance z_c (modulus value) can be calculated by [15]

$$z_c = \sqrt{(R_0 + j\omega L_0)/(G_0 + j\omega C_0)},$$

where j is the imaginary unit ($j^2 = -1$), and ω is angular frequency.

Under the industrial frequency, there are $R_0 \ll \omega L_0$ and $G_0 \ll \omega C_0$, then the wave impedance of a lossless transmission line is obtained as

$$z_c \approx \sqrt{L_0/C_0}. \tag{1}$$

It can also be said that the wave impedance of a lossless transmission line is a pure resistance. In this case, the electric field energy released (or absorbed) from the distributed capacity C_0 is equal to the magnetic field energy absorbed (or released) from the distributed reactance L_0. That is to say, the reactive power of the distributed capacity and the distributed reactance are just balanced. Thus, the transmitted power is completely equal to the load active power. If the rated voltage is U_N and the wave impedance (modulus value) is z_c, the natural power of a transmission line is

$$S_c = U_N^2/z_c. \tag{2}$$

The natural power is often used for analyzing the transmission condition, voltage level, etc. of a transmission line. When natural power is transmitted, the voltages at the two terminals of the line will be the same, and so are the power factors. When the transmitted power is lower than the natural power, because of the decreased transmission line current, the reactive power of the distributed reactance which is proportional to the square of the line current, will be less than the reactive power of the distributed capacity. This leads to a higher voltage at the line terminal. Similarly, when the transmitted power is larger than the natural power, due to the increased transmission line current, the reactive power of the distributed reactance will be larger than that of the distributed capacity and the voltage at the line terminal will be lower. Theoretically, it is ideal to transmit the natural power. This is hard to be achieved in practice, particularly when traction loads fluctuate sharply: the voltage increases at no load and drops at heavy load. In order to maintain a good transmission capacity, sometimes the inductive reactive power compensation is needed, or sometimes the capacitive reactive power compensation is required, even both are demanded at the same time.

By Eq. (1), the wave impedance decreases as the distributed capacitance increases, and increases as the distributed reactance increases. Usually the wave impedance of overhead lines is between 270 and 380 Ω [14]. The mutual inductance approaches to the self-reactance when the cable cores are arranged closely. The equivalent distributed reactance is greatly reduced when the values of currents in two cores are the same with opposite directions. At the same time, when the distance between the cable cores is closer and the insulation material between the cores has a relative dielectric constant up to 4 or 5, the distributed capacitance is greatly increased. So the cable's wave impedance would be much less than overhead line's impedance. The cable's wave impedance would be approximately 1/7 of overhead line's impedance [15]. From Eq. (2), it can be seen that the transmission capability of a cable is approximately 7 times those of overhead lines or catenaries at the same voltage level, since the natural power is inversely proportional to the wave impedance. Empirical data [14] show that the output power at one side of a single loop (head-to-end) of 35 kV overhead lines can reach up to 10 MW with a transmission distance of approximately 20 km, while the output power of cables under the same condition can reach up to 70 MW (for double-loop circuit lines to 140 MW). Thus it can be derived that the cable's one-side transmission distance is up to 100 km or longer. It is able to fully meet the needs of any metro or light rail line.

It should be noted that power transmission may have different needs, and the transmission capacity is affected by many factors. In addition to the natural power, other factors such as economic factor, thermal stability limit, voltage drop limit, system stability, etc. [14], should also be taken into account so as to find out an optimal solution.

In current metro and light rail, 1,500 V DC voltage is adopted for the OCS with an exception of 750 V in particular cases. If the OCS uses an industrial frequency AC system, its voltage grade will be determined by the following factors: (1) AC peak voltage, (2) tunnel clearance and OCS insulation requirement, (3) transformerless AC–DC–AC ET, and so on. Selecting a voltage grade as high as possible will improve the OCS's power supply capacity and reduce the power loss. In general, a feasibility study of the OCS voltage (RMS value) between 2 and 3.5 kV should be carried out.

Figure 2 shows a chain network, which is the equivalent circuit of CTN illustrated in Fig. 1. The OCS-R loop can be imputed to the cable loop side, or the cable loop can be imputed to the OCS-R loop side.

In Fig. 2, the thick solid line Z_{1i} represents the distribution parameter of the i-th segment of the cable loop; the thick solid line Z'_{2i} represents the distribution parameter of the OCS-R loop, which is imputed to the cable loop side; the traction transformer is considered as a lumped parameter element and its leakage reactance imputed to the cable loop side is represented by Z_{Ti}. Taking into account the compactness of TCC and the isometry between TCC loop and OCS-R loop, the coupling between TCC loop and OCS-R loop can be ignored generally in the analysis and calculation. To further simplify the calculation, the lumped parameters can be used to establish the equivalent circuit of OCS-R loop.

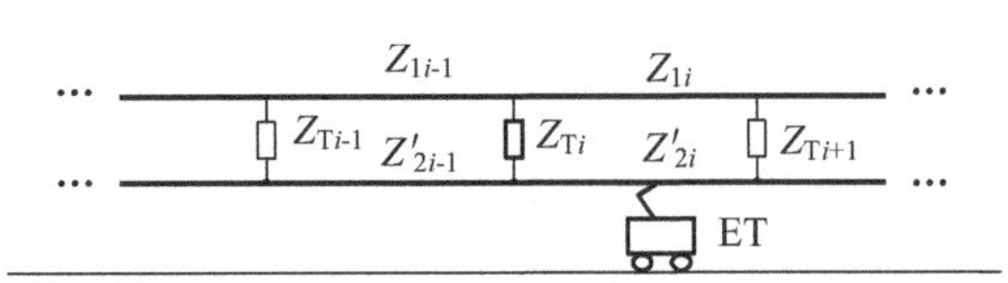

Fig. 2 Equivalent circuit of traction network

The current of TCC loop or OCS-R loop is inversely proportional to the loop impedance. From the above analysis about the wave impedance, it can be inferred that the TCC loop impedance will be much smaller than the OCS-R loop impedance and the current in TCC will be larger. Moreover, if OCS-R loop is imputed to TCC loop side, the imputed impedance modulus value of OCS-R loop will be the square of its turn ratio. For example, when the TT primary voltage is 35 kV, its turn ratio would be no less than 10 if the chosen OCS voltage does not exceed 3.5 kV, and the impedance modulus value of TCC loop will be less than 1 % of the impedance modulus value of OCS-R loop imputed to TCC loop side. The OCS-R loop between two nearest TTs can be called short loop (the i-th segment in Fig. 2). And the parallel connection of the TCC loop and the OCS-R loop from the short loop to the main traction busbar can be called long loop (from the 1st to the $(i-1)$th segments in Fig. 2). If so, the short loop will take the train load while the OCS on the long loop will hardly bear any burden for power supply (less than 1 %). The TCC on the long loop will bear almost all power supply burden (larger than 99 %). That is, the OCS on the short loop is mainly responsible for the power supply of the train in the current segment, and the TCC on the long loop is responsible for the power supply of the entire traction network.

In [16], analyses and calculations of a coaxial-cable traction power supply network are given for reference.

Metro or light rail's upward and downward traction network works in parallel. Besides, auxiliary cables should be set up to work with the main cable at the same time, which means that they are reserved for each other. This can help improving the power supply capacity and reducing the power loss and voltage loss of the traction network.

3 Traction power supply segmentation, status identification, and protection

As mentioned in Sect. 1, the CTN is characterized by strong transmission capacity and long distance transmission ability. However, its construction is complex, and particularly more complex when the upward and downward network together with auxiliary cables are working in parallel. Based on the analyses and experimental studies of the control and monitoring for all-parallel AT (auto-transformer) power supply mode of the high-speed railway in China [11, 17], the segmented power supply has become an effective approach for timely removal of faults, which limit faults to a minimum scope so that failure effects can be reduced to the minimum. The segmented power supply avoids the scenario in which one failure could lead to

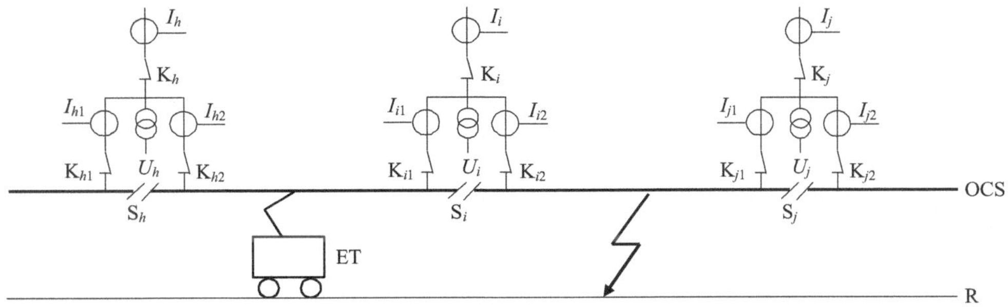

Fig. 3 Diagram of SS feeder and OCS segments

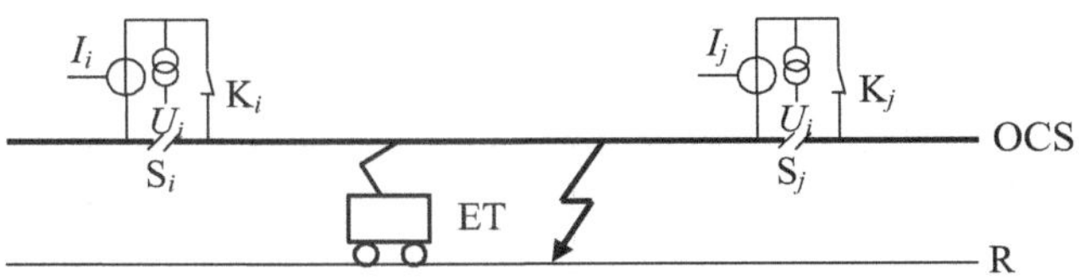

Fig. 4 OCS segments diagram

system failure and thus jeopardize the system's reliability and availability.

Cables and OCSs can be segmented separately or segmented at a common location, for convenient purpose, lumped segment is better to be at the location of a TT where the traction substation (SS) is built. The SS i includes 35 kV entrance cables, a main tie line, a TT, traction busbars, feeder lines, a feeder breaker K_i, a sub-feeder, and its circuit breaker, etc. At the exit of SS i, there should be a nearby OCS slicer S_i which is in series with OCSs, so as to allow a train with uninterrupted supply. The SS i should also include the sub-feeder breakers K_{i1} and K_{i2} connected in parallel with S_i, the current transformer (measured currents denoted by I_{i1} and I_{i2}), and the voltage transformer connected to the OCS slicer S_i (measured voltage denoted by U_i). The segmented power supply at SS feeder lines and OCS segments is shown as Fig. 3.

The distance between two adjacent SSs, in reference to existing metros or light rail lines, is typically 2–4 km. If the OCS's length between SSs is considered too long, it can be more segmented, as shown in Fig. 4.

The power supply of a traction network has two states: normal state and faulty state. The former includes no-load state and normal load state while the latter includes the short circuit and the open circuit (the broken line). These states can be identified by measuring the voltage from voltage transformer or the current from current transformer. Based on the fact that the cable's failure rate is much lower than the OCS's failure rate in a traction network, even the power supply through the OCS is more complex than that through the cable, the main focus in this paper will be given to the OCS segmented power supply, status identification, and protection methods.

Assumptions: When the local OCS voltage (a modulus value, the same applies hereinafter) $U_i \geq U_s$ (a specified value), it is in the normal power supply state on the left and right segments of the OCS. If $U_i < U_s$, then the train on the left and right segments of the OCS is not in a working state; that is, the train load is zero and thus the OCS segment may be in a failure state.

Acquire the OCS voltage U_i of the SS i measured by the voltage transformer, the left-side OCS current I_{i1} and the right-side OCS current I_{i2} of the SS i measured by the current transformer, where $i = 1, 2, \ldots, N$. For a OCS segment between SS i and SS j ($j = i + 1$) as shown in Fig. 3, the operation state of a train can be identified by calculating the arithmetic difference of currents between the two ends of the OCS segment if the following assumptions are satisfied [11, 17]: when the OCS voltages U_i, $U_j \geq U_s$, the current I_{i2} that flows into the segment and the current $\dot{I}_{j1}$ that flows out of the segment are all positive, their power factors are the same, and the OCS-to-ground distributed capacitive current and the measurement error are ignored, who are relatively small.

If $I_{i2} - I_{j1} = 0$, there is no electric train in the segment; when $I_{i2} = I_{j1} = 0$, the segment is on a no-load state; when $I_{i2} = I_{j1} > 0$, the segment acts as a transmission line to pass transmission power.

If $I_{i2} - I_{j1} > 0$, there is a moving train being in traction in the segment.

If $I_{i2} - I_{j1} < 0$, there is a moving train being in regenerative breaking in the segment.

When the OCS voltage $U_i < U_s$, according to the above assumptions, there would be no load within the adjacent OCS segments. So there is only short-circuit current. For the OCS segment between SS i and SS j ($j = i + 1$) as shown in Fig. 3, it is still assumed that the current I_{i2} flowing into the segment is positive, also the current I_{j1} flowing out of the segment is positive, and their traction impedance angles are equal. Hence the differential current is

$$I_{Di2} = \left| I_{i2} - I_{j1} \right|. \tag{3}$$

Clearly, if there is no short circuit, then $I_{i2} = I_{j1}$, and

$I_{Di2} = |I_{i2} - I_{j1}| = 0$. If a short circuit occurs, then I_{j1} is reversed and $I_{Di2} = |I_{i2} + I_{j1}| > 0$. This leads to a low-voltage starting differential current protection for OCS. Specifically, acquire the OCS voltage U_i, the left-side OCS current I_{i1}, the right-side OCS current I_{i2}, the right-side OCS current I_{h2} on the left adjacent substation h ($h = i - 1$) and the left-side OCS current I_{j1} on the right adjacent substation j ($j = i + 1$). The OCS differential current at the right-side of substation i is given by Eq. (3) and the one at the left-side of the substation i is

$$I_{Di1} = |I_{i1} - I_{h2}|. \tag{4}$$

When $U_i < U_s$ at substation i, if the left-side OCS differential current $I_{Di1} > 0$, it can be concluded that there is a short-circuit fault on the OCS between SS i and the left adjacent SS h. Then, open the breaker K_{i1} at SS i and the breaker K_{h2} at the SS h, otherwise both breakers K_{i1} and K_{h2} are still in closed state. Similarly, if the right-side OCS differential current $I_{Di2} > 0$ at SS i, it can be concluded that there is a short-circuit OCS fault between SS i and the right adjacent SS j, then open breakers K_{i2} and K_{j1} for the same reason.

Further analysis reveals that if one of the currents in $\{I_{i1}, I_{h2}\}$, or $\{I_{i2}, I_{j1}\}$ is zero, then at the side where the current is zero, an open-circuit fault has occurred, while at another side where the current is not zero, a short-circuit fault has occurred.

At the same time, the short-circuit impedance can be calculated from the OCS's voltage and current in a faulty state, subsequently the fault location can be identified.

The setup of low-voltage starting differential current protection of OCS at SS i is to meet the quick-action requirement. It is only dependent on the local SS and two adjacent SSs. The train operational state identification is required by the MSS or the dispatching center, and should cover every SS in the entire traction network [5]. The MSS or the dispatching center can set up backup protection based on acquired OCS voltages and currents from each of SSs. Due to these differences, the low-voltage start and differential current protection at each SS is called distributed protection, while the backup protection at MSS or the dispatching center is called centralized protection.

In order to improve the real-time performance of the centralized protection, the sign value method can be utilized [11, 17], which is based on the fault power flow direction. The fault power flow is defined as the current phasor under the unit voltage as the reference phasor. Set the sign value to 1 if the direction of the fault power flows into the segment, to -1 if it flows out of the segment, and to 0 for no-load fault power flow. Calculate the sign value of fault power flow into the current segment and its absolute value p, faults can be identified as follows:

If $p = 0$, there is no short-circuit fault at the current segment.

If $p = 1$, there is a short-circuit fault at one current segment end if the absolute value of the sign value is 1; there is an open-circuit fault at the other end if the absolute value of the sign value is 0.

If $p = 2$, there are short-circuit faults at both ends of the current segment.

Obviously, this result is consistent with the result obtained from the low-voltage starting differential current protection approach mentioned previously. In addition, the method of fault current sign value is easier to further determine the OCS's composite faults.

The segmented power supply and protection for OCSs can greatly improve the reliability of the system [14]. This technique is also applicable to the protection of the cable segmented supply in a similar way.

It is evident that the use of traction network segmentation technique allows more flexibility for the traction network's operation and maintenance.

4 Main substation power supply scheme

The scheme of an MSS power supply is dependent on the number and the relative position of metro or light rail line that requires the power supply.

4.1 One line case

If only one metro or light rail line is needed by the power supply, the main substation (MSS) should be located as close to the middle of the line as possible, as shown in Fig. 5, where MSS provides power for both sides of the line. If the condition doesn't permit, the MSS can be placed at one line end and thus uses unilateral power supply as shown in Fig. 1. In both arrangements, the MSS supplies the power by combining a single-phase main transformer (MTT) with a negative-sequence compensation device (NCD), to realize the centralized negative-sequence management as shown in Fig. 6. This scheme also includes a transformer for three-phase power and single-phase lighting supply (not shown in the diagram, the same applies hereinafter). The NCD consists of a high-voltage matching transformer (HMT), an AC–DC–AC converter (ADA), a traction matching transformer (TMT), and so on. The length of the bilateral power supply doubles that of the unilateral power supply. If the supply's length is the same, the bilateral supply can save cable material and reduce power loss and voltage loss.

It should be noted that the negative-sequence compensation mentioned here is referred to as the symmetrical compensation in a broad sense. As stated in [10], the

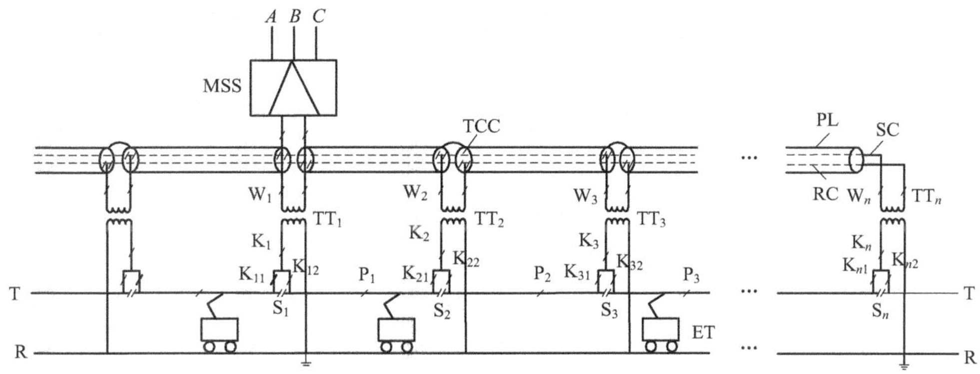

Fig. 5 Traction power supply system diagram

symmetrical compensation includes negative-sequence compensation and reactive power compensation based on three-phase to one-phase symmetrical transformation. Since the power factor of an AC–DC–AC train is close to 1, the reactive power compensation has been achieved, and just the negative-sequence compensation has become the main target of compensation. Thus the symmetrical compensation can be termed as the negative-sequence compensation. In the same principle as applied in a co-phase compensation device [11], the NCD needs to transfer active power to compensate negative sequence. If necessary, functionalities of reactive power compensating and harmonic filtering are able to be realized at these device terminals.

If the short-circuit capacity of a utility grid connected with the MSS is denoted by s_{sc} (MVA), and the three-phase voltage unbalance value (95 % probability value or the maximum) specified by the national standard in China [18] (GB for short, the same applies hereinafter) is denoted by u_ε (%), then the allowable value of the negative-sequence power, that is s_ε, is given by

$$s_\varepsilon = u_\varepsilon \times s_{sc}/100.$$

Using the load process $s_L(t)$ obtained from power supply calculation, we can work out the load power s corresponding to the GB specified value (95 % probability value or the maximum). Then, the calculated MTT capacity s_T and the calculated NCD capacity s_C are obtained as follows [11]:

$$\begin{bmatrix} s_T \\ s_C \end{bmatrix} = \frac{1}{2}\begin{bmatrix} 1 & 1 \\ 1 & -1 \end{bmatrix}\begin{bmatrix} s \\ s_\varepsilon \end{bmatrix}. \tag{5}$$

Further determination of their installed capacities should base on the overload characteristics of MTT and NCD.

The MSS should be equipped with backup devices as demonstrated in Fig. 7. In addition to the devices shown in Fig. 6, a backup MTT (BMTT) and a backup HMT (BHMT) are added. To save space, the MTT and the HMT can be fabricated into one box, so can the BMTT and the BHMT. The NCD adopts an m/n system; that is to say, among n units, m units work while n–m units act as backup units. Figure 7 shows a 3/4 system or a 2/4 system.

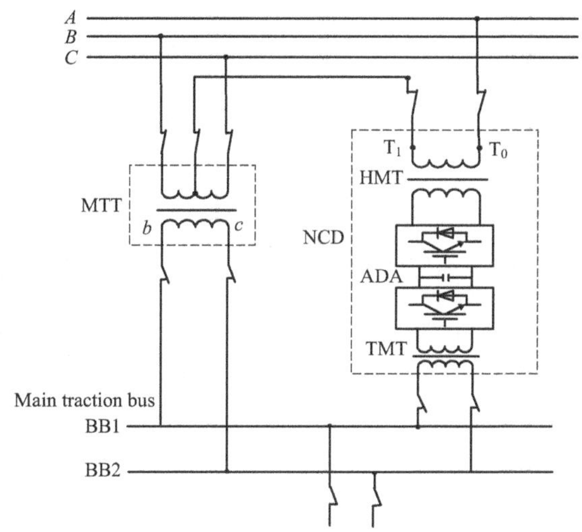

Fig. 6 MSS connection for one line

4.2 Two line case

If an MSS needs to supply power for two independent metro or light rail lines whose trains can't cross each other, the three-phase to two-phase balanced traction transformers [7] should be utilized as in a trunk railway line to reduce the impact of negative sequence. However, considering their scalability and dynamic negative-sequence compensation, it is better to build a combination scheme based on the system shown in Fig. 6. To reduce negative-sequence impact and lower down NCD costs, only two of the three phases are used for power supply, as shown in Fig. 8. In this figure, the primary side of the main traction transformer MTT_{AB} is connected to phase A and phase B of a utility grid, where U_{AB} is the line voltage, and the secondary side feeder F_{ab} supplies power for one line.

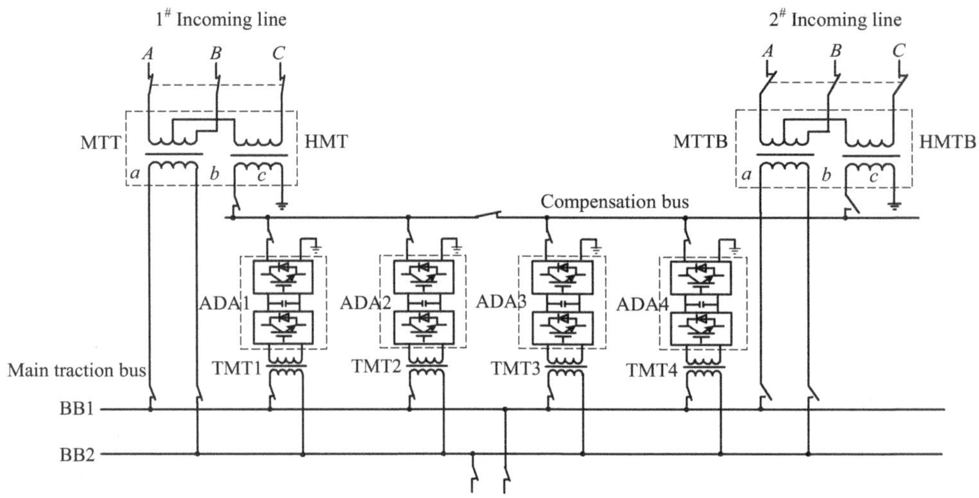

Fig. 7 MSS connection and standby equipments for one line

Similarly, the primary side of the main traction transformer MTT_{BC} is connected to phase B and phase C of a utility grid, where U_{BC} is the line voltage and the secondary side feeder F_{bc} supplies power for another line. The YNd11-connected three-phase high-voltage matching transformer HMT is connected to two NCDs, where AC–DC–AC converters are ADA_a and ADA_c, respectively, ADA_a, MTT_{BC}, and F_{bc} form one group, while ADA_c, MTT_{AB}, and F_{ab} form another group; Δ, $\bullet$, and $*$ represent the dotted terminals.

In general, the traction power is larger than the regenerative power on one line; that is, regenerative power will be absorbed by the train in traction. For this reason, pure regenerative power will be ignored in calculating capacities of the MTT and the negative-sequence compensation. Denote the traction load process $s_{Lab}(t)$ of F_{ab} and the traction load process $s_{Lbc}(t)$ of F_{bc} as s_{ab} and s_{bc} correspondingly to the GB-specified values (95 % probability value or the maximum). Assuming that their load power factors are the same, if the power of two lines is supplied with the same line voltage as shown in Fig. 6, the synthesis of negative sequence can be expressed as Eq. (6).

$$s_1^- = s_{ab} + s_{bc}. \tag{6}$$

If the power of the two lines is supplied with two different line voltages as shown in Fig. 8, the negative-sequence modulus value based on 120 degree synthesis will be less than or equal to the larger one between s_{ab} and s_{bc}, that is,

$$s_2^- \leq \max\{s_{ab}, s_{bc}\}. \tag{7}$$

Comparing Eq. (6) with Eq. (7), we can see that the negative-sequence power and the NCD capacity are lowered down greatly in Eq. (7). The maximum value is achieved at $s_{ab} = s_{bc}$ so that the NCD capacity can be decreased by 50 %.

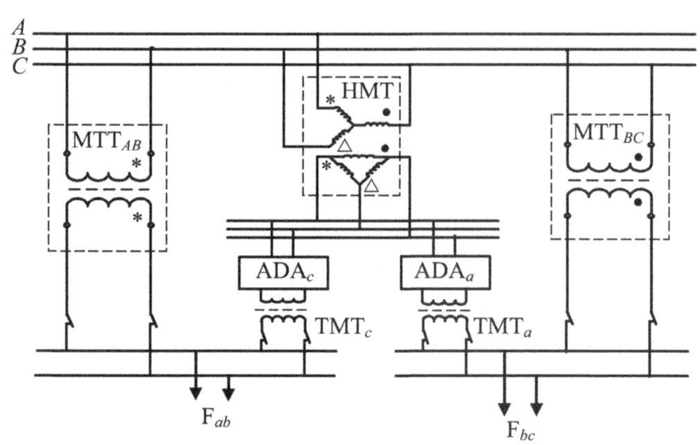

Fig. 8 MSS connection for two lines

According to the GB [18], the MSS with two lines as shown in Fig. 8 is assessed through synthesis of negative-sequence components. The compensation power portions are decomposed into the two NCDs according to proportions of load power s_{ab} and s_{bc}, respectively. Each of the NCDs only works in accordance with its feeder's traction load. That is to say, the ADA_a is a current source controlled by $s_{Lbc}(t)$ of F_{bc}. Similarly, the ADA_c is a current source controlled by $s_{Lab}(t)$ of F_{ab}. The calculated capacities of MTT_{BC} and ADA_a can be obtained by substituting $s = s_{bc}$ into Eq. (5). Likewise, the calculated capacities of MTT_{AB} and ADA_c can be obtained by substituting $s = s_{ab}$ into Eq. (5).

4.3 Three lines case

In order to minimize the impact of negative sequence and reduce the NCD capacity and costs, apparently three metro

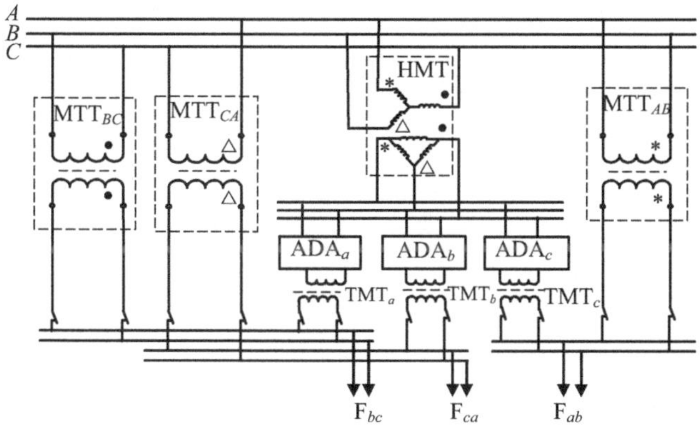

Fig. 9 MSS connection for three lines

or light rail lines should be powered with three different line voltages as shown in Fig. 9. Here, the primary side of MTT_{AB} is connected to phase A and phase B of a utility grid, and F_{ab} in the secondary side supplies the power for the first line. The primary side of MTT_{BC} is connected to phase B and phase C, and F_{bc} in the secondary side supplies the power for the second line. The primary side of MTT_{CA} is connected to phase C and phase A, and F_{ca} in the secondary side supplies power for the third line. The YNd11-connected three-phase HMT is used to link three NCDs, where AC–DC–AC converters are ADA_a, ADA_b, and ADA_c, respectively. Thereinto, ADA_a, MTT_{BC}, and F_{ab} form the first group. ADA_b, MTT_{CA}, and F_{CA} form the second group. ADA_c, MTT_{AB}, and F_{ab} form the third group.

As to GB-specified values (95 % probability value or the maximum), s_{ab}, s_{bc}, and s_{ca} are corresponding the load power of the load processes $s_{Lab}(t)$, $s_{Lbc}(t)$ and,$s_{Lca}(t)$ for feeders F_{ab}, F_{bc}, and F_{ca}, respectively. The minimum among them is denoted by

$$s_{\min} = \min\{s_{ab}, s_{bc}, s_{ca}\}.$$

The synthesis of three negative-sequence components is determined by differences between two larger load powers and the minimum load power. For instance, if the minimum load power $s_{\min} = s_{ca}$, the synthesis of negative-sequence components can be synthesized from $s_{ab} - s_{ca}$ and $s_{bc} - s_{ca}$ at 120°. In this case, the ADA_b in the group of feeder F_{ca} is not functioning, while the ADA_a in the group of feeder F_{bc} and the ADA_c in the group of feeder F_{ab} are functioning. Similar to the two line case, the calculated capacity of ADA_a can be obtained by substituting $s = s_{bc} - s_{ca}$ into Eq. (5) and that of ADA_c can be obtained by substituting $s = s_{ab} - s_{ca}$ into Eq. (5). The calculated capacity of the MTT should be added in $s_{\min} = s_{ca}$ after obtained from Eq. (5). Following a similar design process, all results can be obtained in the cases of $s_{\min} = s_{ab}$ and $s_{\min} = s_{bc}$. The final calculated capacities of ADA_a, ADA_b, and ADA_c should take the maximum among three cases of $s_{\min} = s_{ca}$, $s_{\min} = s_{ab}$, and $s_{\min} = s_{bc}$. It is notable that the closer the three lines' load power s_{ab}, s_{bc}, and s_{ca} are, the smaller the NCD capacity required is. When s_{ab},s_{bc}, and s_{ca} are equal, the NCD capacity is zero, which means no negative-sequence compensation is required.

When more than three lines are considered, the rule would be that the load to each of lines should be shared by three line voltages as equally as possible. The capacity of MTTs or the number of feeders can be increased, but the number of MTTs should be unchanged. Likewise, the number of backup MTTs should be kept unchanged.

5 System's economy and reliability

Merits in terms of reliability, economy, and technical performance of an AC traction power supply system (AC system, in short) proposed in this paper can be obtained by comparing with the existing DC traction power supply system (DC system, in short).

5.1 Economy

The DC system shares the same 35 kV cable for the traction power supply, three-phase power, and single-phase lighting supply [4, 6] while the AC system uses the separate cables for them. But this should not add much extra cost. Compared with the DC system, the AC system can save the anti-stray current devices along the metro line, save rectifier units and DC switches at each substation, and save expensive energy storage devices or renewable energy feedback devices. Consequently the metro's underground equipment can be simplified, thus occupied area is reduced. In terms of integrated automatic control devices [4–6], both the DC and AC systems have the similar scale and costs. However, the AC system needs to have the NCDs, the cost for an AC–DC–C train is higher than that for a DC–AC train. Generally, one-time costs of an AC system may be close to that of a DC system, but the MSS installed capacity of an AC traction system is significantly lower than the capacity sum of all individual SSs. Thus power resources and costs of the basic electricity are saved. In an AC system, the renewable energy can be directly used in an efficient way, resulting in further saving of power cost. All of these merits to an AC system are beneficial to the economy and in line with the national policy of energy conservation and emission reduction.

5.2 Reliability

An AC system obviates rectifier units of a DC system and reduces the number of elements in series, so as to help improving reliability. The AC circuit breaker (switch), compared to the DC breaker, is less expensive and more reliable and durable. Moreover, the segmented power supply technology, the status identification, and the protection technology are well applicable to any complex power supply system and able to improve system's reliability significantly [11, 17]. NCDs can have their reasonable backup. NCD's reliability is not less than that of the MTT in an MSS. Even in the case that all NCDs are out of service for a short period of time, the MTT can take advantages of its overload ability and continue to provide power supply. In short, the AC system has a higher reliability.

5.3 Applicability

The integrated automation system built upon the segmented power supply technology, status identification, and protection technology can allow more convenience for change of traction power supply mode, routine maintenance, or repair.

For an urban rail transit system powered by the AC system, because two lines or more than two lines can share one main transformer, each line only requires one MSS. Such a centralized power supply system can not only cut down the interface to the utility grid substantially but also facilitate daily management.

6 Conclusions

This paper has presented and discussed an AC traction power supply system that has the following salient features:

- An AC system avoids the stray current occurring in a DC system.
- Using the industrial frequency AC, the given traction power supply system can connect to a utility grid directly and thus gains economic, reliable, and strong power supply.
- The use of cables meets the requirements of long distance, high capacity, without phase-splits for power supply.
- A train with an AC–DC–AC drive system and the AC traction motor is able to achieve better traction and more direct use of renewable energy.

Above features can solidify the application foundation of the proposed system in the urban rail transit, such as metro and light rail systems.

Several key technologies employed in the proposed system are discussed. The main work and the conclusion can be summarized as follows:

(1) The power supply structure, principle, equivalent circuit, and transmission ability of the CTN are described. The 35 kV cable-OCS matching technique, and the OCS voltage grade selection are also discussed. Analysis shows that the CTN can meet not only the requirements of power supply capacity but also the distance demands for any metro or light rail lines.

(2) A segmentation technology for the traction network is discussed. Particular attentions are given to OCS segmentation, status identification, and protection methods. Using the segmentation technology for traction network, its operation and maintenance can be conveniently changed or scheduled. Faults can be detected and isolated timely so that their impacts can be limited to the minimum extent. As a result, the system's reliability has been improved greatly.

(3) In addition to the power supply scheme for one line railway, schemes for two or more lines railway where one MSS is shared are also given. For each of the schemes, the method for calculating the minimum capacity of the NCD is presented, which ultimately leads to an optimal power supply solution.

In summary, the proposed system, with good technical performance, good economy, and high reliability, provides a new promising alternative solution for urban rail transit, such as metro and light rail transit systems. Therefore, it is worthy of further research and investigation for the future.

Acknowledgement The author is grateful for numerous support from faculty members, colleagues, and students in School of Electrical Engineering, Southwest Jiaotong University, who participated or helped in this research work.

References

1. Cao JY (1956) The way of electrified railway in China, People's Daily, China (26 Nov, 1956) **(in Chinese)**
2. Cao JY (1983) Power supply system of electrified railway. Press of Chinese Railway, Beijing, pp 106–109 **(in Chinese)**
3. Mark Walter GK (1989) Power supply of electrified railway (trans: Yuan ZF, He QG). Press of Southwest Jiaotong University, Chengdu, pp 141–145 **(in Chinese)**
4. He WJ, Gao SB, Zhang SQ, Wang X (1998) Electric traction power supply and transforming technology. Press of Southwest Jiaotong University, pp 94–104 and 188–224 (**in Chinese**)

5. Qian QQ (2000) Electrified railway computer monitoring technology. Press of Southwest Jiaotong University, Chengdu **(in Chinese)**
6. Yu SW, Yang XS, Han LX, Zhang W (2008) Urban rail transit power supply system design principles and applications. Press of Southwest Jiaotong University, Chengdu, pp 225–242 **(in Chinese)**
7. Lian JS (2001) Introduction to Electric Drive Locomotive. Press of Southwest Jiaotong University, Chengdu, pp 127–151 **(in Chinese)**
8. Li QZ, Zhang JS, He WJ (1988) Research on new power system for heavy haul traction. J China Railw Soc 4:23–31 **(in Chinese)**
9. Li QZ, Lian JS, Gao SB (2006) Electrified engineering of high speed railway. Press of Southwest Jiaotong University, Chengdu, pp 155–166 **(in Chinese)**
10. Li QZ (2006) Electrical analysis of traction substation and technology of comprehensive compensation. Press of Chinese Railway, Beijing, pp 72–93 **(in Chinese)**
11. Li QZ (2014) On new generation traction power supply system and its key technologies for electrified railways. J Southwest Jiaotong Univ 49(4):559–568 **(in Chinese)**
12. Tang XH (2006) Protection setting of loop power grid with voltage of 110 kV and below. Rural Electr 27(9):28–30 **(in Chinese)**
13. Wang W, Gao K (2010) Analysis of loop problem of Beijing power grid. Water Resour Hydropower Northeast China 12:58–61 **(in Chinese)**
14. Encyclopedia of China's Power (Utility Grid Volume) (1995) China Electric Power Press, 1995, Beijing, pp 309–310 **(in Chinese)**
15. Qiou GY (1989) The circuit (2), 3rd edn. Higher Education Press, Beijing, pp 185–198 **(in Chinese)**
16. Li QZ, Yi D, He JM (2013) Analysis of traction power supply system with cables for electrified railways. J Southwest Jiaotong Univ 48(1):81–87 **(in Chinese)**
17. Li QZ, Chen MM, Huang YY (2014) Research of power supply state monitoring and control method of AT traction network for high-speed railway, Electric Railway, Supplement, pp. 11–15 **(in Chinese)**
18. Standardization Administration of the People's Republic of China (2008) GB/T15543—2008 power quality three-phase voltage unbalance **(in Chinese)**

9

Impacts of cell phone use on driving safety and drivers' perception of risk

Dazhi Sun[1] · Anxi Jia[1]

Abstract In this study, the factors influencing phone-related driving safety and drivers' perceptions of cell phone usage were analyzed. A representative sample of more than 500 licensed drivers in Texas who own a cell phone was interviewed based on a well-design questionnaire. Logistic regression model showed that the impact of using cell phone on driving safety varies depending on the characteristics of drivers, such as gender, age, driving experience, and use intensity. Additionally, the results indicated that the strong determinants of phone-related hazard are different from that of phone-related accidents. Regarding the drivers' perception of cell phone usage, there are two key findings. First, there is no explicit belief among the drivers about whether cell phone usage impairs driving safety regardless of the drivers' age, gender, driving education experience etc. Second, most of drivers have not realized that cell phone use while driving would increase their perception reaction time. Based on the analysis of these results, implications of cell phone use on driving safety along with some safety countermeasures, such as selective bans and non-cell phone zones, are discussed in the paper.

Keywords Cell phone · Phone-related hazard · Phone-related accident

1 Introduction

National Highway Traffic Safety Administration [1] reported that almost 80 % of crashes and 65 % of near-crashes involved some form of driver inattention, among which cell phones use is the most common distraction. Cellular Telecommunications & Internet Association estimated that in the United States over 236 million people subscribed to such wireless communication devices as cell phones as of May 2007, compared with approximately 4.3 million in 1990. Increased reliance on cell phones has inevitably led to a rise in the number of people who use the devices while driving. NHTSA [2] had estimated approximately 974,000 drivers on the road nationwide at any time during the day using a handheld cell phone. In recent years, cell phone use while driving has led to an obvious concern about driving safety.

1.1 Current research approaches and results

In general, there are three major research methods currently employed to study the effects of cell phone use on driving safety. The first is to establish a statistical association between cell phone use and accidents using survey data [3–6]. The second is simulator based or on-road controlled experiments [7, 8]. The third method mainly depends on accident statistics from police reports, aggregate crash, and cell phone statistics [9, 10]. Hahn and Dudley [11] and McCartt et al. [12] examined all of the methods and found that each approach has its own shortcomings. The key findings and their shortcomings of each method are summarized in the following.

1.1.1 Survey

The survey data are commonly collected through personal interview, telephone interview, and internet-based questionnaires. Redelmeier and Tibshirani [10] concluded that the risk of a collision when using a cell phone was four times higher than the risk when a cell phone was not being

✉ Dazhi Sun
dazhi.sun@tamuk.edu

1 Texas A&M University-Kingsville, Kingsville, TX, USA

used. Pöysti et al. [13] reported that some factors, such as drivers' age, amount of phone usage, and occupational position, were related to phone-related hazards. Hahn and Prieger [14] found that there is no significant effect of hands-free or handheld cell phone use on accidents. Regarding the survey method, however, the sample selection and the dishonest answers to the sensitive questions, such as whether they were involved in accidents because of cell phone use, may bias the result.

1.1.2 Simulation and experiment

In recent years, driving simulation and an on-road controlled experiment are quite prevalent methods to study the impacts of cell phone usage while driving. Reed and Green [7] provided experimental evidence of negative effects on lane keeping, steering performance, and accelerator control. Strayer and Johnston [15] reported that drivers engaged in cell phone conversations missed twice as many traffic signals and had slower reaction times. More recently, through a new study, Strayer confirms that the reaction time of cell phone users increases dramatically, increasing the risk of accidents and tying up traffic in general. The driving simulation or on-road controlled experiments could also bias the result due to the experimental nature of data collection effort, which has less accuracy than real-world situations. Moreover, it is difficult to obtain representative samples of drivers in conducting such simulation or experiment. Haigney and Westerman [16] insist on the need to specify the conditions in which the experiments are performed and to restrict their conclusions to those conditions.

1.1.3 Accident statistics

It is inevitable to study the drivers' crash history for evaluating the crash risk associated with phone use. Unfortunately, most states in U.S. do not provide data elements in the crash report forms recording drivers' cell phone use. Although the characteristics of cell phone-related crash were examined in some research (Goodman et al. [17], Huang and Stutts [18]), McCatt [12] concluded that police reports would not be appropriate in assessing the role of phone use in crashes because the data on phone use are unreliable. In order to fix this problem, some research used the cell phone company billing records to verify cell phone use. The most representative study was conducted by McEvoy et al. [19] in Australia. They found that a driver's use of a cell phone up to 10 min before a crash was associated with a fourfold increased likelihood of a serious crash. However, another problem is that accident statistic reports may overwhelmingly limit the sample to drivers who had accidents and may overestimate the impact of usage of cell phones on accidents. A recent research, by Hahn and Prieger [14], implies that previous estimates of the impact of cell phone use on risk for the population may be overstated by about one third.

1.2 The objective of this study

The objective of this study is to evaluate the impacts of cell phone use on driving safety and drivers' perception in Texas. Although it is very clear that cell phone use degrades driving performance and safety based on previous studies, it is not clear whether its impacts will vary with the characteristics of drivers, such as gender, driving experience, driving behavior, and cell phone use intensity. Bailey [20] and Nunes and Recarte [21] indicate that cell phone use while driving increases the overall level of cognitive, and sometimes physical, demand of drivers. This result is consistent with the finding that, in this circumstance, both heart rate and self-report workload increase which implies that safety margins are reduced when talking on a mobile phone, because drivers have less spare processing resources to allocate to driving operation. The hypothesis of this paper is that cell phone use may generate different impacts on different driver groups, because driving workload among different groups could be varied. Another difference from previous studies is that drivers' perceptions of the impacts of cell phone use will be examined and compared in this study.

2 Method and data collection

This study was conducted in Texas, with no regulations against using cell phone while driving. Based on careful comparison of the three prevalent approaches discussed in previous section, personal interview survey is finally selected as the data collection method. The driving simulation and on-road experiment are not suitable for evaluating the accident risk of cell phone use due to its experimental nature and unrepresentative sample. Regarding the accident statistics method, although the cell phone company's billing records could help verify the phone use related to crashes, such records have been unavailable in the United States. Among the three options of survey methods, personal interview and internet-based questionnaires could be conducted in an anonymous manner. Chang and Krosnick [22] indicate that anonymous survey can impressively reduce the effect of "dishonest" answers on the survey's accuracy, compared with telephone surveys. Although the personal interview survey may involve more cost, it offers three major advantages compared with internet-based survey. First, people will usually respond when confronted in person. Second, the sample may be more representative than internet-based surveys because the latter limits the sample to the internet-users. In addition, the interviewer can note specific reactions and eliminate misunderstandings about the questions asked.

A well-designed interview questionnaire was developed to collect expected data. This questionnaire (see Table 1) consists of four question categories: basic personal information, cell phone use information, crash or dangerous driving history, and drivers' perceptions. Among the questionnaire, the items concerning that an accident or a hazard situation happened when there's cell phone use are designed as the dependant variables. Marshall and Rodney [23] suggest dependence on the observation that if the cell phone usage increases accident risk, then the driver is more likely to be on the phone at the time of the crash than during the earlier reference period. The drivers' age, driving experience, gender, frequency of cell phone use, text messaging, and driving speed are selected as the independent variables. The items concerning perception are designed to reveal the motorists' beliefs or attitude on what impact, if any, cell phone usage has on their driving behavior and safety.

The interviewers are composed of 50 engineering students, who are thoroughly trained to correctly state and explain the question. In order to obtain the sample as representative as possible, the student went to gas stations and public parking lots at the library and shopping marts to conduct the interview and record the responses. The respondents are randomly selected by the interviewers from 8:00 a.m to 8:00 p.m. More than 1,000 questionnaire sheets were completely finished by the respondents and about 300 drivers were not willing to respond or finish the interview at last. The response rate of this study is about 75 %. After the deletion of problematic questionnaires, the sample finally consisted of 942 participants (57 % male and 43 % female) who have valid driver license. Their ages ranged from 16 to 55 (25 % in 16–20, 24 % in 21–25, 22 % in 26–30, and 29 % above 30). The sample contains drivers who had accidents and who did not, drivers who use a cell phone as well as drivers who do not.

In this study, two separate logistic regression analyses were made. The first is to study the hazard situation experienced while using the cell phone, while the second is for the accidents that happened when they were using cell phone. Chi-square test was used to examine the independence between the dependant and independent variables. For the analysis of drivers' perceptions, unpaired t test was used to compare the mean of the drivers' perceptions. All results are reported at 0.05 confident intervals.

3 Results

The result of this study indicates that there are 862 drivers (87.7 %) who ever use a cell phone while driving and 116 (12.3 %) who do not. Among the 862 drivers who use cell phone while driving, 466 are male drivers (54 %) and 396 are female drivers (46 %).

Table 1 Items of questionnaire

Domain	Subcategory	Questions
Basic information	Age	What is your age?
	Gender	What is your gender?
	Driving experience	How many years driving experience do you have?
	Driving habits	Do you typically drive under, at, or above the posted speed limit?
Cell phone use	Usage	Do you ever drive and use cell phone simultaneously?
	Intensity	Estimate the percentage of time you spend on cell phone use while driving?
	Text message	Do you ever text message while driving?
Crash history	# of Accidents	How many accidents have you had in last year?
	Cell phone related	Were any of these accidents related to your cell phone use?
		If YES, how many of them related to cell phone use?
	Dangerous driving	Have you ever found yourself failing to yield or stop due to using cell phone?
		Have you ever had to make a sudden evasive maneuver to avoid being in an accident while you were driving and using cell phone?
Perception		Most people can carry on a conversation on their cell phone and still drive safely
		Cell phones are more beneficial to drivers than they are harmful
		Using a hands-free device with a cell phone is much safer than using a handheld cell phone
		Using a cell phone while driving increases the perception reaction time of the driver

The respondents were asked to rate their opinions on the perception part on a scale of 0–10, where 0 means "Completely disagree" and 10 means "completely agree"

3.1 Phone-related hazards

About 60 % of cell phone-using drivers admitted having experienced hazard traffic situations while using a cell phone in the last 12 months. The hazard situation experienced while using the cell phone, which is defined as sudden evasive maneuver to avoid being in an accident and failing to stop at signal or stop sign, had been clearly explained to the interviewers by the students. The logistic regression (Table 2) showed that whether the drivers experienced a hazard situation while using a cell phone depends on gender, using frequency, texting message, and driving speed. Female drivers reported the hazard experience two times more than their male counterpart. The more frequently the drivers use their cell phone, the more likely they have experience hazard situations. For instance, the drivers who spend more than 60 % of driving time on cell phone use reported the hazard experience about 2.5 times more than the ones who only spend <20 %. The drivers who use the service of text message reported the hazard situation three times more than the ones who do not use the service but use cell phone while driving. As expected, the drivers who typically drive at a higher speed reported more hazard situations experienced while using a cell phone. This is in line with previous results, which suggest that cell phone use increases the overall level of cognitive, and sometimes physical, demand experienced by drivers [24], while driving at a higher speed also increases the resources demanded for the overall driving task [25–27].

Table 2 Summaries of logistic regression analyses for hazard situation while using cell phone

Predictor		*B*	Odds ratio	Significance
Gender	Male Ref.			
	Female	0.6058	1.833	0.0011
Age	40+ Ref.			
	16–20	0.1404	1.151	0.6956
	21–25	0.0529	1.054	0.1418
	26–30	0.0605	1.062	0.2162
	31–35	−0.2118	0.809	0.8285
	36–40	0.1050	1.111	0.5422
Driving experience	10+ Ref.			
	0–2	0.1500	1.162	0.5927
	3–5	0.1022	1.108	0.4571
	6–10	−0.1578	0.854	0.1226
Speeding	Above 10+ Ref.			
	Below	−0.5253	0.591	0.0039
	At	−0.3369	0.714	0.0372
	Above 0–5	−0.1461	0.864	0.9683
	Above 5–10	−0.0516	0.950	0.2755
Frequency	<20 % Ref.			
	<40 %	0.6532	1.922	<.0001
	<60 %	0.7293	2.074	<.0001
	60 %+	0.9462	2.576	<.0001
Text	0 Ref.			
	1	1.0403	2.830	<.0001

3.2 Phone-related accidents

The logistic regression (Table 3) showed that whether the drivers experienced an accident while using a cell phone depended on age, driving experience, driving speed, and using frequency. Young and inexperienced drivers are more likely involved in accidents when they were using a cell phone than the older and experienced drivers. Similar to the analysis of hazard situation, the drivers who typically drive at a higher speed and use the cell phone more frequently reported more accidents experienced while using a cell phone. There is no significant difference between male and female drivers in terms of accidents experienced while using a cell phone.

The above analysis suggests several variables could be used to predict the hazard situation or accidents experienced while using the cell phone. However, it is possible that not of all them are the strong determinants of phone-related hazards or accidents, if there were correlations between these variables. Table 4 shows the correlations between the six variables. Age has strong correlations with driving experience, driving speed, and text messaging: young drivers definitely have less driving experience, typically drive at higher speed, and are more likely to use the cell phone frequently and the text message service. Female drivers are more likely to use the cell phone frequently and the text message service than males. The drivers, who typically travel at higher speed, reported more use frequency and use of text message service. Based on the forward logistic regression model, the final determinants are use frequency (P value = 0.000) and text message (P value = 0.015) for phone-related hazards. For the phone-related accidents, driving experience (P value = 0.001) and use frequency (P value = 0.038) are selected as the strong determinants.

3.3 Drivers' perception of the effects of cell phone use

This research also studied the differences in the perception of the effects of cell phone use on safety. The respondents were asked to rate their opinion on four cell phone use related driving safety statements (see Table 1). After that, the respondents are categorized into five different groups: (1) drivers taking driver education class or not; (2) male and female drivers; (3) drivers using cell phone while driving and not; (4) drivers involved in accidents or not; and (5) novice drivers (0–2) and experienced drivers (3+) drivers. In each group, the differences of the perception are analyzed using unpaired t test, which is shown in Table 5.

Table 3 Summaries of logistic regression analyses for accidents while using cell phone

Predictor		*B*	Odds ratio	Significance
Gender	Male Ref.			
	Female	0.1074	1.113	0.5573
Age	40+ Ref.			
	16–20	0.5319	2.169	0.0196
	21–25	0.4489	1.567	0.0009
	26–30	0.7745	1.702	0.0339
	31–35	0.2343	1.264	0.1626
	36–40	0.2183	1.244	0.1229
Driving experience	10+ Ref.			
	0–2	0.8035	2.233	0.0070
	3–5	0.4420	1.556	0.0026
	6–10	0.5369	1.711	<.0001
Speeding	Above 10+ Ref.			
	Below	−0.9075	0.404	0.0271
	At	−0.5082	0.602	0.0491
	Above 0–5	−0.3261	0.722	0.0620
	Above 5–10	−0.2676	0.765	0.0857
Frequency	<20 % Ref.			
	<40 %	0.0803	1.084	0.4157
	<60 %	0.3405	1.406	0.0040
	60 %+	0.3595	1.433	0.0256
Text	0 Ref.			
	1	−0.0745	0.928	0.6855

3.3.1 Drivers taking driver education class versus drivers not

In Texas, all new drivers under 18 must complete an approved driver education course before they can be licensed. Moreover, the drivers who have been involved in accidents or issued tickets could also take the driver education course to keep their insurance rates reasonable or reduce the fines. According to the results of the survey, however, there are no statistically significant difference of perception between the drivers who had taken driving education classes and the drivers who had not. The two groups have the same neutral attitudes about the effects of cell phone use on their driving safety. For the questions regarding PRT, neither of the two groups is sure whether cell phone use impairs their PRT. Both of them are inclined to agree that a hands-free device is much safer than a handheld device.

3.3.2 Novice driver versus experienced driver

No statistically significant difference has been found about the perception between the novice drivers and their experienced counterparts, according to the P values of the t test shown in Table 5. Moreover, the results show that both of them are not very clear whether cell phone use has affected their driving safety. According to the mean rates, the novice drivers are more likely to underestimate the effect of cell phone use on PRT than the experienced drivers.

3.3.3 Male versus female

The t test results show that there is no statistically significant difference of perception by gender. For statement 3 (hands-free vs. handheld), the mean rates of female drivers

Table 4 Correlations between the variables

	Age	Gender	Drive experience	Speeding	Frequency	Text
Age	1.00000	−0.14348	0.65881	−0.13944	−0.03482	−0.23233
		0.0014	<.0001	0.0020	0.4413	<.0001
Gender	−0.14348	1.00000	−0.20080	−0.03672	0.19819	0.12680
	0.0014		<.0001	0.4169	<.0001	0.0049
Drive experience	0.65881	−0.20080	1.00000	−0.01773	0.02929	−0.12186
	<.0001	<.0001		0.6952	0.5174	0.0069
Speeding	−0.13944	−0.03672	−0.01773	1.00000	0.21592	0.17643
	0.0020	0.4169	0.6952		<.0001	<.0001
Frequency	−0.03482	0.19819	0.02929	0.21592	1.00000	0.30159
	0.4413	<.0001	0.5174	<.0001		<.0001
Text	−0.23233	0.12680	−0.12186	0.17643	0.30159	1.00000
	<.0001	0.0049	0.0069	<.0001	<.0001	

are still greater than male by 7.7 %, although these differences are not statistically significant.

3.3.4 *Drivers using cell phone while driving versus drivers not*

Statistically significant differences have been found about the perception between the drivers who use a cell phone while driving and the ones who do not. The results indicate that the drivers who do not use a cell phone are much more inclined to agree that cell phone use while driving could impair the driving safety than the drivers who use. However, the former group is more likely to underestimate the effect of cell phone use on the PRT than the later group. For the hands-free and handheld issue, the drivers who use a cell phone are more inclined to agree that the hands-free device is much safer.

3.3.5 *Drivers involved in cell phone-related accidents versus drivers not*

There is no statistically significant difference of perception between the drivers who have been involved in cell phone-related accidents and the ones who have not. Although the former group has been involved in cell phone-related accidents, they have the similar attitude as the latter group about the effects of cell phone use on driving safety. Paradoxically, the mean rates of statement 1 and statement 2 of the former group are greater than the rates of the latter group by 9.2 % and 22.6 %, respectively. This finding indicates that the drivers who have been involved in some cell phone-related accidents usually are those who are more likely to underestimate the effect of talking on cell phone on driving safety.

4 Discussion and recommendation

The results of this study show that frequency use and text messaging are two strong determinants of a phone-related hazard. When it comes to the phone-related accident level, however, the determinant variable is driving experience and use frequency. Cell phone usage while driving could increase the likelihood of both traffic hazards and accidents; however, whether the phone-related accidents happen also depends on the drivers' driving experience. Moreover, driving experience contributes more than use frequency for the phone-related accidents according to the regression model. That means that using a cell phone is not threatening the safety of all driver groups equally: cell phone usage will result in phone-related accident more easily for novice drivers than experienced ones. Even for novice drivers, about 50 % of them have experienced phone-related hazards, while only 6 % of them have phone-related accidents. This indicates that the safety situation with a cell phone is not extremely serious. The potential phone-related accident risks have been controlled at the acceptable level and have not increased in line with the rapid growth of cell phone usage. Some authors have also suggested that most drivers can manage with their mobile phone while driving (Mikkonen and Backman [28]). However, it does not mean the phone-related accidents should be overlooked, considering that driving the car is the major task, while using cell phones is just a

Table 5 Perceptions of different groups of the effects of cell phone use

Category	Rate	Statement 1	Statement 2	Statement 3	Statement 4
Take class	Mean (R)	5.00	4.48	5.09	4.20
Not take class	Mean (R)	5.12	4.36	4.70	3.92
	P value	0.670	0.670	0.197	0.398
Male	Mean (R)	5.02	4.40	4.72	3.99
Female	Mean (R)	5.08	4.49	5.24	4.22
	P value	0.828	0.747	0.067	0.474
Use cell phone	Mean (R)	5.25	4.70	5.89	4.25
Not use cell phone	Mean (R)	3.64	2.55	2.90	2.95
	P value	0.000	0.000	0.005	0.008
Have cell phone-related accidents	Mean (R)	5.48	5.37	4.44	4.44
Not have cell phone-related accidents	Mean (R)	5.02	4.38	4.98	4.07
	P value	0.448	0.091	0.442	0.565
Novice drivers	Mean (R)	5.17	4.41	4.80	3.80
Experienced driver	Mean (R)	4.94	4.46	5.07	4.35
	P value	0.369	0.852	0.352	0.083

secondary task and may result in additional accident risk. Therefore, traffic safety communities should still enhance the management and education to reduce or prevent the phone-related accidents, especially for the specific risk groups, novice drivers.

According to the results, novice drivers are more likely to be involved in phone-related accidents than experienced ones. Unfortunately, there is no significant difference between novice drivers and experienced drivers about their perception of the impact of cell phone use on travel safety. This is an obvious inconsistency between their actual performance and their subjective perception. The cell phone usage effects on gender only appeared at the phone-related hazard level. Compared to female drivers, the male drivers are found to be more likely to experience cell phone-related hazards. Concerning the phone-related accidents, however, there is no significant difference between male and females.

Another interesting finding is that current driving education classes do not show significant improvement on the drivers' perception of the effects from cell phone use. The drivers who have taken a driving class do not exhibit expected right perception of the impact of cell phone use on their driving safety as well as PRT. Their perception of effects is quite similar to the drivers who have never taken these classes. To reduce the cell phone-related accident risk, current driving education programs should be enhanced to improve the awareness of impairing effects of using a cell phone while driving, especially for the specific risk groups, i.e., young drivers or novice drivers.

Besides negative effects, it is undoubted that cell phone use could also provide some benefits to drivers, such as emergency calls to report accidents immediately. Moreover, in most cases, the legislation only restricts the use of handheld phones while driving. In the U.S., New York, New Jersey, Washington, D.C. and many municipalities fully outlawed in-vehicle handheld phone use. However, this restriction may be problematical itself. For example, Strayer et al. [8] and Ishigami [29] concluded that allowing hands-free cell phone use will have very little effect on reducing the driver distractions and accident risk. Therefore, it seems unreasonable to ban all drivers from using their cell phones while driving. It could be a feasible solution to selectively ban vulnerable drivers to use cell phones. For instance, drivers with <2 years of driving experience, and the drivers who have been involved in cell phone-related accidents in the last 1 or 2 years could be considered as the vulnerable drivers. In addition, non-cell phone zone would be another reasonable, possible remedy to address the cell phone-related safety issues. The qualified locations for the non-cell phone zone could include some roadway segments requiring heavy driving workload, such as highway work zones, busy intersections, and high crash risk locations with constrained sight distance. In all these cases, there will obviously be enforcement related issues.

Besides the legislative efforts, insurance companies could play a vital role to alleviate the increasing impairing effects of cell phone use on driving safety. For example, in the United Kingdom and Germany, insurance companies are allowed to cancel the drivers' insurance coverage if they are involved in a crash while talking on phone. The impact of such measures by insurance companies will have important policy implications in the country too.

References

1. National Highway Traffic Safety Administration (2006) The impact of driver inattention on near-crash/crash risk: an analysis using the 100-car naturalistic driving study data. Department of Transportation, NHTSA, Washington
2. National Highway Traffic Safety Administration (2004) Traffic safety facts 2002: a compilation of motor vehicle crash data from the fatality analysis reporting system and the general estimates system. U.S. Department of Transportation, NHTSA, National Center for Statistics and Analysis, Washington
3. Dreyer NA, Loughlin JE, Rothman KJ (1999) Cause-specific mortality in cellular telephone users. J Am Med Assoc 282(19):1814–1815
4. Redelmeier DA, Tibshirani RJ (1997) Association between cellular-telephone calls and motor vehicle collisions. N Engl J Med 336(7):453–458
5. Violanti JM, Marshall JR (1996) Cellular phones and traffic accidents: an epidemiological approach. Accid Anal Prev 28(2):265–270
6. Violanti JM (1998) Cellular phones and fatal traffic collisions. Accid Anal Prev 30(4):519–524
7. Reed MP, Green PA (1999) Comparison of driving performance on road and in a low-cost simulator using a concurrent telephone dialing task". Ergonomics 42(8):1015–1037
8. Strayer DL, Crouch D, Drews FA (2006) A comparison of the cell-phone driver and the drunk driver. Hum Factors 48:381–391
9. Laberge-Nadeau C, Maag U, Bellavance F, Lapierre S, Desjardins D, Messier S, Saidi A (2003) Wireless telephones and the risk of road crashes. Accid Anal Prev 35:649–660
10. Redelmeier DA, Tibshirani RJ (1997) Interpretation and bias in case-crossover studies. J Clin Epidemiol 50(11):1281–1287
11. Hahn RW, Dudley PM (2002) The disconnect between law and policy analysis: a case study of drivers and cell phones. Adm Law Rev 55(1):127–185
12. McCartt Anne T, Hellinga Laurie A, Bartiman Keli A (2006) Cell phones and driving: review of research. Traffic Inj Prev 7:89–106
13. Pöysti L, Rajalin S, Summala H (2005) Factors influencing the use of cellular (mobile) phone during driving and hazards while using it. Accid Anal Prev 37:47–51

14. Hahn RW, Prieger JE (2006) The impact of driver cell phone use on accidents. J Econ Anal Policy Adv 6:1–37
15. Strayer DL, Johnston WA (2001) Driven to distraction: dual-task studies of simulated driving and conversing on a cellular phone. Psychol Sci 12:462–466
16. Haigney D, Westerman SJ (2001) Mobile cellular phone use and driving: a critical review of research methodology. Ergonomics 44(2):132–143
17. Goodman MJ, Tijerina L, Bents FD, Wierwille WW (1999) Using cellular phones in vehicles. Transp Hum Factors 1:3–42
18. Huang HF, Stutts JC (2003) The characteristics of cell phone-related motor vehicle crashes in North Carolina. Transportation Research Board Preprints [CD-ROM]: Washington, D.D., TRB
19. McEvoy SP, Stevenson MR, McCartt AT, Woodward M, Haworth C, Palamara P, Cercarelli R (2005) Role of mobile phones in motor vehicle crashes resulting in hospital attendance: a case-crossover study. Br Med J 331:428–430
20. Bailey TJ (1994) The road safety implications of mobile phone use whilst driving: a short literature review. Report Series 4/94 (Brisbane: Office of Road Safety, South Australian Department of Transport)
21. Nunes L, Recarte MA (2002) Cognitive demands of hands-free-phone conversation while driving. Transp Res Part F 5:133–144
22. Chang LC, Krosnick JA (2003) National surveys via RDD telephone interviewing vs. the internet: comparing sample representativeness and response quality. Ohio State University. http://www.psy.ohio-state.edu/social/CSR-KN-HI.doc
23. Marshall RJ, Rodney TJ (1993) Analysis of case-crossover designs. Stat Med 12:2333–2341
24. Fairclough SH, Ashby MC, Ross T, Parkes AM (1991) Effects of handsfree telephone use on driving behavior. In Proceedings of the ISATA conference, Florencia, Italia, ISBN 0 947719458
25. Wickens CD (1991) Processing resources and attention. In: Damos DL (ed) Multiple-task performance. Taylor & Francis, London
26. Wilde GJS (1982) The theory of risk homeostasis: implications for safety and health. Risk Anal 2:209–225
27. Wilde GJS (1988) Risk homeostasis theory and traffic accidents: propositions, deductions and discussion of dissention in recent reactions. Ergonomics 31:441–468
28. Mikkonen V, Backman M (1988) Use of the car telephone while driving. Technical Report # A39. Department of Psychology, Univeristy of Helsinki
29. Ishigami Y (2007) Is a hands-free phone safer than a handheld phone? In: BBCS/SCCC 17th annual meeting

10

Residential relocation in response to light rail transit investment: case study of the Hudson–Bergen Light Rail system

Xiaobo Liu[1] · Yi Deng[2] · Scott Le Vine[1,3,4]

Abstract It is widely acknowledged that the improved accessibility enabled by investment in public transport services can, under favorable market conditions, impact the local real estate market within the zone of influence of the service's stations. The motivation for this study is to establish the nature of two such impacts, specifically the spatial and socio-economic patterns of residential relocations that are driven by the new light rail transit (LRT) service. Using empirical data ($n = 1{,}023$) from the Hudson–Bergen Light Rail system in New Jersey (US), we report findings regarding the impacts of the introduction of the new LRT service. We investigate two linked dimensions; the first is the distinctive socio-economic profile of LRT passengers who self-report having relocated to the new transit corridor due, at least in part, to the new transit service. The second is their proximity (following their residential relocation) to the new LRT line's stations. We present a novel analysis that accounts for endogeneity between these two dimensions of residential relocation. Of light rail passengers who engaged in a residential relocation in the 5 years prior to the survey, two-thirds (69 %) indicate that proximity to the light rail service was a 'somewhat' or 'very' important consideration. Via the multivariate analysis, we demonstrate that small household size, low income, youth (as opposed to older age), and low car ownership are each positively linked, *ceteris paribus*, with having engaged in a residential relocation motivated by the new transit service. Finally, higher household income is found to be associated with distance (after relocation) to the nearest transit station, which is consistent with bid-rent theory.

✉ Xiaobo Liu
Xiaobo.liu@swjtu.cn

1 School of Transportation and Logistics, Southwest Jiaotong University, Chengdu 610031, People's Republic of China

2 Parsons Transportation Group, 100 Broadway, New York, NY 10005, USA

3 State University of New York at New Paltz, New Paltz, NY 12561, USA

4 Imperial College London, South Kensington, London SW7 2AZ, England, UK

Keywords Light rail transit · Residential relocation · Transit-oriented development

1 Introduction

Light rail transit (LRT) has attracted increasing interest from policymakers in recent decades, with new services being introduced across a range of metropolitan regions. The improved accessibility engendered by LRT is seen by many observers as a catalyst for transit-oriented developments (TODs), and indeed in many cases residential development accompanies the introduction of new LRT services [1, 2]. The general issue encapsulated by the TOD concept is that the stimulus of introducing LRT is thought to have non-trivial impacts on the real estate market of the service area.

The motivation for this study was to investigate the dynamics of residential relocation associated with the inauguration of new LRT service. Empirical data are sourced from a 2008 survey of passengers ($n = 1{,}023$) of the Hudson–Bergen Light Rail (HBLR) system in New Jersey (US). Fieldwork for the survey was undertaken eight years after service inauguration, which implies that this study's results should be interpreted as mid-term effects after the residential property market has had the chance to initially re-equilibrate in response to the stimulus of the new LRT system. The HBLR system (see Fig. 1) feeds heavy rail and ferry services

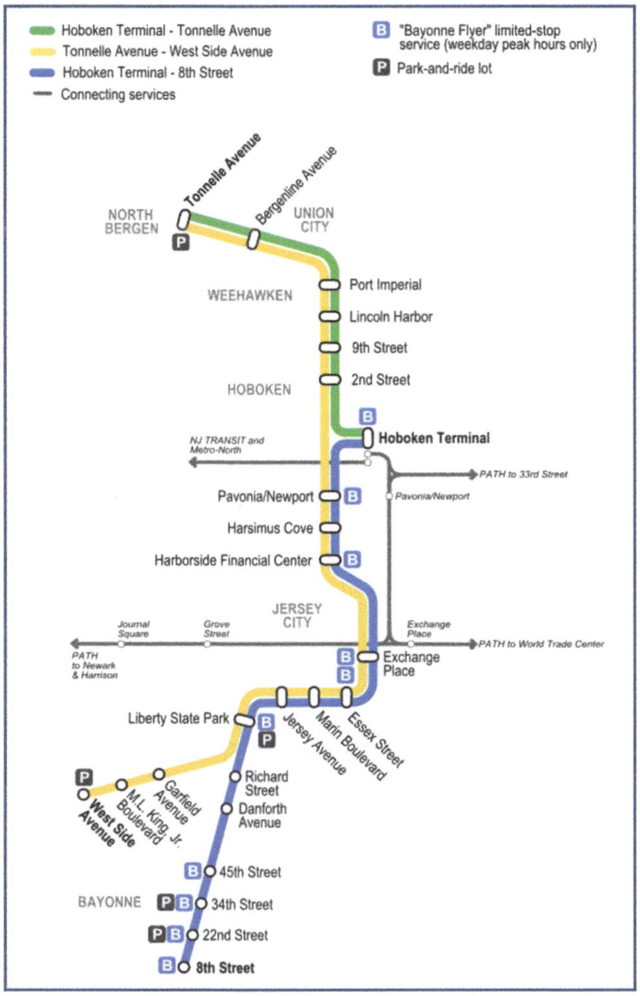

Fig. 1 Schematic representation of the Hudson–Bergen Light Rail (HBLR) system

that connect to midtown and downtown Manhattan, and also provides local connectivity between pre-automobile urban neighborhoods of Hudson County, New Jersey (which is characterized by the sixth-highest residential density among counties in the US). In a study of the land use impacts of the HBLR system, Robins and Wells [3] estimate approximately 10,000 newly built residential units, representing a gross investment of $5.3 billion (in 2008 prices).

The results reported in this paper are twofold. First, we investigate the distinctive socio-economic profile of LRT passengers who self-report having relocated to the new transit corridor due, at least in part, to the new transit service. Second, we document the spatial patterns (proximity to LRT stations) of residential relocations which were self-reported to have been influenced by the introduction of the LRT system, as a function of socio-economic characteristics. Both descriptive statistics and results from multivariate analyses are presented.

Although the characteristics of transit-oriented development and the impact of transit on property values are well documented (cf. [4]), few studies have explicitly investigated the impact of public transport investment on residential relocation behavior. Table 1 contains a summary listing of the key characteristics of relevant earlier studies. Specifically, the contribution of the present paper is to provide new results regarding two research questions that remain under-researched, while accounting for the possibility of endogeneity between them:

1. *What is the socio-economic profile of people most likely to relocate in response to the stimulus of public transport investment? (a new LRT service, in the case study reported in this paper)*
2. *Conditional on an LRT passenger having relocated in the prior 5 years, how do socio-economic characteristics relate with the proximity of their residential choice to the public transport service?*

The rest of this paper is structured as follows. Section 2 introduces the empirical data employed on this study, and Sect. 3 outlines the analytical framework. Section 4 then summarizes and concludes this paper.

2 Empirical data

The empirical data for this study were collected via an intercept-survey undertaken of passengers waiting on platforms at the seven HBLR stations between the Hoboken Terminal and Tonnelle Avenue stations (see Fig. 1) on 1 May, 2008. The questionnaire instrument contained 39 questions organized around three themes:

- travel, employment, and residential patterns and changes,
- customer satisfaction (not considered for the purposes of the present study), and
- socio-economic and demographic characteristics.

5,384 questionnaires were distributed; the overall response rate was 19 %. Following data processing, a total of 1,023 complete responses for which the respondent's residential address could be successfully geocoded were taken forward for statistical analysis in this study.

Survey respondents reported the length of time that they had lived at their current address, and those who indicated that they lived at their current address for five or fewer years were subsequently asked to indicate how important the availability of the HBLR service was to their residential relocation decision.[1]

[1] The question wording was: *How important was the availability of the Hudson–Bergen Light Rail service in your decision to move?*

Table 1 Summary of the previous literature

Reference	Empirical context	Dataset	Key findings
Lund [5]	Los Angeles, San Diego, and San Francisco, California	2003/2004 hand-delivered mail-back survey ($n = 605$)	Among respondents that had recently moved to a Transit-Oriented Development (TOD), access to transit was ranked fifth (out of nine factors) in importance as a factor (after housing quality/type, housing cost, neighborhood quality, and access to shops/services)
Ketraungroch [6]	Bangkok, Thailand	Household Socio-Economic Survey [SES] ($n = 1{,}445$ households in 1998; $n = 1{,}512$ households in 2004)	Higher income workers were found to exhibit sharper bid-rent curves than lower income workers, consistent with economic theory
Olaru et al. [7]	Perth, Western Australia (newly opened commuter rail service)	2007 personal-interview survey ($n = 1{,}034$)	Olaru and colleagues report an estimate that, *ceteris paribus*, each additional minute of distance from a rail station is associated with a decrease in property value of US $3,770
Cao and Schoner [8]	Light rail transit in Minneapolis	2011 postal survey ($n = 267$)	Among respondents that had recently moved to the LRT corridor, access to public transport was ranked fourth (of 30 candidate factors) in importance as a factor in residential location choice (after affordability, walkability, and proximity to parks/open space)
Dai and Weinzimmer [9]	San Francisco Bay Area (Silicon Valley)	2013 online survey ($n = 130$) of employer shuttle bus passengers	Among shuttle passengers, proximity to the service was ranked fifth (of 11 candidate factors) in importance as a factor in residential location choice (after Walkability, proximity to cultural amenities, proximity to traditional public transport, and urban-neighborhood amenities)
This study	Hudson–Bergen Light Rail System, New Jersey, US	2008 passenger platform survey ($n = 1{,}023$)	See Sect. 4

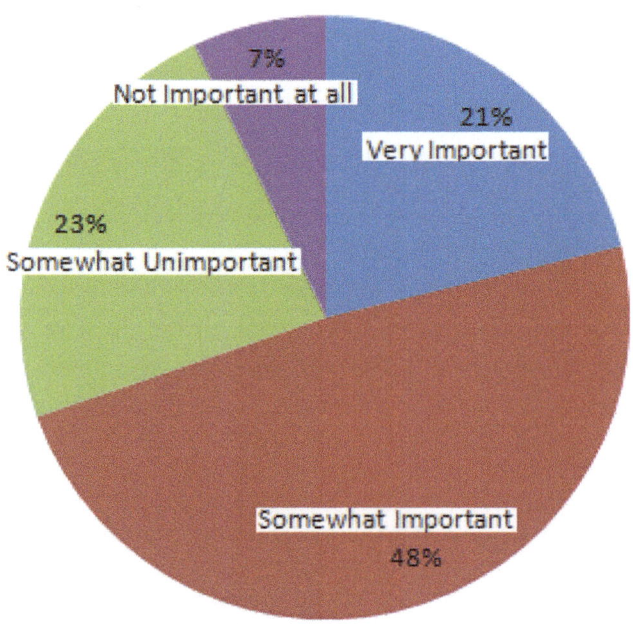

Fig. 2 Among LRT passengers that report having relocated within the past 5 years, the self-reported 'importance' of the LRT service in the relocation decision. Adapted from [10]

Two-fifths (42 %) of respondents reported having relocated within the past 2 years, with a further quarter (24 %) indicating that they relocated more than two but fewer than five years prior. Figure 2 shows the distribution of responses to the importance-of-LRT-in-relocation decision question; 21 % of movers indicated that the LRT service was very important to their relocation decision, and a further 48 % reported that it was somewhat important.

3 Analytical framework

A Type II Tobit discrete–continuous specification [11] was employed in the subsequent quantitative analysis, with the discrete dimension specified to be whether or not a person had relocated to the HBLR corridor within the prior 5 years and the continuous dimension (conditional on a respondent having relocated) the street-network distance of their new residence to the nearest LRT station.

Formally, we denote these dimensions as follows:

$$d_i^* = \alpha' z_i + e_i, \quad i = 1, \ldots, N, \tag{1}$$

$$d_i = 1, \quad \text{if } d_i^* > 0; \quad d_i = 0, \quad \text{if } d_i^* \leq 0, \tag{2}$$

$$y_i^* = \boldsymbol{\beta}'\boldsymbol{x}_i + \varepsilon_i, \quad i = 1, \ldots, n, \quad n < N, \tag{3}$$

$$y_i = y_i^*, \text{ if } d_i = 1; \quad y_i = 0, \text{ if } d_i = 0, \tag{4}$$

where d_i^* and y_i^* are the latent, continuous dependent variables for individual worker i; N, and n are the numbers of observations in the full dataset (including respondents who did not relocate) and the subset of respondents who did relocate, respectively. z_i and $\boldsymbol{x}_i$ are the vectors of observed variables that are treated as independent variables in the discrete and continuous models, respectively. Note that the variables within these two vectors are not mutually exclusive; the same variable can appear in both of the modeled dimensions. $\boldsymbol{\alpha}'$ and $\boldsymbol{\beta}'$ are corresponding vectors of parameters to be estimated, and e_i and ε_i are the disturbance terms which may be correlated through a correlation coefficient ρ that is independent of z_i and $\boldsymbol{x}_i$. e_i and ε_i are bivariate normally distributed, with zero mean and unknown covariance matrix, denoted in Eq. (5) as follows:

$$\begin{bmatrix} e_i \\ \varepsilon_i \end{bmatrix} \sim N\left(\begin{bmatrix} 0 \\ 0 \end{bmatrix}, \begin{bmatrix} 1 & \rho\sigma_\varepsilon \\ \rho\sigma_\varepsilon & \sigma_\varepsilon^2 \end{bmatrix} \right), \tag{5}$$

with σ_ε^2 normalized to one for purposes of model identification.

The standard Heckman two-step estimator is employed [11], which is based on the conditional mean expressions and the truncated bivariate normal distribution of the error terms. The expected value of the observed dependent variable y_i is

$$\begin{aligned} \boldsymbol{\beta}'\boldsymbol{x}_i + \rho\sigma_\varepsilon\lambda_i\left(\frac{\boldsymbol{\alpha}'z_i}{\sigma_e}\right) E(y_i|\boldsymbol{x}_i, z_i) &= E(y_i^*|\boldsymbol{x}_i, z_i, d_i = 1) \\ &= \boldsymbol{\beta}'\boldsymbol{x}_i + E(\varepsilon_i|e_i \succ -\boldsymbol{\alpha}'z_i), \end{aligned} \tag{6}$$

which simplifies to

$$\boldsymbol{\beta}'\boldsymbol{x}_i + \rho\sigma_\varepsilon \frac{\phi(\boldsymbol{\alpha}'z_i)}{\Phi(\boldsymbol{\alpha}'z_i)}, \tag{7}$$

where σ_ε is fixed at one; $\lambda_i(\boldsymbol{\alpha}'z_i) = \phi(\boldsymbol{\alpha}'z_i)/\Phi(\boldsymbol{\alpha}'z_i)$ is the inverse Mills ratio; and ϕ and Φ are standard normal probability distribution function and cumulative distribution function, respectively. Equation (7) implies that ignoring the term $\rho\sigma_\varepsilon\lambda_i$ would in effect omit a variable from Eq. (3) under censoring. Thus, the ordinary least squares (OLS) regression of Eq. (2) will yield unbiased estimates of β only if $\rho = 0$ or if correlation between λ_i and $\boldsymbol{x}_i$ is zero. Equation (7) also demonstrates that we could estimate $\boldsymbol{\beta}$ consistently using the "relocators-only" subset of the survey data by an OLS regression of y_i on $\boldsymbol{x}_i$ and λ_i ($\boldsymbol{\alpha}'z_i$, if $\boldsymbol{\alpha}$ were known). Based on this observation, Heckman's two-step estimator is calculated by the following procedure:

1. First, the full sample is employed to estimate a binary probit model using standard maximum-likelihood techniques, to obtain estimates of $\boldsymbol{\alpha}$, i.e., $Pr(d_i = 1) = \Phi(\boldsymbol{\alpha}'z_i)$ and $Pr(d_i = 0) = 1 - \Phi(\boldsymbol{\alpha}'z_i)$.
2. Next, λ_i ($\boldsymbol{\alpha}'z_i$) is estimated for each survey respondent that reported having relocated i.

Finally, the sub-sample of only respondents that relocated is used to estimate $\boldsymbol{\beta}$ and $\beta_\lambda = \rho\sigma_\varepsilon$, by OLS of regressing y_i on $\boldsymbol{x}_i$ and the estimated λ_i.[2]

The OLS standard error estimates calculated by this estimation procedure require correction, as the error term in Eq. (7) may be heteroskedastic and we use fitted rather than actual values of λ_i. Furthermore, the resulting estimates are consistent, but not asymptotically efficient (i.e., not minimum variance) under a standard assumption of normality. More efficient estimates can be obtained using the full information maximum-likelihood (FIML) approach, which can be expressed as follows:

$$\ln L = \sum_{d_i=0} \ln \Phi(-\boldsymbol{\alpha}'z_i) + \sum_{d_i=1} \left[-\ln\sigma_\varepsilon + \ln\phi\left(\frac{y_i - \boldsymbol{\beta}'\boldsymbol{x}_i}{\sigma_\varepsilon}\right) + \ln\Phi\left(\frac{\boldsymbol{\alpha}'z_i + \rho\sigma_\varepsilon^{-1}(y_i - \boldsymbol{\beta}'\boldsymbol{x}_i)}{\sqrt{1-\rho^2}}\right)\right].$$

Maximizing this likelihood function produces simultaneous estimates of the parameters of both the discrete and continuous dimensions ($\boldsymbol{\alpha}$, $\boldsymbol{\beta}$, ρ, and σ_ε). If $\rho = 0$, the log likelihood function reduces to the sum of a probit and a standard OLS regression, which can each be estimated separately. In comparison to the two-step Heckman procedure described above, the FIML estimator is computationally intensive to numerically identify optimal values. Reasonable starting values for FIML that are close to the true parameter values are therefore required. In this study, the final values of the Heckman two-step estimation procedure were used as the starting values for the FIML procedure.

4 Estimation results

Prior to undertaking estimation of the statistical model, the degree of correlation between each pair of the candidate independent variables was calculated. All such correlation coefficients were found to be smaller than 0.40; it was therefore determined that explicit correction for multicollinearity was not necessary.

A structured specification search was then undertaken, using the standard Akaike Information Criterion (AIC) to select between alternative candidate specifications. AIC is

[2] Note that the estimator of σ_ε can be computed using the procedure described in Step 1, which yields an estimate of the correlation coefficient is obtained: $\hat{\rho} = \hat{\beta}_\lambda / \hat{\sigma}_\varepsilon$.

Table 2 Estimation results

	Move/do not move dimension		Distance between LRT station and new residence	
	Coefficient	t statistic	Coefficient	t statistic
Constant	−1.44	8.72	−1.45	7.81
Household income				
Up to \$35 K	Reference class	N/A	Reference class	N/A
\$35–100 K	−0.185	2.25	−0.391	7.15
\$100 K+	−0.458	4.01	−1.231	5.12
Age				
Up to age 34	Reference class	N/A	Reference class	N/A
35–55	−0.135	1.48	0.213	1.64
55+	−0.506	3.15	0.439	7.32
Household Size				
One or two members	Reference class	N/A	Reference class	N/A
Three or more members	−0.185	1.67	0.432	6.40
Mode used to access LRT				
Drive	Reference class	N/A	Reference class	N/A
Transit	0.496	3.94	1.231	4.35
Walk	0.427	4.03	−0.644	5.33
Number of cars owned by household	−0.103	2.33	1.219	3.12
Number of children (under age 18) in household	−0.268	2.31	2.321	2.43
Length of time LRT has been used (months)	0.556	1.51	0.149	2.41
		σ_ε	0.81	2.34
		ρ	0.92	2.12
			Wald Chi square	56.73 ($p < 0.001$)

a metric of global goodness-of-fit which penalizes added parameters, and is widely used as to determine objectively whether the improved goodness-of-fit due to adding additional free parameters is warranted on the grounds of information theory [12].

The preferred model specification and resulting parameter estimates are presented in Table 2.[3] Positive coefficients in the *move/do not move* dimension indicate that the relevant variable has a positive effect, *ceteris paribus*, on the likelihood of a survey respondent having relocated to the HBLR corridor within the prior 5 years and likewise positive coefficients in the *distance between LRT station and new residence* dimension are interpreted to mean that this distance increases, *ceteris paribus*, with the value of that variable.

As shown in Table 2, the majority of parameter estimates are statistically significant at the $p = 0.05$ level of confidence. The signs of parameter estimates are consistent with a priori expectations, as discussed in the remainder of this section. Further, the structural parameters and are both estimated to be statistically different than 1.0 ($t = 2.12$, 2.34, respectively), indicating the presence of statistically significant correlation between the error terms of the discrete and continuous models.

Household income was found to relate negatively, net of confounding effects, with the likelihood of having relocated in the prior 5 years; in other words having a lower household income was associated with a lower level of residential stability. Likewise, having a higher household income was associated with shorter distances between the new residences of residential-movers and the nearest LRT station. The fact that higher income groups chose to live closer to the stations of the LRT network is consistent with access to the LRT being a normal good, though this analysis cannot determine this definitively. Likewise, this result is also consistent with bid-rent theory, in which locations with higher accessibility are allocated to the land use that is willing/able to pay the most to occupy them [13].

Being under age 35 was associated with the highest propensity to have relocated within the prior 5 years, and also with the greatest propensity to relocate near to the access points to the LRT system. These findings are consistent with a more general pattern of younger adults

[3] Variables that were tested and excluded from the preferred model specification, on the basis of AIC values, were as follows: dwelling type, race, employment status, occupation type, destination-type (of the surveyed journey), and whether or not the surveyed passenger receives a fare subsidy through their employer.

having greater residential mobility (cf. [14]), and also being more sensitive to the accessibility afforded by the LRT system. Similarly intuitive effects were found with respect to household size and the presence of children.

Automobile ownership was, by contrast, negatively associated with propensity to have relocated and, conditional on having moved, the proximity to the nearest LRT station. Having driven to access the LRT system for the surveyed journey was also negatively linked with having relocated (relative to having walked or taken another form of public transit). However, conditional on having relocated the largest distance between new home and nearest LRT station was associated with having used transit to access the LRT system—longer distance than both driving and walking.

Finally, Table 2 shows that the length of time that one has been using the LRT system is not significantly associated with the propensity to have relocated within the prior 5 years ($t = 1.51$), though is positively associated with the distance between [new] home and LRT station. Put another way, this suggests that people who started using the LRT service more recently are likely to live closer to the LRT system's stations than people who have been using it for a longer period of time (all else equal).

5 Conclusions

This paper contributes to the body of literature regarding the effects of public transport investments on residential property markets. Specifically, we investigated the effects of a new LRT system (the HBLR system in New Jersey, US) on: (1) the likelihood of LRT passengers having relocated in the prior 5 years and (2) the proximity of the 'new' residential location to the LRT system's stations. Give our a priori expectation of correlation in unobserved effects across these two dimensions of analysis, we employed a simultaneous discrete–continuous specification (a Type II Tobit). The empirical results suggest that the residual error terms were indeed correlated across these two dimensions; failing to take this correlation into account would have yielded biased and inefficient parameter estimates (i.e., effects).

The substantive results indicate that, among the sample of surveyed LRT passengers ($n = 1{,}023$), small household size, low income, being a young adult (under age 35) and low household car ownership are each independently associated with heightened propensity to have residentially relocated during the prior 5 years. Subsequently, conditional on having relocated, these same characteristics (with the exception of low income) are associated with relocating in close proximity to the LRT network's access points (stations).

Further research will be required to establish whether the empirical findings reported here are indicative of generally applicable relationships, or are idiosyncratic to the HBLR system and/or the dynamics of the local real estate market in which it is located. For instance, the HBLR serves in part as a connecting service to heavy rail and ferry services to Manhattan's labor market and cultural destinations. It would be worthwhile, for instance, to establish whether or not similar functional relationships hold for LRT systems that comprise the core of their urban area's transit network (e.g., Portland Oregon's MAX system or Calgary's C-Train network), and for urban areas with characteristics different than the New York/New Jersey metropolitan region.

References

1. Cervero R, Landis J (1997) Twenty years of the bay area rapid transit system: land use and development impacts. Transp Rev A 31(4):309–333
2. Guiliano G (1989) New directions for understand transportation and land use. Environ Plan A 21:145–159
3. Robins ME, Wells JS (2008) Land development at selected Hudson–Bergen Light Rail Stations. Alan M. Voorhees Transportation Center (VTC) at Rutgers. April, 2008
4. Cervero R, Ferrell C (2002) Murphy S Transit-oriented development and joint development in the United States: a literature review. TCRP Report #52. Transportation Reserch Board, October 2002
5. Lund H (2006) Reasons for living in a transit-oriented development, and associated transit use. J Am Plan Assoc 72:357–366
6. Ketraungroch P (2008) The effect of mass rapid rail transit on residential location pattern in Bangkok. Master Thesis, Thammasat University, Bangkok, Thailand
7. Olaru D, Smith B, Taplin JHE (2011) Residential location and transit-oriented development in a new rail corridor. Transp Res Part A Policy Pract 45(3):219–237
8. Cao J, Schoner J (2013) Transportation impact of transitways: a case study of Hiawatha light rail transit in Minneapolis. CTS Report 13-13, Report #7 in the series Transitway Impacts Research Program
9. Dai D, Weinzimmer D (2014) Riding first class: impacts of silicon valley shuttles on commute and residential location choice. Working Paper, UCB-ITS-WP-2014-01, February 2014
10. Liu R (2009) Hudson Bergen Light Rail customer impact and retention study. Final Report
11. Heckman JJ (1979) Sample selection bias as a specification error. Econometrica 47:153–161
12. Greene WH (2012) Econometric analysis, 7th edn. Prentice Hall, New York
13. Alonso W (1964) Location and land value. Harvard University Press, Cambridge
14. Schachter J (2001) Geographical mobility population characteristics March 1999–March 2000. US Census Bureau

11

A data mining approach to characterize road accident locations

Sachin Kumar[1] · Durga Toshniwal[2]

Abstract Data mining has been proven as a reliable technique to analyze road accidents and provide productive results. Most of the road accident data analysis use data mining techniques, focusing on identifying factors that affect the severity of an accident. However, any damage resulting from road accidents is always unacceptable in terms of health, property damage and other economic factors. Sometimes, it is found that road accident occurrences are more frequent at certain specific locations. The analysis of these locations can help in identifying certain road accident features that make a road accident to occur frequently in these locations. Association rule mining is one of the popular data mining techniques that identify the correlation in various attributes of road accident. In this paper, we first applied *k*-means algorithm to group the accident locations into three categories, high-frequency, moderate-frequency and low-frequency accident locations. *k*-means algorithm takes accident frequency count as a parameter to cluster the locations. Then we used association rule mining to characterize these locations. The rules revealed different factors associated with road accidents at different locations with varying accident frequencies. The association rules for high-frequency accident location disclosed that intersections on highways are more dangerous for every type of accidents. High-frequency accident locations mostly involved two-wheeler accidents at hilly regions. In moderate-frequency accident locations, colonies near local roads and intersection on highway roads are found dangerous for pedestrian hit accidents. Low-frequency accident locations are scattered throughout the district and the most of the accidents at these locations were not critical. Although the data set was limited to some selected attributes, our approach extracted some useful hidden information from the data which can be utilized to take some preventive efforts in these locations.

Keywords Road accidents · Accident analysis · Data mining · *k*-Means · Association rule mining

✉ Sachin Kumar
sachinagnihotri16@gmail.com

Durga Toshniwal
durgatoshniwal@gmail.com

1 Centre for Transportation Systems (CTRANS), Indian Institute of Technology Roorkee, Roorkee, Uttarakhand 247667, India

2 Computer Science & Engineering Department, Indian Institute of Technology Roorkee, Roorkee, Uttarakhand 247667, India

1 Introduction

Road and traffic accidents (RTA) are one of the important problems in India. MORTH [1] mentioned in its report that every year there are 0.4 million accidents reported in India, which makes India a country with large accident rate. This report shows that there is a negative trend of accidents from 2012 to 2013; however, as accidents are unpredictable and can occur in any type of situation, there is no guarantee that this trend will sustain in future also. Therefore, the identification of different geographical locations where most of the accidents have occurred and determining the various characteristics related to road accidents at these locations will help to understand the different circumstances of accident occurrence. Kannov and Janson [2] stated that systematic relationship between accident frequency and other variables such as geometry of road, road side features, traffic information and vehicle information can help to develop effective accident prevention measures.

Lee et al. [3] indicated that statistical models were a good choice in the past to analyze road accidents to identify the correlation between accident and other traffic and geometric factors. However, Chen and Jovanis [4] determined that analyzing large dimensional datasets using traditional statistical techniques may result in certain problems such as sparse data in large contingency tables. Also, statistical models have their own model specific assumptions and violation of these can lead to some erroneous results. Due to these limitations of statistical techniques, data mining techniques are being used to analyze road accidents. Data mining is a set of techniques to extract novel, implicit and hidden information from large data. Barai [5] discussed that there are various applications of data mining in transportation engineering such as road roughness analysis, pavement analysis and road accident analysis. Various data mining techniques [6] such as association rule mining, classification and clustering are widely used for the analysis of road accidents.

Accident cases in India are usually recorded by police officer of the region in which the accident has occurred. Also, the area covered by a police station is limited and they keep record of accidents that have occurred in their regions only. Ponnaluri [7] discussed that the report prepared by police only contains the basic information that are not much useful for the research purpose. He suggests that data collection method used by police needs a lot of improvement. However, Indian researchers used these data and analyzed it for some highway portions using statistical methods [8, 9]. Data mining can be described as a novel technique to extract hidden and previously unknown information from the large amount of data. Several data mining techniques such as clustering, classification and association rule mining are widely used in the road accident analysis by researchers of other countries. Geurts et al. [10] used association rule mining technique to understand the various circumstances that occur at high-frequency accident locations on Belgium road networks. Tesema et al. [11] used adaptive regression tree model to build a decision support system for the road accidents in Ethiopia. Abellan et al. [12] developed various decision trees to extract different decision rules for different trees to analyze two-lane rural highway data of Spain. They found that bad light conditions and safety barriers badly affect the crash severity. Depaire et al. [13] used clustering technique to analyze road accident data of Belgium and suggest that cluster-based analysis of road accident data can extract better information rather analyzing data without clustering. Kashani et al. [14] used classification and regression tree (CART) to analyze road accidents data of Iran and found that not using seat belt, improper overtaking and speeding badly affect the severity of accidents. Kwon et al. [15] used Naïve Bayes and decision tree classification algorithm to analyze factor dependencies related to road safety. Severity of accident is directly concerned with the victim involved in accidents, and its analysis only targets the type of severity and shows the circumstances that affects the injury severity of accidents. Sometime accidents are also concerned with certain locations characteristics, which makes them to occur frequently at these locations. Hence identification of these locations where accident frequencies are high and further analyzing them is very much beneficial to identify the factors that affect the accident frequency at these locations.

In this paper, we are making use of data mining techniques to identify high-frequency accident locations and further analyzing them to identify various factors that affect road accidents at those locations. We first divide the accident locations into k groups based on their accident frequency counts using k-means clustering algorithm. Then association rule mining algorithm is applied on these to reveal the correlation between different attributes in the accident data and understand the characteristics of these locations. Hence, our main emphasis will be the interpretation of the outcomes.

2 Methodology

2.1 k-Means clustering

Clustering is an unsupervised data mining technique whose main task is to group the data objects into different clusters such that objects within a group are more similar than the objects in other clusters. k-means algorithm [16] is very popular clustering technique for numerical data. It groups the data objects into k clusters. There are various clustering algorithms existing but selection of suitable clustering algorithm depends on type and nature of data. Our prime motive of this paper is to discriminate the accident location based on their frequency count. We have two choices to do this: First, we can decide a threshold level for each of the category of accident locations and group them in some categories. The problem with this approach is that it is very difficult to identify the number of categories of accident locations and decide a threshold level for each category. The other way is to use k-means algorithm which can divide the accident locations into different groups. The number of groups can be identified using some cluster selection criteria such as gap statistic.

Initially, we have frequency counts of 87 locations with 7,027 road accidents. In order to divide the location into different groups we used k-means algorithm. A brief formal description of k-means algorithm is given below [18]:

Input: $\{D, k\}$ // $D\rightarrow$ Data set consists of n data objects, $k\rightarrow$ Number of clusters
Output: k clusters
Method:

1. Choose the k objects at random from D, as initial cluster centers.
2. **repeat**
3. Assign each data object to the cluster to which its distance is closest
4. Update the cluster means, i.e., calculate the mean value of the objects for each cluster.
5. **Until** no data object changes its cluster membership or any other convergence criteria is met.

2.2 Number of cluster selection

The major problem with a clustering algorithm is to identify the number of clusters to be made. The weakness of k-means clustering is that the user has to provide the value for k. An inappropriate value for k may lead to wrong clustering results. In this paper, we have used the gap statistics [17] in order to find the value of k that can be supplied to divide the accident locations into different groups based on their frequency counts. Gap statistics can be used with any type of clustering technique, but they have been scarcely used to determine the number of clusters in road accident analysis.

Consider a data set D_{ij}, $i = 1,2,\ldots,m$, $j = 1,2,\ldots,n$, consisting of m data objects with values of n attributes. Assuming d_{xy} is the squared Euclidean distance between objects X and Y given by $d_{xy} = \sum (X_j - Y_j)^2$. If the data set has been clustered into k clusters, $c_1, c_2,\ldots,c_k$, where c_i indicates the ith cluster, then $n_i = |c_i|$.

Let $D_i = \sum d_{xy}$, (where $x, y \in c_i$) is the sum of pair-wise distances for all points in cluster i and W_k is the collective within cluster sum of squares around the cluster means and is given by Eq. (1). Gap$_n(k)$ can be defined as the difference between expected and observed values of $\log(W_k)$ and given in Eq. (2). K can be taken for the value maximizing Gap$_n(k)$.

$$W_k = \sum_{i=1}^{k}\left(\frac{1}{2n_i}\right)D_i, \qquad (1)$$

$$\text{Gap}_n(K) = E_n^*\{\log(W_k)\} - \log(W_k), \qquad (2)$$

where E_n^* denotes the expectation under a sample size n from the reference distribution.

2.3 Association rule mining

Association rule mining is a very popular data mining technique based on market basket analysis that extracts interesting rules between various attributes in a large data set [18]. Association rule mining produces a set of rules that define the underlying patterns in the data set. Given a data set D of n transactions where each transaction $T \in D$. Let $I = \{I_1, I_2,\ldots, I_n\}$ be a set of items. An item set A will occur in T if and only if $A \subseteq T$. $A \rightarrow B$ is an association rule, provided that $A \subset I$, $B \subset I$ and $A \cap B = \emptyset$. In case of road accident data, an association rule can identify the various attribute values which are responsible for an accident occurrence.

In association rule mining, various interesting measures are there to assess the quality of a rule. These interesting measures for a rule $A \rightarrow B$ are discussed as follows:

2.3.1 Support (S_p)

The support of a rule $A \rightarrow B$ defines the percentage how often A and B occur together in a data set and can be calculated using Eq. (3). Support is also known as frequency constraint. A set of items satisfying certain support threshold is known as frequent item set. These frequent item sets are further used to generate association rules based on other measures.

2.3.2 Confidence (C_f)

Confidence of a rule $A \rightarrow B$ defines the ratio of the occurrence of A and B together to the occurrence of A only and can be calculated using Eq. (4). Higher the confidence values of a rule $A \rightarrow B$, higher the chances of occurrence of B with the occurrence of A. Sometimes, only confidence values are not sufficient enough to evaluate the descriptive interest of a rule.

2.3.3 Lift (L_t)

Lift for a rule $A \rightarrow B$ measures the occurrence of A and B together than expected. In other words, lift is the ratio of the confidence and the expected confidence of a rule. Expected confidence can be defined as the occurrence of A and B together with the occurrence of B. A lift value ranges from 0 to ∞. Lift values greater than 1 make a rule potentially useful for predicting the consequent in future data sets. Lift determines how far from independence are A and B. Lift measures co-occurrence only and is also symmetric with respect to A and B. Lift can be calculated using Eq. (5).

2.3.4 Leverage (L_v)

Leverage for a rule $A \rightarrow B$ measures the difference of A and B appearing together in the data set and the expectation if A and B are statistically dependent [19]. It can be calculated using Eq. (6). The values for leverage range from [−0.25 to +0.25]. A leverage value 0 indicates that the variables are statistically independent. It will increase towards +1 if the variables occur more often together and will decrease towards −1 if one of the variables alone occurs more often.

2.3.5 Conviction (C_v)

Conviction is another measure that undertakes some of the weaknesses of confidence and lift [20]. Conviction of a rule $A \rightarrow B$ compares the probability that A occurs without B if they are dependent with the actual frequency of the appearance of A without B. Conviction is not symmetric i.e. conviction $(A \rightarrow B) \neq$ conviction $(B \rightarrow A)$. Conviction is rather inspired in the logical definition of implication and attempts to calculate the degree of implication of any rule. The value for conviction ranges within $[0.5, \infty]$. The values which are distant from 1 indicate interesting rules. In conviction, the supports of both antecedent and consequent are taken into account. It can be calculated using Eq. (7):

$$S_p = \frac{P(A \cap B)}{N}, \quad (3)$$

where N is the total number of accident records.

$$C_f = \frac{P(A \cap B)}{P(A)}, \quad (4)$$

$$L_t = \frac{P(A \cap B)}{P(A) \times P(B)}, \quad (5)$$

$$L_v = P(A \cap B) - P(A) \times P(B), \quad (6)$$

$$C_v = \frac{P(A) \times P(B)}{P(A \cap \bar{B})}, \quad (7)$$

For a better understanding of the above concept, consider the following short example in Table 1, from road accident domain in which the set of items is $I = \{$Fog, High Traffic, Speed > 100, Low Traffic, Fatal Accident$\}$. In Table 1, 1 shows the presence of the item and 0 indicates the absence of the item.

An example rule {Fog, Speed > 100} → {Fatal Accident} specifies that if Fog and Speed > 100 occur together, Fatal Accident will also take place. To select the interesting rules, minimum threshold value of support is provided.

In the above example, the support and confidence value for the item set {Fog, Speed > 100, Fatal Accident} can be computed using Eqs. (3) and (4) as follows:

Table 1 Example data set

Accident id	Fog	Speed > 100	High traffic	Low traffic	Fatal accident
1	1	1	0	1	1
2	1	1	1	0	0
3	0	0	1	0	0
4	0	0	1	0	0
5	1	1	1	0	1

$$S_p = P(\text{Fog} \cap \text{Speed} > 100 \cap \text{Fatal Accident}) / \text{Total Number of Accidents} = 2/5 = 0.4.$$

A support value of 0.4 indicates that in 40 % of accident records, Fog, Speed > 80 and Fatal Accident take place together.

$$C_f = \frac{P(\text{Fog} \cap \text{Speed} > 100 \cap \text{Fatal Accident})}{P(\text{Fog} \cap \; > 100)} = \frac{(2/5)}{(3/5)} = 0.66.$$

The confidence value of 0.66 means that in 66 % of accident cases when Fog and Speed > 100 occur together, then Fatal Accident also occurs. This value indicates that there are 66 % chances of fatal accident if the weather is foggy and vehicle speed is greater than 100.

The lift of the above rule can be calculated using Eq. 5 as follows:

$$L_f(\{\text{Fog, Speed} > 100\} \rightarrow \{\text{Fatal Accident}\}) = \frac{P(\text{Fog} \cap \text{Speed} > 100 \cap \text{Fatal Accident})}{P(\text{Fog}) \times P(\text{Speed} > 100) \times P(\text{Fatal Accident})} = \frac{0.4}{0.144} = 2.77,$$

The above lift value indicates that fog, speed more than 100 and fatal accidents are strongly correlated with each other and this rule can be used to predict future accident with good accuracy.

The leverage for the above rule can be calculated using Eq. (6) as follows:

$$\text{Leverage}(\{\text{Fog, Speed} > 100\} \rightarrow \{\text{Fatal Accident}\}) = P(\text{Fog} \cap \text{Speed} > 100 \cap \text{Fatal Accident}) - P(\text{Fog}) \times P(\text{Speed} > 100) \times P(\text{Fatal Accident}) = 0.4 - 0.144 = 0.256.$$

The conviction for the above rule can be calculated using Eq. (7) as follows:

$$C_v(\{\text{Fog, Speed} > 100\} \rightarrow \{\text{Fatal Accident}\}) = \frac{P(\text{Fog} \cap \text{Speed} > 100) \times P(\text{Fatal Accident})}{P(\text{Fog} \cap \text{Speed} > 100 \cap \text{Fatal Accident})} = \frac{4}{1} = 4.0.$$

2.4 Data set

In India, 108 is an emergency ambulance service which provides help to the accident victims. This emergency service is being operated in several states of India. Uttarakhand is an Indian state where this service is running. This service is operated and handled by Emergency Management Research Institute (GVK-EMRI) [21], which provides emergency service to the accident victims and also keeps track of record of every accident served. Also, this information is stored at their central server located at one particular place in the state. Hence, these data provide information about accidents that have happened in the road network of entire city, district and state. The data for this study have been obtained from GVK-EMRI, Dehradun. The data set consists of 15,574 road accidents for 6 years period from 2009 to 2014, in Dehradun District of Uttarakhand State. After preprocessing, 9,640 accident records have been considered for this research.

The attributes of the data are mostly categorical in nature. A brief description of the data is given in Table 2 as follows:

3 Results and discussion

3.1 Categorization of accident location

In order to categorize accident locations on the basis of their accident frequency count. We extracted the frequency count of accident locations from the data. In total there were around 158 locations found with 9,640 road accidents. In order to provide better analysis, we excluded those locations which have frequency count of less than 20 over 6 years of duration, which results in 87 locations with 7,327 road accidents. These locations are plotted with location id in the increasing number of frequency, as shown in Fig. 1.

Number of clusters to be made is identified using gap statistic as discussed in Sect. 2.3. The values obtained for W_k, log (W_k), E_n^* (log (W_k)) and gap statistic are plotted against the number of clusters in Fig. 2a, b, c respectively. The graphs clearly indicate that there are three clusters for the accident locations. Figure 2a shows a knee curve at cluster 3, Fig. 2b plots the observed and expected values for log(W_k) and Fig. 2c shows that gap value at cluster 3 that maximizes the Gap$_n(k)$.

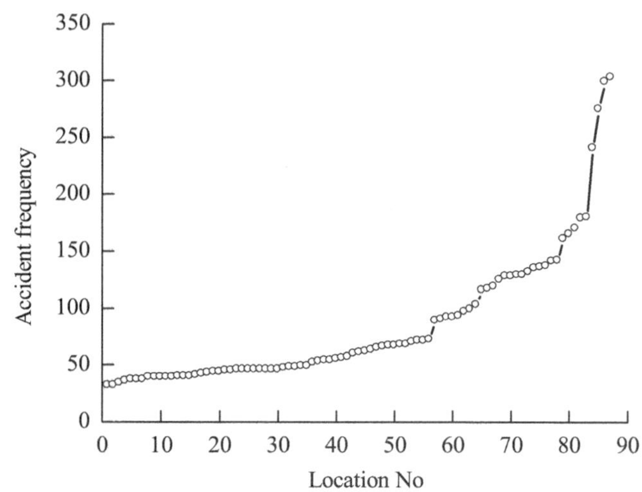

Fig. 1 Accident frequency count of locations in ascending order

Table 2 Road accident attributes used for analysis

Attribute name	Type	Values
Number of victims	Nominal	1, 2, >2
Victim age group	Nominal	Children, Young, Adult, Senior
Victim gender	Binary	Male/Female
Accident category	Nominal	2 wheeler, 3 wheeler, Vehicle_Fall_height, Pedestrian_hit, Multi_Vehicular_Incident (MVI), Vehicle_rollover/skid
Time of accident	Nominal	1,…,12
Day	Nominal	Monday, Tuesday, Wednesday, Thursday, Friday, Saturday, Sunday
Month	Nominal	1, 2….,12
Location_No	Ratio	1,……,158
Lighting on road	Binary	Daylight, street light, no light
Roadway feature	Nominal	intersection, curve, slope, other
Accident severity	Nominal	Critical/non-critical
Area around	Nominal	Hospital, colony, market, forest, hills, agriculture land
Road type	Nominal	Highway/non-highway

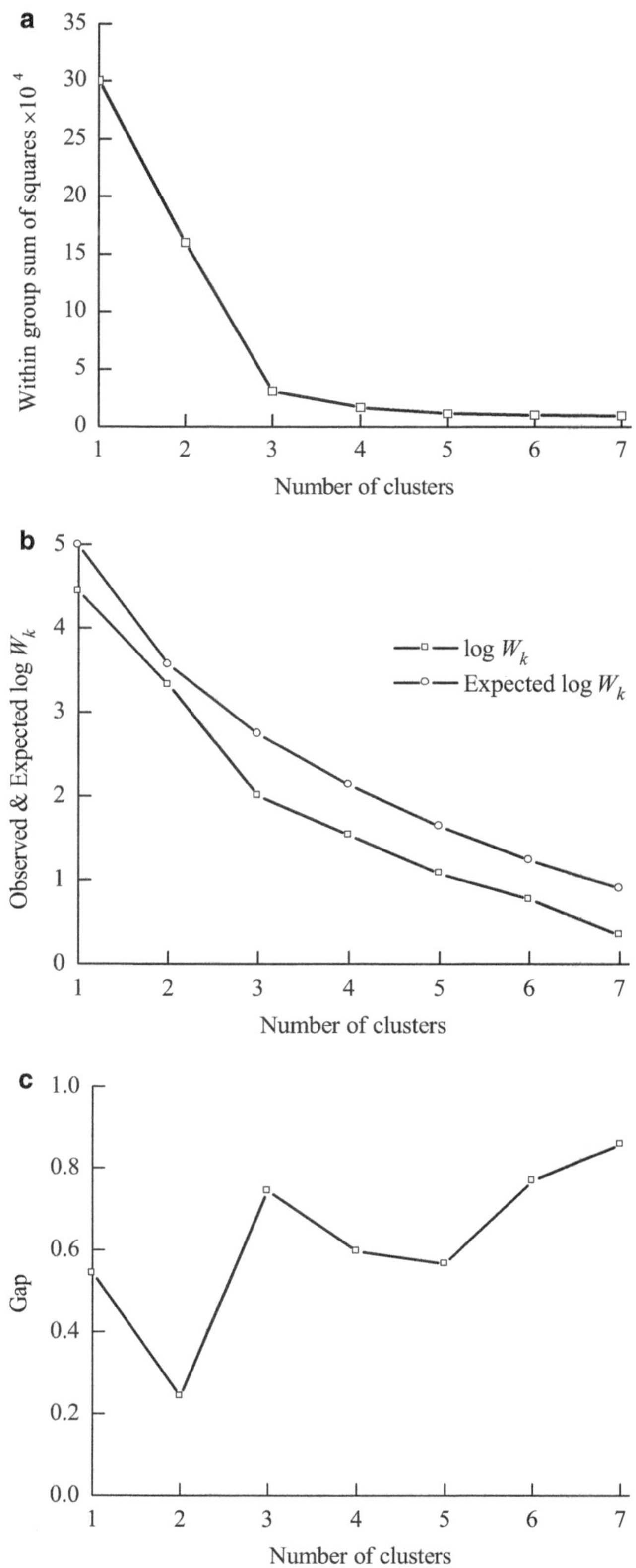

Fig. 2 **a** Within group sum of squares function W_k against the number of clusters. **b** Observed log (W_k) and Expected $E_n^*\{\log(W_k)\}$. **c** Gap statistic curve

Further, *k*-means clustering technique was applied on the accident frequency counts of accident locations to get three clusters of 87 locations as given in Table 3. We renamed three clusters as high-frequency accident location (HFAL), moderate-frequency accident locations (MFAL) and low-frequency accident locations (LFAL) based on their frequency count.

In LFAL cluster, there are 56 locations with a frequency count of less than 75, where a total of 2,646 (36.11 %) accidents occurred. Similarly, 22 locations with accident count in the range of 75–150 are in MFAL cluster which consists of 35.06 % of accidents. There are eight locations in HFAL cluster where around 28.82 % of the total accidents have occurred.

3.2 Mining association rules

Initially, we divided the locations into three categories namely HFAL, MFAL and LFAL. We used Apriori [22] algorithm in Weka 3.6.12 in order to generate rules. In order to identify strong rules, a minimum support of 5 % has been taken. The following tables show the association rules generated for each category.

Tables 4, 5 and 6 show the association rule generated for HFAL, MFAL and LFAL, respectively. Association rules provide an understanding of correlation between different attribute values that occur together when an accident occurs. Several rules are generated for each category but only some interesting rules (based on lift value) have been chosen to show in this paper. The rules for HFAL, MFAL and LFAL are discussed as follows:

3.3 Association rules for HFAL

The association rules reveal that HFAL locations are primarily hill roads towards hill stations with continuous traffic flow. Most of the accidents at these locations are severe accidents. These locations are highly sensitive for two-wheeler accidents. Other type of accidents such as pedestrian hit, multi-vehicular accidents are very less as compared to two-wheeler accidents. The major accidents have happened at curve and slope on hilly roads. The other locations with high-frequency accidents are markets with high traffic volume. Strong rules with high lift value disclose that intersections at market on highway roads are the main locations where accidents have occurred. The rules suggest that highways are the high-frequency accident locations, and intersections on highways which come across market locations are more prone to severe accidents.

Table 3 Accident frequency distribution and categorization

Cluster id	Number of locations	Accident frequency	Total accidents	% Accidents	Category name
1	56	30–75	2,646	36.11	LFAL
2	22	75–150	2,569	35.06	MFAL
3	8	150–303	2,112	28.82	HFAL

Table 4 Association rules for HFAL

Rule no.	Rule body	Conf.	Lift	Lev.	Conv.
1.	{Hills} → {Curve, Highway}	0.56	5.33	0.05	2.03
2.	{Curve, Highway} → {Hills}	0.6	5.33	0.05	2.03
3.	{Time = 12, Agriculture Land} → {Others, No light}	0.9	4.27	0.05	7.34
4.	{Intersection, Highway} → {Daylight, Market}	0.48	3.07	0.05	1.6
5.	{Market, 2 wheeler} → {Intersection, Highway}	0.42	2.71	0.04	1.45
6.	{Intersection, Highway} → {Market}	0.72	2.61	0.07	2.53
7.	{Others, Agriculture Land} → {No light, 2 wheeler}	0.39	2.07	0.06	1.32
8.	{No light, Non-highway} → {Agriculture Land}	0.97	1.98	0.06	13.35
9.	{Pedestrian, Market} → {Intersection}	0.56	1.81	0.02	1.54
10.	{Intersection} → {Non-Highway, 2 wheeler}	0.39	1.46	0.04	1.2
11.	{Others, Vechicle_Rollover} → {Agriculture Land}	0.71	1.45	0.02	1.7
12.	{Intersection, Market, 2 wheeler} → {Highway}	0.81	1.44	0.02	2.19
13.	{Intersection, Market} → {Highway}	0.79	1.42	0.03	2.1
14.	{Daylight, Highway} → {Multi_Vehicular_Incident}	0.56	1.18	0.01	1.03
15.	{Highway, Agriculture Land → 2 wheeler}	0.45	1.07	0.01	1.02

Table 5 Association rules for MFAL

Rule no.	Rule body	Conf.	Lift	Lev.	Conv.
16.	{Colony} → {Non-highway, Pedestrian}	0.48	3.22	0.01	1.14
17.	{Intersection, Market} → {Highway, Pedestrian}	0.44	3.11	0.01	1.11
18.	{Curve, Agriculture Land} → {Multi_Vehicular_Incident}	0.36	2.31	0.01	1.29
19.	{Nolight, Non-highway → {Agriculture Land, Multi_Vehicular_Incident}	0.71	2.11	0.01	1.1
20.	{Role_over} → {Other, Agriculture Land}	0.64	2.08	0.03	1.89
21.	{Daylight, Highway} → {Curve, Fall_height}	0.47	1.93	0.01	1.03
22.	{Daylight, 3 wheeler} → {Other}	0.97	1.87	0.01	9.09
23.	{Street Light, 2 wheeler} → {Intersection}	0.55	1.8	0.02	1.52
24.	{Agriculture Land, 2 wheeler} → {Other, No light, Non-highway}	0.68	1.78	0.02	1.52
25.	{Intersection, Non-highway} → {Daylight, Agriculture Land}	0.42	1.77	0.03	1.3
26.	{Intersection, Daylight, 2 wheeler} → {Agriculture Land, Non-highway}	0.46	1.7	0.02	1.34
27.	{Daylight, 2 wheeler} → {Time = 9}	0.41	1.6	0.02	1.1
28.	{2 wheeler} → {Intersection, Agriculture Land}	0.49	1.55	0.04	0.49
29.	{Market} → {Daylight, Highway}	0.45	1.46	0.04	1.25
30.	{Pedestrian, Colony} → {Intersection}	0.58	1.43	0.02	1.54

3.4 Association rules for MFAL

In MFAL locations, colonies on non-highway roads and intersections in market on highway roads are dangerous for pedestrians. Curve or bends on roads surrounded with agriculture land on non-highway roads are dangerous for multi-vehicle accidents. Roads that pass through agriculture land are also found dangerous to vehicle role over

Table 6 Association rules for LFAL

Rule no.	Rule body	Conf.	Lift	Lev.	Conv.
31.	{Intersection, Market} → {Street Light}	0.8	5.53	0.03	4.06
32.	{Curve, Highway} → {Hills, 2 wheeler}	0.26	4.47	0.02	1.26
33.	{Month = 10, Highway} → {Market}	0.57	2.06	0.02	1.63
34.	{Other, Daylight} → {Colony}	0.43	2.04	0.03	1.15
35.	{Agriculture Land} → {Time = 10,Other, Non-highway, 2 wheeler}	0.34	1.85	0.01	1.02
36.	{Intersection} → {Pedestrian}	0.38	1.81	0.02	1.1
37.	{Pedestrian_hit} → {Hospital, Highway}	0.37	1.77	0.02	1.24
38.	{Curve,Fall_height} → {Highway}	0.97	1.73	0.01	8.96
39.	{Day = Monday, Non-highway, 2 wheeler} → {Agriculture Land}	0.67	1.38	0.01	1.49
40.	{Daylight} → {Curve, Hills, Highway, Fall_height}	0.54	1.12	0.2	1
41.	{Agriculture Land, Non-highway} → {Day = Saturday, 2 wheeler}	0.59	1.02	0	1
42.	{Time = 11, Other, 2 wheeler} → {Highway}	0.56	1.01	0	0.99
43.	{Time = 9, Daylight, Agriculture Land} → {Non-highway}	0.44	1.01	0	0.99
44.	{Day = Wednesday} → {Daylight, Agriculture Land, 2 wheeler}	0.45	1.01	0	1
45.	{Month = 11, 2 wheeler} → {Non-highway}	0.44	1.01	0	0.98

accidents but road feature are unknown. Fall height accidents are associated with curve on roads. Three-wheeler accidents have happened at day time but the road features are unknown. Two-wheeler accidents are scattered on non-highway roads where area around is mostly agriculture land and road feature is intersection.

3.5 Association rule for LFAL

In LFAL locations, rule shows that pedestrians are more prone to accidents at intersections on road. In these locations, most of the accidents have occurred after sunset and before sunrise. Two-wheeler accidents mostly occur in night time. In LFAL locations, October and November month have more accidents than any other month. In India, these months have more national festivals such as Vijaydashmi and Diwali, and these festival celebrations cause congested traffic and rush on roads. This congestion and rush could be a cause of more accident rate in these months. Monday and Saturday that are first day of working week and last day of working week have more two-wheeler accidents in comparison to other week days.

The association rules for HFAL, MFAL and LFAL are slightly similar but have different interesting measures. Hence we tried to identify the reasons why the accidents mostly occurred at HFAL locations and identified that most of the HFAL locations are on the highway towards the tourist hill stations where traffic movement are high every day of the week. Another HFAL location was found to be non-highway roads connecting two or more cities which contain all type of transport activities like highways and also non-highways features such as local colonies, markets. Hence these locations contain highway and non-highway traffic on these roads making them vulnerable to road accidents. MFAL locations are mainly highways that go through agriculture land, forest areas and non-tourist hill areas. Usually, these roads are free from side road traffic such as pedestrians on the road. So we can expect vehicles with high speed at these locations. LFAL locations consist of both highway and non-highway roads in the city premises and contain colonies, congested market, hospitals etc. As these locations are scattered throughout the city area and similarly city traffic, pedestrians are also distributed as per their localities in the city, there the accident counts are relatively low as compared to HFAL and MFAL. However, the association rule concludes the following information related to the circumstances of the accident occurrence:

The association rule reveals that curvy roads in hilly regions are more prone to accidents for every type of vehicles. So, one should carefully drive on these locations in order to prevent the chance of accidents. Government officials should also do some prevention work such as putting hoardings about sudden bend and slopes on road. The roads that pass through the agricultural land are risky at night as there are no road side lights. Various intersections are not visible on these roads at night time, which create the chances for an accident as intersection is a place where several vehicles can encounter with each other at the same time. Road lightning in these locations can help in overcoming the accident rates at night time.

Pedestrian hit cases are mostly found near intersections in the market and colonies. Market and colonies are highly congested areas where mostly pedestrians are found. The reason of pedestrian hit cases is attributed to the parked vehicles in the market roads, which makes the pedestrian walk difficult. Hence, there must be different vehicle parking zones so that vehicles cannot enter the congested area and thereby pedestrian hit accidents can be prevented in these areas.

Vehicle rollover accidents are mainly found in roads through agriculture land. Although the road type feature is unknown for this type of accidents but based on authors personal knowledge, potholes and bad road conditions can be the reason for these accidents. Potholes are usually rare on highways but it can cause severe accidents when a high speed vehicle comes into its contact. So, government should periodically monitor the conditions on road and provide proper maintenance to overcome these accidents. Several other rules indicate that in the night time highways become more sensitive to road accidents for all type of accidents. Other road accident features such as the speed of the vehicle at the time of road accident, condition of the road, driver characteristics and weather information are very much helpful to extract other useful information which can better explain the circumstances of accident occurrence.

Figure 3 illustrates the time-wise distribution of road accidents in HFAL, MFAL and LFAL locations. It indicates that accident pattern in these locations are almost similar except that for MFAL locations there is a downfall at 14:00–16:00 h while an there is an increase in the LFAL and MFAL locations. Figure 4 illustrates the month-wise distribution of road accidents in HFAL, MFAL and LFAL

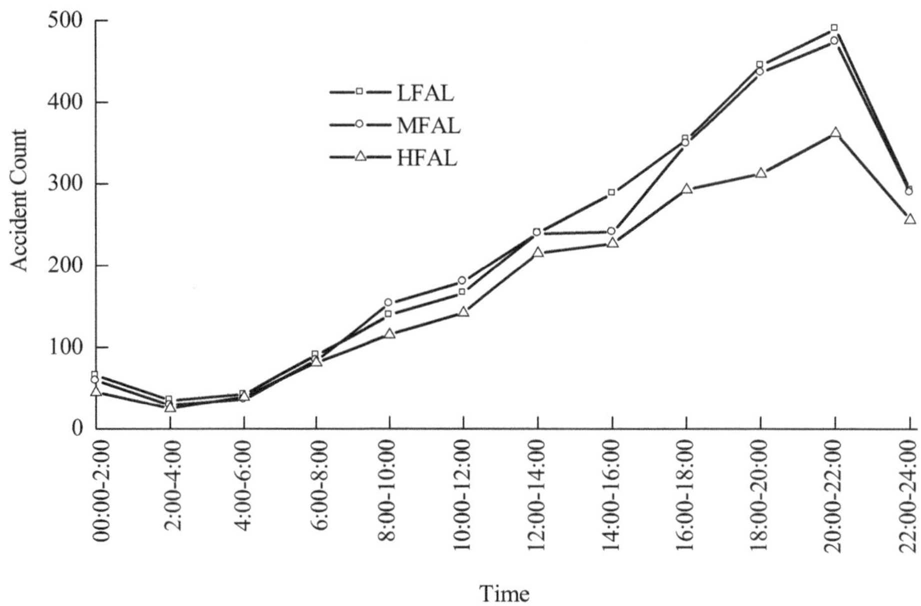

Fig. 3 Time-wise distribution of accidents

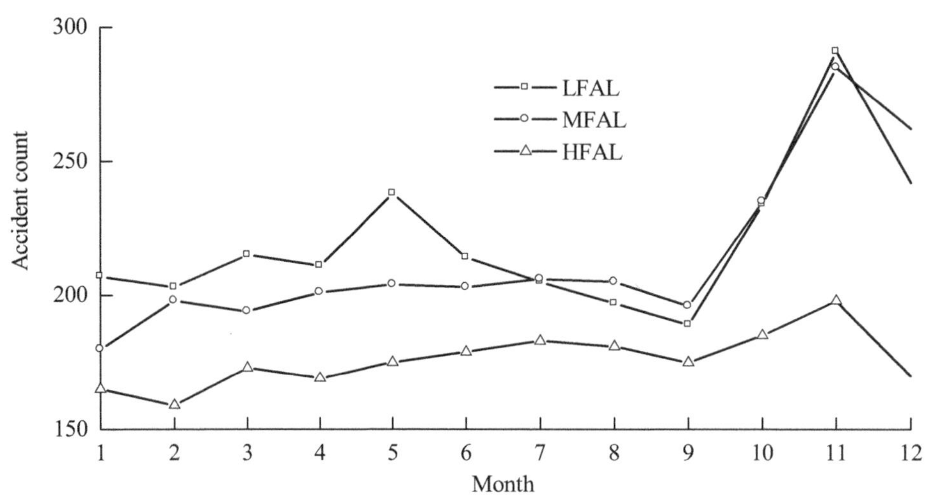

Fig. 4 Month-wise distribution of accidents

locations. Month-wise illustration of accidents shows that there is a different pattern of road accidents. As mentioned that HFAL locations contain tourist hill stations, Fig. 4 illustrates that there is an increase in accident rate from April to July. The reason can be most of the people visit tourist places in summer duration. LFAL locations on September show the minimum peak of accident count. As most of the rainfall occurs in September month, the people are happier to stay at home and hence very few accidents are reported at that duration.

Though authors try to extract the best characteristics of accident occurrence in the available data through data mining techniques, other accident features such as speed of the vehicle at the time of accident, weather information, traffic volume information can surely strengthen our results, which are not available with the data.

4 Conclusion

Data mining has been proven as a reliable technique in analyzing road accident data. Various authors used data mining techniques to analyze road accident data of different countries. Various data mining techniques such as clustering, classification and association rule mining are widely used in the literature to identify reasons that affect the severity of road accidents. It is the first time that *k*-means algorithm is used to identify high- and low-frequency accident locations based on accident count as it provides some technical measures to divide the accident locations based on threshold values. Association rule mining is a very popular technique that can be used to identify the relationship among different sets of attributes that frequently occur together when an accident takes place. In our study, we applied association rule mining algorithm on different groups of accident locations. The rules generated for every group exposed the various factors associated with road accidents in these locations. Although certain rules are quite similar in each group, they have different interest scores for each cluster. The road accident dataset and its analysis using *k*-means clustering and association rule mining algorithm illustrate that this approach can be reused on other accident data with more attributes to identify various other factors associated with road accidents. Although this data mining approach is quite sufficient to uncover reasonable information from the selected data set, the results remain at very general level as source data does not contain other accident related information such as the speed of vehicle at the time of accident, weather information, road surface condition. The data with more number of attributes can reveal more information using our approach.

Acknowledgments We are thankful to GVK-EMRI Dehradun for providing data for our research.

References

1. MORTH (2014) Road Accidents in India 2013. New Delhi: Ministry of Road Transport and Highways Transport Research Wing, Government of India, August 2014. http://morth.nic.in/showfile.asp?lid=1465. Accessed 20 May 2015
2. Kononov J, Janson BN (2002) Diagnostic methodology for the detection of safety problems at intersections. Transp Res Rec. doi:10.3141/1784-07
3. Lee C, Saccomanno F, Hellinga B (2002) Analysis of crash precursors on instrumented freeways. Transp Res Rec. doi:10.3141/1784-01
4. Chen W, Jovanis P (2000) Method for identifying factors contributing to driver-injury severity in traffic crashes. Transp Res Rec. doi:10.3141/1717-01
5. Barai S (2003) Data mining application in transportation engineering. Transport 18:216–223. doi:10.1080/16483840.2003.10414100
6. Tan PN, Steinbach M, Kumar V (2006) Introduction to data mining. Pearson Addison-Wesley, Boston
7. Ponnaluri RV (2012) Road traffic crashes and risk groups in India: analysis, interpretations, and prevention strategies. IATSS Res 35:104–110. doi:10.1016/j.iatssr.2011.09.002
8. Kumar CN, Parida M, Jain SS (2013) Poisson family regression techniques for prediction of crash counts using Bayesian inference. Proc Soc Behav Sci 104:982–991. doi:10.1016/j.sbspro.2013.11.193
9. Parida M, Jain SS, Kumar CN (2012) Road traffic crash prediction on national highways. Indian Highw Indian Road Congr 40:93–103
10. Geurts K, Wets G, Brijs T, Vanhoof K (2003) Profiling of high frequency accident locations by use of association rules. Transp Res Rec. doi:10.3141/1840-14
11. Tesema TB, Abraham A, Grosan C (2005) Rule mining and classification of road accidents using adaptive regression trees. Int J Simul 6:80–94
12. Abellan J, Lopez G, Ona J (2013) Analysis of traffic accident severity using decision rules via decision trees. Expert Syst Appl 40:6047–6054. doi:10.1016/j.eswa.2013.05.027
13. Depaire B, Wets G, Vanhoof K (2008) Traffic accident segmentation by means of latent class clustering. Accid Anal Prev 40:1257–1266. doi:10.1016/j.aap.2008.01.007
14. Kashani T, Mohaymany AS, Rajbari A (2011) A data mining approach to identify key factors of traffic injury severity. Promet-Traffic Transp 23:11–17. doi:10.7307/ptt.v23i1.144
15. Kwon OH, Rhee W, Yoon Y (2015) Application of classification algorithms for analysis of road safety risk factor dependencies. Accid Anal Prev 75:1–15. doi:10.1016/j.aap.2014.11.005
16. MacQueen J (1967) Some methods for classification and analysis of multivariate observations. In: Proceedings of the fifth Berkeley symposium on mathematical statistics and probability, volume 1:

Statistics, pp 281–297, University of California Press, Berkeley, 1967. http://projecteuclid.org/euclid.bsmsp/1200512992
17. Tibshirani R, Walther G, Hastie T (2001) Estimating the number of clusters in a data set via the gap statistic. J R Statist Soc B 63:411–423. doi:10.1111/1467-9868.00293
18. Han J, Kamber M (2001) Data mining: concepts and techniques. Morgan Kaufmann Publishers, Burlington
19. Piatetsky-Shapiro G (1991) Knowledge discovery in databases. In: Piateski G, Frawley W (eds) Discovery, analysis, and presentation of strong rules. AAAI/MIT Press, Menlo Park, pp 229–248
20. Brin S, Motwani R, Ullman JD, Tsur S (1997) Dynamic itemset counting and implication rules for market basket data. In: Proceedings of the 1997 ACM SIGMOD international conference on management of data, Tucson, Arizona, pp 255–264. doi:10.1145/253260.253325
21. http://www.emri.in
22. Agrawal R, Srikant R (1994) Fast algorithms for mining association rules in large databases. In: Proceedings of the 20th international conference on very large data bases, pp. 487–499

12

Investigation into rail corrugation in high-speed railway tracks from the viewpoint of the frictional self-excited vibration of a wheel–rail system

G. X. Chen[1] · X. L. Cui[1] · W. J. Qian[1]

Abstract A finite element vibration model of a multiple wheel–rail system which consists of four wheels, one rail, and a series of sleepers is established to address the problem of rail corrugation in high-speed tracks. In the model, the creep forces between the wheels and rail are considered to be saturated and equal to the normal contact forces times the friction coefficient. The oscillation of the rail is coupled with that of wheels in the action of the saturated creep forces. When the coupling is strong, self-excited oscillation of the wheel–rail system occurs. The self-excited vibration propensity of the model is analyzed using the complex eigenvalue method. Results show that there are strong propensities of unstable self-excited vibrations whose frequencies are less than 1,200 Hz under some conditions. Preventing wheels from slipping on rails is an effective method for suppressing rail corrugation in high-speed tracks.

Keywords Rail corrugation · Wear · Friction-induced vibration · Self-excited vibration · Elastic vibration mode · Wheel-rail system · High-speed railway

1 Introduction

The total mileage of Chinese high-speed railway has exceeded 16,000 km since the Beijing–Tianjin high-speed intercity railway line with an overall length of 120 km was put into operation in 2008. With the operation speed of high-speed trains arriving at 300–350 km/h, the operation safety of trains has become a main concern to the railway administrators and passengers. Rail corrugation and polygonalization of high-speed railway wheels are two emergent problems influencing the operation safety and comfort of trains in the Chinese high-speed railway industry. For instance, the Beijing–Shanghai high-speed railway line has an overall length of 1,318 km, and was first put into operation in 2011. After 2–3 months' operation, rail corrugation occurred in some track sections. Polygonalization of high-speed railway wheels was also found to be serious after the high-speed trains were used for 1–2 years. Rail corrugation and polygonalization of high-speed railway wheels can not only lead to a poor passenger ride comfort and a high environment noise level but also decrease the service life of some key parts such as wheelsets, bearings, and rails. Therefore, it is significant to study the rail corrugation and polygonalization of high-speed railway wheels in high-speed railway lines. This paper mainly concerns the problem of rail corrugation.

Rail corrugation is an undulatory wear which occasionally occurs on the rail rolling surface. The research history of rail corrugation can be traced back to 1900s. Since then, many researchers have made gigantic contributions to the understanding of the generation mechanism of rail corrugation. It is well known that the generation mechanism of rail corrugation consists of both the material damage and wavelength fixing mechanisms [1]. Several sagacious concepts have been proposed in the literature, which include the wear law of the frictional power causing rail working surface wear [2], the feedback mechanism for corrugation development [3], the contact filter [4], the pinned–pinned resonance [5], and so on. Specifically, Hempelmann et al. [2] studied rail corrugation, holding an

✉ G. X. Chen
chen_guangx@163.com

1 State Key Laboratory of Traction Power, Tribology Research Institute, Southwest Jiaotong University, Chengdu 610031, China

opinion that rail working surface wear is dominated by the wear law of the frictional power. Hempelmann and Knothe [3] extended a linear model to study the short pitch corrugation. Müller [6] studied rail corrugation from the instability of the wheel–rail system. Igeland and Ilias [7] studied rail corrugation from the viewpoint of non-linear interaction between wheel and rail. Knothe and Ripke [8] studied parameter sensitivity of rail corrugation. Andersson and Johansson [9] studied rail corrugation using a three-dimensional wheel–rail interaction model. Johansson and Nielsen [10] studied the influence of powered wheelsets with wheel tread irregularities on rail corrugation. Clark [11] and Brockley [12] studied rail corrugation from the viewpoint of stick–slip vibration of the wheel–rail system. Wu and Thompson [13] attributed rail corrugation to the micro-slip between wheel and rail. Xie and Iwnicki [14] analyzed the railhead wear using a three-dimensional non-Hertzian and non-steady contact model. Daniel and Coworkers [15] discussed the effect of the wheel pass time delay on rail corrugation. Matsumoto [16] and Suda [17] all believed that rail corrugation was probably due to the stick–slip vibration of the wheel–rail system. Sun and Simson [18] studied rail corrugation using an integrated model including rail surface roughness and stick–slip vibration of the wheel–rail system. Jin [19] developed a comprehensive three-dimensional train-track model to study rail corrugation. Contributions from these researchers constitute the theoretical groundwork of rail corrugation.

As concluded by Grassie [20], wear is a main material damage mechanism of rail corrugation. Most of the above investigations on rail corrugation were focused on the wavelength fixing mechanism. Research [1, 21, 22] shows that the wavelength fixing mechanism includes the P2 resonance of the interaction between wheel and rail, the pinned–pinned resonance of tracks, the second torsional resonance of driven axles, and the transient dynamic interaction between wheel and rail due to initial rail surface roughness. Among these proposed mechanisms, the transient dynamic interaction between wheel and rail is widely accepted. Now, most simulation works on rail corrugation are all based on this mechanism.

However, it should be noted that the existing rail corrugation knowledge cannot lead to a good solution to rail corrugation other than rail grinding. Rail corrugation is severer than ever before. In Chinese metro tracks, almost all tracks whose curve radii are less than 350 m suffer from rail corrugation. Rail corrugation in high-speed railway lines and polygonalization of high-speed railway wheels are puzzling the high-speed railway industry. We cannot predict where rail corrugation will occur in the design stage of a high-speed railway line. We know that rail corrugation is inevitable in tight curved tracks, but it is strange that most rail corrugations usually occur on the low rail and seldom occur on the high rail although there is no apparent difference between the receptances of the low and the high rails. To date, it appears that a satisfactory and objective explanation to the reason why most rail corrugations occur on the low rail of tight curved tracks cannot be available in the literature.

Severe and uncontrolled rail corrugation suggests that there is still a need to study rail corrugation from a new perspective. Recently, the authors proposed the viewpoint that rail corrugation is caused by friction-induced vibration of the wheel–rail system [23]. At the same time, Kurzeck and Hecht [24] also proposed the similar viewpoint based on a simple model with 3 degrees of freedom. In their works, the material damage mechanism is wear and the wavelength fixing mechanism is the friction-induced self-excited vibration of the wheel–rail system. Different from the second torsional resonance of driven axles, the saturated creep force-induced mode coupling between wheel and rail leads to self-excited vibration of the wheel–rail system to create corrugation. From the new insight of friction-induced vibration causing rail corrugation, one can objectively explain many rail corrugation phenomena such as most rail corrugations occurring on the low rail of tight curved tracks and little rail corrugation occurring in the curved track whose radius is larger than 600–700 m [25, 26].

In this paper, rail corrugation in high-speed tracks is studied from the viewpoint of self-excited vibration of the wheel–rail system causing corrugation. An elastic vibration model of a multiple wheel–rail system is established for a vehicle traveling on a high-speed track. It is found that when the traction force or brake force is not well controlled, the slip friction between the wheel and rail can lead to self-excited vibration of the wheel–rail system to create rail corrugation.

2 Self-excited vibration model of a multiple wheel–rail system

2.1 Rail corrugation phenomena in high-speed railway lines

Figure 1 shows a picture of rail corrugation in a high-speed track [27]. Based on field investigations of rail corrugation in high-speed tracks, the following corrugation phenomena are found: (1) Rail corrugation in high-speed tracks most probably occurs in curved tracks, and the smaller the curve radius, the more easily rail corrugation takes place. (2) There is an apparent difference between the rail corrugation occurrence propensities in the train-up and train-down directions. In one direction, train brake is applied and in the other direction, no train brake is applied. Rail corrugation most probably occurs in track sections where trains are braked.

Fig. 1 Rail corrugation in a high-speed track [27]

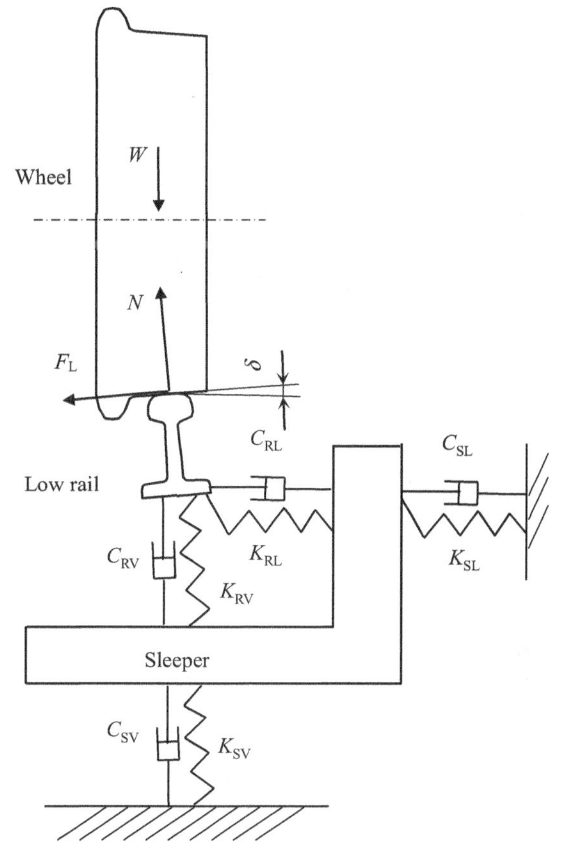

Fig. 2 Position of the contact point between four wheels and rail

2.2 Contact and friction forces acting on a wheel

The minimum radius of curved tracks in high-speed railway lines is larger than 7,000 m. According to the authors' investigation, rail corrugation seldom occurs in metro tracks whose radii are larger than 600–700 m. As mentioned above, rail corrugation in high-speed tracks most probably arises in the train brake section of tracks. Since rails are exposed to rain, snow, and dust, the friction coefficient between the wheel and rail is changed in a large range. Based on these facts, the authors can reasonably assume that the wheels slip on the rail in some brake applications. In the present paper, the self-excited vibration model of the wheel–rail system is based on this assumption. Wheels are generally always kept in contact with the rail when a vehicle travels on a track. According to the vehicle system dynamics, the contact points between the wheels and rail on a tangential track fall in the central area of the rail top surface. The corresponding contact angles are in the range of about 1–3°. The interaction forces between the wheels and rail in this case is shown in Fig. 2. The friction forces between the wheels and rail are along the direction of rail. In Fig. 2, δ is the contact angle; N is the normal force; F_L is the lateral creep force; w is the vertical suspension force applied by the axle box; K_{RV} and K_{RL} are vertical stiffness and lateral stiffness of the rail fastener spring, respectively; C_{RV} and C_{RL} are vertical and lateral damping coefficients of the rail fastener dampers, respectively; K_{SV} and K_{SL} are vertical and lateral stiffness of the sleeper support springs, respectively; and, C_{SV} and C_{SL} are vertical and lateral damping coefficients of the sleeper support dampers, respectively.

2.3 Finite element oscillation equations of the multiple wheel–rail system

In the present work, the wheel–rail system consists of four wheels, one rail, and a series of sleepers. The creep force between the wheels and rail is considered to be saturated and is approximately equal to the normal force times the coefficient of friction. In the finite element oscillation equations, the friction coupling between the wheel and rail vibrations is very important for the occurrence of self-excited oscillation of the multiple wheel–rail system. The friction force is expressed as follows:

$$F_{wr,i} = -F_{rw,i} = \mu k\left(u_{w,i} - u_{r,i}\right), \tag{1}$$

where $F_{wr,i}$ and $F_{rw,i}$ are the friction force components acting on the ith node of the wheel and its corresponding ith node of the rail, respectively; μ is the coefficient of friction; k is the contact stiffness between the wheel and rail; and, $u_{w,i}$ and $u_{r,i}$ are the displacements of the ith contact points of the wheel and rail in the normal direction, respectively. All components of the friction forces on the contact surfaces can be written in matrix form as follows:

$$\boldsymbol{F}_{ff} = \boldsymbol{K}_{ff}\boldsymbol{u}, \tag{2}$$

where $\boldsymbol{F}_{ff}$ is the friction force vector of all the contact nodes, $\boldsymbol{K}_{ff}$ is the contact stiffness matrix, and $\boldsymbol{u}$ is the displacement vector of all nodes of the multiple wheel–rail system. In the absence of friction between the sliding surfaces, the motion equations of the wheel–rail system are written as follows:

$$\boldsymbol{M}\ddot{\boldsymbol{u}} + \boldsymbol{C}\dot{\boldsymbol{u}} + \boldsymbol{K}\boldsymbol{u} = \boldsymbol{0}, \tag{3}$$

where $\boldsymbol{M}$, $\boldsymbol{C}$, and $\boldsymbol{K}$ are respectively the mass, damping and stiffness matrixes of the system; and they are all symmetric matrices. Equation (3) is a stable system. In the presence of friction, the friction coupling between the wheels and rail cannot be ignored. The motion equations of the wheel–rail system with friction coupling are rewritten as follows:

$$\boldsymbol{M}\ddot{\boldsymbol{u}} + \boldsymbol{C}\dot{\boldsymbol{u}} + (\boldsymbol{K} - \boldsymbol{K}_{ff})u = \boldsymbol{0}. \tag{4}$$

The eigenvalue equation of Eq. (4) is

$$\left(\boldsymbol{M}\lambda^2 + \lambda\boldsymbol{C} + (\boldsymbol{K} - \boldsymbol{K}_{ff})\right)\boldsymbol{\varphi} = \boldsymbol{0}, \tag{5}$$

where λ is the eigenvalue, and $\boldsymbol{\varphi}$ is the eigenvector.

The general solution of Eq. (4) is

$$\boldsymbol{u}(t) = \sum \boldsymbol{\varphi}_i \exp(\lambda_i t) = \sum \boldsymbol{\varphi}_i \exp((\alpha_i + \mathrm{j}\omega_i)t), \tag{6}$$

where t is time; $\boldsymbol{\varphi}_i$ is the ith eigenvector of Eq. (5); $\lambda_i = \alpha_i + \mathrm{j}\omega_i$ is the ith eigenvalue of Eq. (5), in which α_i and ω_i are the real and imaginary parts of the ith eigenvalue, and j is the imaginary unit. From Eq. (6), one can see that when a real part of eigenvalues is larger than zero, the displacement will increase with time, and the vibration displacements of the wheel–rail system will become larger and larger. In this case, the wheel–rail system becomes an unstable system. The self-excited vibration governing equations of frictional systems are composed of Eqs. (1–6). Generally, there are more than one million displacement degrees of freedom in a multiple wheel–rail system. A simulation analysis of the self-excited vibration of such a frictional system needs a high-efficient computer and consumes much computational time. More recently, the self-excited vibration theory of frictional systems has been accepted generally. Both ABAQUS and ANSYS finite element software packages provide a capability of analyzing the motion stability and transient dynamics of frictional systems [28]. The procedures of applying ABAQUS to evaluate the self-excited vibration occurrence propensity of the wheel–rail system are presented as follows:

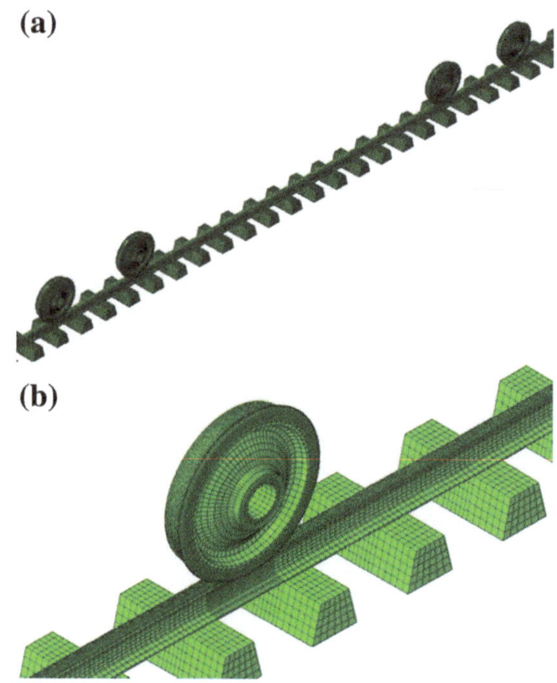

Fig. 3 Multiple wheel–rail model. **a** Overall layout. **b** Details

(1) Contact non-linear analysis to extract contact forces without friction.
(2) Contact non-linear analysis when the wheels are imposed to slide on the rail to extract contact forces with friction.
(3) Mode analysis to extract mode parameters when the friction coupling is ignored.
(4) Complex eigenvalue analysis to extract unstable mode parameters in the presence of friction coupling.

Figure 3 shows the finite element self-excited vibration model of the wheel–rail system. In the model, the rail is supported on a series of sleepers and the sleepers are supported on a monolithic track bed. The connections between the rail and sleepers and the connections between the sleepers and the monolithic track bed comprise a group of lateral springs and dampers, a group of vertical springs and dampers, and a group of longitudinal springs and dampers at each location of sleepers. The number of each group of springs and dampers between the rail and sleepers is the same as the number of the rail nodes on the contact area between the rail and each sleeper. The number of each group of springs and dampers between the sleepers and the monolithic track bed is the same as the number of the sleeper nodes on the contact area with the monolithic track bed.

Both the self-excited vibration of the wheel–rail system in the present paper and the traditional stick–slip vibration of the wheel–rail system fall into the category of the friction-induced vibration, but there is also some difference between them. The generation mechanism of the former vibration is attributed to the mode coupling induced by friction, while the latter vibration is attributed to the negative friction-speed slope. The feedback energy of the self-excited vibration of the wheel–rail system is provided by the vibration displacement of the system, while the feedback energy of the traditional stick–slip vibration is provided by the vibration speed of the system.

2.4 Nominal parameters of the wheel–rail system

The Young's modulus of both the wheel and rail materials is $E = 2.06 \times 10^{11}$ N/m. The density of the wheel and rail materials is $\rho = 7{,}800$ kg/m^3. The Poisson's ratio of both the wheel and rail materials is $v = 0.3$. The contact angle is $\delta = 2.5°$. The vertical force applied on the wheel is $w = 75{,}000$ N. The wheel is 840 mm in nominal diameter. The rail is the 60 kg/m type, and 46,000 mm in length. The distance between two sleepers is 600 mm. The width of sleepers is 160 mm. The coefficient of friction between wheel and rail is $\mu = 0.4$. In the present work, the lateral spring stiffness value K_{RL} of the rail fastener between the rail and each sleeper is assumed to be a half of the vertical spring stiffness value K_{RV}; specifically $K_{RV} = 5.0 \times 10^7$ N/m, and $K_{RL} = 2.5 \times 10^7$ N/m. The vertical and lateral damping coefficients of each rail fastener are $C_{RV} = 5{,}000$ Ns/m and $C_{RL} = 5{,}200$ Ns/m, respectively. The vertical and lateral spring stiffness values between each sleeper and the monolithic track bed are $K_{SV} = 8.9 \times 10^7$ N/m and $K_{SL} = 5.0 \times 10^7$ N/m, respectively. The vertical and lateral damping values between each sleeper and the monolithic track bed are $C_{SV} = 8.98 \times 10^4$ Ns/m and $C_{SL} = 4.0 \times 10^4$ Ns/m, respectively.

3 Results

3.1 Unstable vibration characteristics of the wheel–rail system

Rail corrugation is generally attributed to the vibration with a frequency in the range from 20 to 1,200 Hz. Therefore, the simulation results are restricted to this frequency range. The effective damping ratio is a useful parameter for measuring the occurrence tendency of self-excited vibration. The larger the absolute value of the effective damping ratio, the more easily the corresponding vibration becomes unstable. The effective damping ratio is defined as $\zeta = -2\mathrm{Re}(\lambda)/|\mathrm{Im}(\lambda)|$, where $\mathrm{Re}(\lambda)$ is the positive real part of the eigenvalue λ, and $|\mathrm{Im}(\lambda)|$ is the absolute value of the corresponding imaginary part of λ. Figure 4 demonstrates unstable vibration characteristics of the wheel–rail system in a frequency range from 20 to 1,200 Hz. From this figure, we can see that in the prescribed frequency range, there are two unstable vibration modes, in which the wheel vibrates on the rail in the lateral direction. Since the effective damping ratio corresponding to the unstable vibration of 508.89 Hz is much larger than that corresponding to the unstable vibration of 331.44 Hz, the unstable vibration of 508.89 Hz takes place more easily. It needs to be noted that the unstable vibration of 508.89 Hz is very close to the measured vibration frequency of corrugated rail, which is about 533.26 Hz [27].

3.2 Influence of friction coefficient on rail corrugation

Figure 5 demonstrates the change of the effective damping ratio of unstable vibration with the friction coefficient. From this figure, it is seen that when the friction coefficient between the wheels and rail is less than 0.28, there is no unstable vibration, which suggests that no rail corrugation arises. We can also see that with the increasing of friction coefficient, the effective damping ratio is decreased, which suggests that rail corrugation takes place more easily in case of larger friction coefficient.

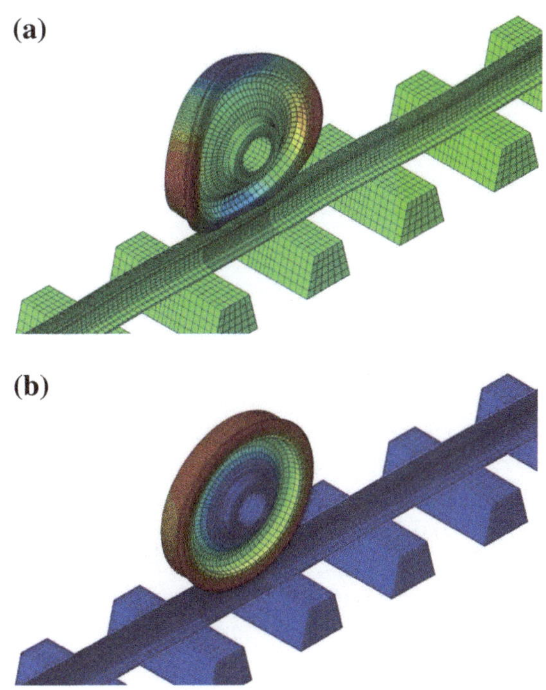

Fig. 4 Mode shapes of unstable vibrations of the wheel–rail system. **a** Unstable vibration frequency $f_R = 331.44$ Hz and corresponding effective damping ratio $\xi = -0.001$. **b** $f_R = 508.89$ Hz and $\xi = -0.0094$

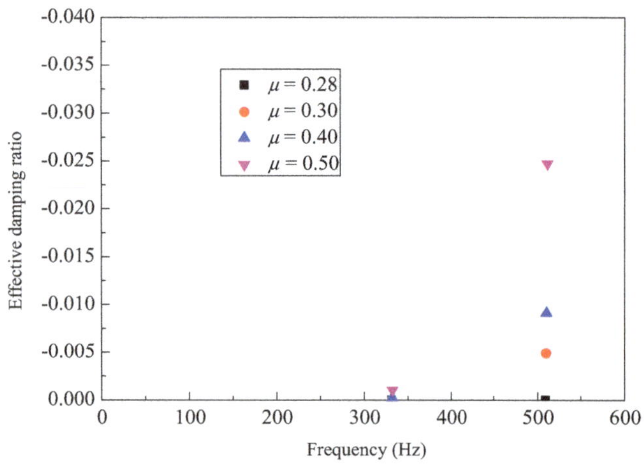

Fig. 5 Change of the effective damping ratio with the friction coefficient

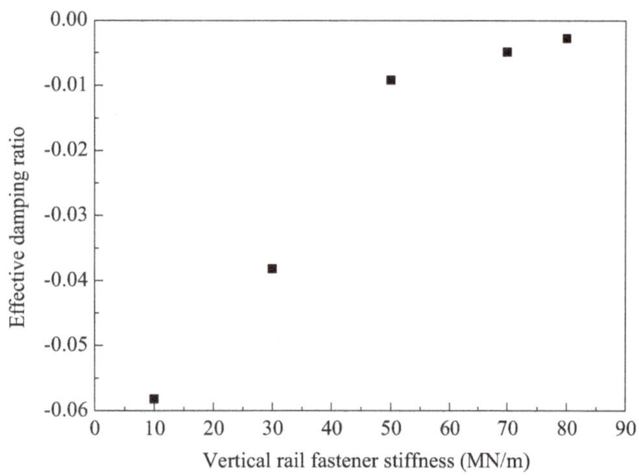

Fig. 6 Change of the effective damping ratio with the rail support stiffness. In the simulation, the lateral rail fastener stiffness is equal to a half of the vertical rail fastener stiffness

3.3 Influence of rail fastener stiffness on rail corrugation

Rail fastener stiffness is the connection stiffness between the rail and each sleeper. The rail is supported on sleepers by a series of rail fasteners. Figure 6 demonstrates the change of the effective damping ratio of unstable vibration with the vertical rail fastener stiffness. From Fig. 6, it is seen that with the decreasing of the vertical rail fastener stiffness, the effective damping ratio is decreased. Therefore, decreasing the vertical rail fastener stiffness will deteriorate rail corrugation.

In some cases, the lateral rail fastener stiffness is not equal to a half of the vertical rail fastener stiffness. Figure 7 shows the change of the effective damping ratio with the lateral rail fastener stiffness. We can see that the

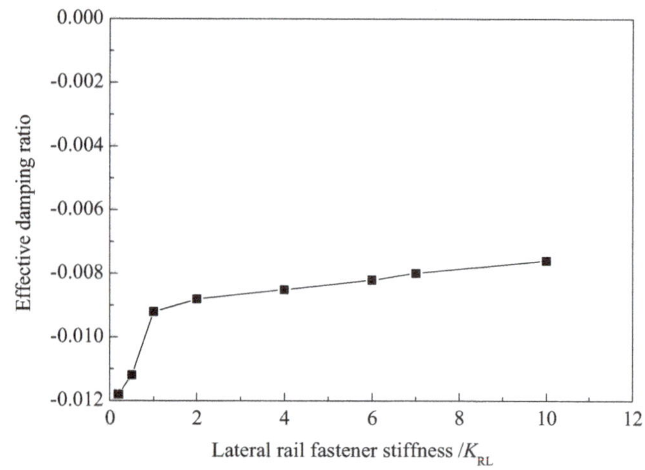

Fig. 7 Change of the effective damping ratio with lateral rail fastener stiffness, $K_{RV} = 5.0 \times 10^7$ N/m

effective damping ratio increases with the lateral rail fastener stiffness but the magnitude of change is small when the lateral rail fastener stiffness is in the range of $(0.2\text{–}10)K_{RL}$.

3.4 Influence of rail fastener damping on rail corrugation

Now we study the influence of rail fastener damping on rail corrugation from the viewpoint of self-excited vibration of the wheel–rail system. Figure 8 shows the change of the effective damping ratio of the wheel–rail system with the rail fastener damping. As can be observed, when the rail fastener damping coefficient is six times the nominal rail fastener damping coefficient C_{RV}, the absolute value of the effective damping ratio achieves its minimum value, which suggests that the rail corrugation can be suppressed as largely as possible in this case. When the rail fastener

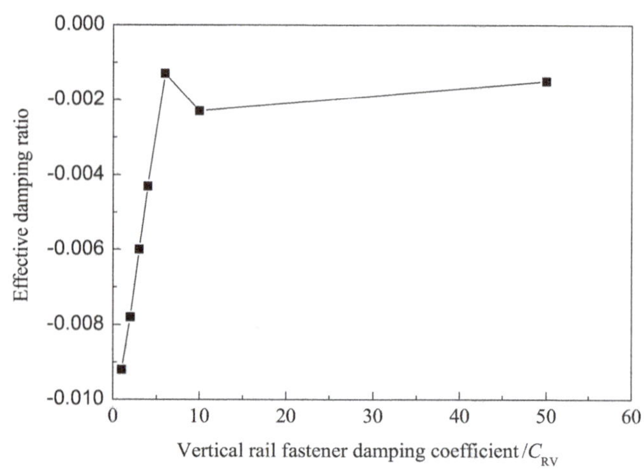

Fig. 8 Change of the effective damping ratio with the rail fastener damping, $C_{RL} = 5{,}200$ Ns/m

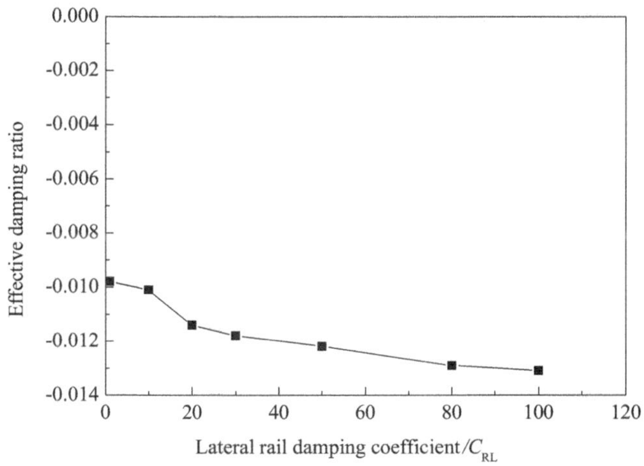

Fig. 9 Change of the effective damping ratio with the lateral rail fastener damping, C_{RV} = 5,000 Ns/m

damping coefficient increase to 10–50 times the nominal rail fastener damping coefficient C_{RV}, the effective damping ratio is close to each other, implying that a larger vertical rail fastener damping coefficient can alleviate rail corrugation.

Figure 9 shows the change of the effective damping ratio with the lateral rail fastener damping coefficient. It is found that when the lateral rail fastener damping coefficient is in the range of 1–100 times the nominal lateral rail fastener damping coefficient C_{RL}, the effective damping ratio only has a small change, implying that the lateral rail fastener damping coefficient has a little influence on rail corrugation.

4 Discussion

From the above simulation results, it is found that when wheels slip on the rail and the friction coefficient between the wheel and rail is large enough, unstable vibration of the wheel–rail system will occur, signifying that rail corrugation will arise. Therefore, the effective method for controlling and suppressing rail corrugation in high-speed tracks is to prevent wheels slipping on rails. If the wheel does not slip on the rail, self-excited vibration of the wheel–rail system will not take place. Thus, rail corrugation will not occur. The present result explains why rail corrugation most probably takes place in train brake sections of tracks. This is because the wheel is easy to slip on the rail due to a low coefficient of friction when a train is braked at a high speed. This result can also be used to explain why most of rails in high-speed tracks are not subjected to corrugation, where wheel slipping on rail rarely takes place.

5 Conclusions

In this paper, a numerical study on rail corrugation in high-speed railway lines is performed from the viewpoint of friction-induced self-excited vibration causing rail corrugation. The following conclusions can be drawn.

(1) When a wheel slip on a rail, the friction force can induce unstable vibration of the wheel–rail system. This unstable vibration leads to rail corrugation. Numerical simulation reveals that unstable vibration of frequency 508.89 Hz most probably takes place.
(2) Preventing wheels from slipping on rails is an effective method for suppressing rail corrugation.
(3) With vertical rail fastener stiffness increasing, the occurrence propensity of rail corrugation will decrease.

Acknowledgments This work was supported by the National Natural Science Foundation of China (No. 51275429).

References

1. Grassie SL, Kalousek J (1993) Rail corrugation: characteristics, causes and treatments. J Rail Rapid Transit 207:57–68
2. Hempelmann K, Hiss F, Knothe K, Ripke B (1991) The formation of wear patterns on rail tread. Wear 144:179–195
3. Hempelmann K, Knothe K (1996) An extended linear model for the prediction of short pitch corrugation. Wear 191:161–169
4. Remington PJ (1976) Wheel/rail noise, parts I–IV. J Sound Vib 46:359–451
5. Hiensch M, Nielsen JCO, Verheijen E (2002) Rail corrugation in the Netherlands-measurements and simulations. Wear 253:140–149
6. Müller S (1999) A linear wheel-track model to predict instability and short pitch corrugation. J Sound Vib 227:899–913
7. Igeland A, Ilias H (1997) Railhead corrugation growth predictions based on nonlinear high frequency vehicle/track interaction. Wear 213:90–97
8. Knothe K, Ripke B (1989) The effects of the parameters of wheelset, track and running conditions on the growth-rate of rail corrugations. Veh Syst Dyn 18:345–356
9. Andersson C, Johansson A (2004) Prediction of rail corrugation generated by three-dimensional wheel–rail interaction. Wear 257:423–434
10. Johansson A, Nielsen JCO (2007) Rail corrugation growth-influence of powered wheelsets with wheel tread irregularities. Wear 262:1296–1307
11. Clark RA (1984) Slip-stick vibrations may hold the key to corrugation puzzle. Railw Gaz Int 7:531–533

12. Brockley CA (1988) An investigation of rail corrugation using friction-induced vibration theory. Wear 128:99–106
13. Wu TX, Thompson DJ (2005) An investigation into rail corrugation due to micro-slip under multiple wheel/rail interactions. Wear 258:1115–1125
14. Xie G, Iwnicki SD (2008) Calculation of wear on a corrugated rail using a three-dimensional contact model. Wear 265(9–10):1238–1248
15. Meehan PA, Daniel WJT, Campey T (2005) Prediction of the growth of wear-type rail corrugation. Wear 258:1001–1013
16. Matsumoto A, Sato Y et al (1996) Study on the formation mechanism of rail Corrugation on curved track. Veh Syst Dyn 25:450–465
17. Suda Y, Hanawa M, Okumura M, Iwasa T (2002) Study on rail corrugation in sharp curves of commuter line. Wear 253:193–198
18. Sun YQ, Simson S (2008) Wagon-track modeling and parametric study on rail corrugation initiation due to wheel stick-slip process on curved track. Wear 265:1193–1201
19. Jin XS, Wen ZF, Wang KY et al (2006) Three-dimensional train-track model for study of rail corrugation. J Sound Vib 293: 830–855
20. Grassie SL (2005) Rail corrugation: advances in measurement, understanding, and treatment. Wear 258:1224–1234
21. Sato Y, Matsumoto A, Knothe K (2002) Review on rail corrugation studies. Wear 253:130–139
22. Oostermeijer KH (2008) Review on short pitch rail corrugation studies. Wear 265:1231–1237
23. Chen GX, Zhou ZR, Ouyang H et al (2010) A finite element study on rail corrugation based on saturated creep force-induced self-excited vibration of a wheelset-track system. J Sound Vib 329(22):4643–4655
24. Kurzeck B, Hecht M (2010) Dynamic simulation of friction-induced vibrations in a light railway bogie while curving compared with measurement results. Veh Syst Dyn 48(supl):121–138
25. Qian WJ, Chen GX, Ouyang H et al (2014) A transient dynamic study of the self-excited vibration of a railway wheel set–track system induced by saturated creep forces. Veh Syst Dyn 52(9): 1115–1138
26. Cui XL, Chen GX, Yang HG, Zhang Q, Ouyang H, Zhu MH (2015) Effect of the wheel/rail contact angle and the direction of the saturated creep force on rail corrugation. Wear 330-331:554–562
27. Wang LQ (2013) Research and control on railroad rail corrugation. Railw Constr Technol 12:101–117 **(in Chinese)**
28. Ouyang H, Nack W, Yuan Y, Chen F (2005) Numerical analysis of automotive disc brake squeal. Int J Veh Noise Vib 1:207–231

13

Overview and outlook on railway track stiffness measurement

Ping Wang[1,2] · Li Wang[1,2] · Rong Chen[1,2] · Jinhui Xu[1,2] · Jinmang Xu[1,2] · Mingyuan Gao[1,2]

Abstract Stiffness is one of the basic performance parameters for railway track. The efficient and accurate stiffness measurement has been considered as the foundation for further development of railway engineering, and therefore has great theoretical and practical significance. Based on a summary of the connotation and measurement of track stiffness, the state of the art of measurement methods for track stiffness was analyzed systematically. The standstill measurement of track stiffness can be performed with the traditional jack-loading method, impact hammer method, FWD (falling weight deflectometer) method, and track loading vehicle method. Although these methods can be adopted in stiffness measurement for a section of railway track, they are not desirable owning to small range and low efficiency. In the recent 20 years, researchers have proposed many methods like unbalanced-loading laser displacement method, deflection basin deformation rate method, and eccentricity excitation method to continuously measure track stiffness; however, these methods have drawbacks like poor accuracy, low speed, and insufficient data analysis. In this work, the merits and demerits of these methods were summarized, and optimization suggestions were presented. Based on the wave transmission mechanism and principle of vibration energy harvesting, an overall conception on continuous measurement of stiffness and long-term stiffness monitoring for special sections was proposed.

Keywords Railway · Track structure · Track stiffness · Measurement · Maintenance · Vibration

✉ Rong Chen
chenrong@home.swjtu.edu.cn

1 Key Laboratory of High-speed Railway Engineering, Ministry of Education, Southwest Jiaotong University, Chengdu 610031, China

2 School of Civil Engineering, Southwest Jiaotong University, Chengdu 610031, China

1 Introduction

Track structure, a load bearing and transmitting structure of railway system, has stiffness to provide necessary elasticity. Track structure comprises several layers and each layer has its own stiffness, namely the track component stiffness. The overall track stiffness is usually adopted to characterize the total stiffness of all layers and reflect the stiffness characteristics of the whole track structure [1, 2]. The stiffness measurement of a track, generally, refers to measuring the overall stiffness. In laboratory tests, the stiffness of certain track component can be measured, such as fastener stiffness and ballast stiffness [3, 4].

Discrete rail support model and continuous rail support model are two classical models for the traditional mechanical analysis of railway track [1, 2]. In the discrete rail support model, the stiffness of fastener supporting track is a spring constant, while in the continuous rail support model, it is denoted by track modulus, which is calculated with the spring constant divided by the spacing between centers of discrete supports. The stiffness measurement method based on the Winkler elastic foundation model uses track modulus to characterize the overall track stiffness.

The track stiffness can also be classified into dynamic stiffness and static stiffness [2, 5, 6]. The resistance of track structure to deformation under a static load is called as static stiffness, which, usually, is determined by the deformation degree of track structure under the static load.

The dynamic stiffness refers to the resistance of track structure to deformation under certain dynamic excitation, which is determined by the natural frequencies of track structure. The track dynamic stiffness reflects the supporting performance of the vibrating track structure and therefore is an important factor influencing the wheel–rail interaction and train running performance. However, the static stiffness is always adopted instead for the traditional mechanical analysis of the wheel–rail system. The rapid development of high-speed railway passenger transportation, heavy-loaded train transportation, and railway engineering technologies makes the theoretical analysis and practical measurement of dynamic stiffness significant.

Suppose that under load P the rail beam will deform. If the deformation amount is y, the track static stiffness K will be

$$K = \frac{P}{y}. \tag{1}$$

Track structure is characterized by material nonlinearity and geometrical nonlinearity [7]. For example, the rubber pad in the fastening system is of material nonlinearity, and voids are distributed between the sleeper and ballast. Thus, the track static stiffness is also nonlinear, as shown in Fig. 1.

In order to take into account the nonlinearities of track structure, the track static stiffness is usually calculated by the slope of the linear part of the loading history curve [2, 8]. In this sense, the tangent stiffness and secant stiffness on the loading curve or unloading curve can be defined:

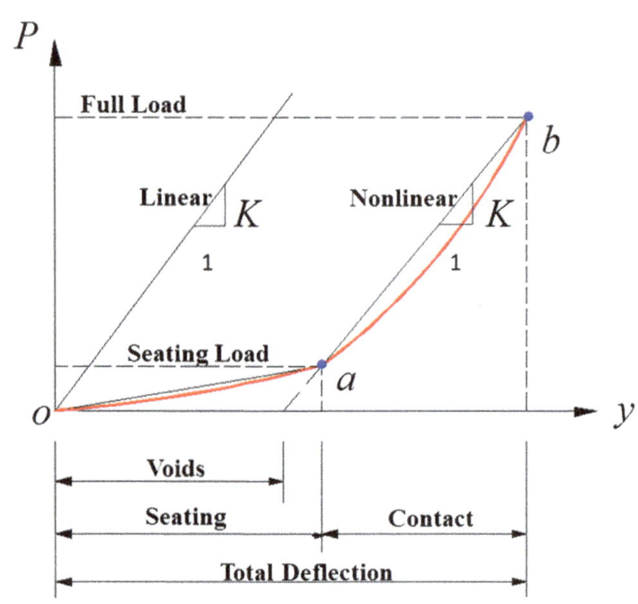

Fig. 1 Load–deflection diagram illustrating nonlinearities

$$K_{a-b} = \frac{P_b - P_a}{y_b - y_a}, \tag{2}$$

where a and b can be defined according to the specific requirement. For example, define b as the final loading state and a as the exact release state of the fastener clamping force.

The dynamic stiffness is influenced by many factors, such as the applied load and the frequency of excitation [9–11]. In order to conduct a quantitative study on the influence of the factors like loading frequency and amplitude on the track dynamic stiffness, a vertical harmonic force $P = P_0 e^{j\omega t}$ is exerted on the rail. If the dynamic displacement at the force acting point is $Z = Z_0 e^{j(\omega t+\varphi)}$, the track dynamic stiffness can be defined as the ratio of the acting force amplitude to the dynamic displacement amplitude:

$$k_d = \frac{|P|}{|Z|}. \tag{3}$$

Under this definition, the track dynamic stiffness can be adopted to reflect the stiffness characteristics in steady forced vibration at a certain frequency.

In mechanical engineering, the dynamic stiffness is always studied in frequency domain. Similarly, the dynamic force and dynamic displacement can be processed through Fourier transform or to use the transfer functions by assuming that the stiffness is linear to some extent. Thus, the track dynamic stiffness can be defined as the ratio of dynamic load to dynamic displacement in frequency domain:

$$k(f) = \frac{F(f)}{z(f)}. \tag{4}$$

Now, the track dynamic stiffness is a complex value, with the amplitude and phase [8, 12]. The track dynamic stiffness can also be measured and analyzed through its reciprocal, namely the track dynamic flexibility.

2 Significance of track stiffness measurement

Track stiffness is an important factor influencing the safety and stability of train operation and the vibration and deformation of track structure as well as the dynamic response of substructures like subgrades and bridges [13–22].

Researches in relation to track stiffness mainly focus on three aspects: ① the analysis of the influence of track stiffness on the vehicle-track-substructure coupled dynamics; ② the nonlinear characteristics of track component stiffness, such as the nonlinearities of the rubber pad; and ③ the evaluation methods of track stiffness and the reasonable stiffness for track structure and its components.

2.1 Stiffness design of new railway lines

Stiffness design of new railway lines is to determine overall track stiffness and track component stiffness to meet the operation conditions of new railways. The measurement of track stiffness can help evaluate the stiffness design of new railways and present the relevant modification opinions.

2.2 Daily maintenance of railway lines

Track stiffness is also an important parameter for maintenance of railway lines in addition to the track geometric irregularity [23]. For most maintenance methods, the track geometric irregularity of railway lines is always adopted as the diagnostic parameter. The track geometry can be adjusted by ballast tamping to keep it within the limited range. However, the performance of ballasts, mostly, is dependent on the performance of bottom subgrade or other substructures. Ballast tamping cannot help improve the performance of substructures, such that track geometric irregularity repeatedly occurs.

The measurement of track stiffness for maintenance purpose covers four aspects:

(1) *Low track stiffness* The railways built in soft soil areas and bad soil zones tend to experience serious rail displacement and fast cracking of track components after a period of operation. These diseases are mainly caused by poor stiffness of soil subgrade and therefore it should be strengthened. It is very important for checking the railway lines in time to keep safe train operation.
(2) *Variable track stiffness* The stiffness always changes abruptly at the railway–bridge, railway–tunnel, and bridge–tunnel sections, and therefore transition zones shall be set [24]. Besides, when track geometric irregularity is deteriorated, support for the track will be uneven.
(3) *Virtual track stiffness* Diseases, such as dirty ballast, hanging sleepers, and loose fasteners, can cause deformation and holes, leading to virtual stiffness for the rail support. The virtual stiffness represents potential risk to safe train operation.
(4) *Assortative stiffness* The unmatched supporting stiffness of the right and left rails will result in uneven rail displacement, which can cause center bending of sleepers as well as the other unreasonable stress and cracking problems of track components.

Fig. 2 Jack-loading diagram and vertical rail deflection measurement

The measurement of track stiffness can help direct the routine maintenance of railway lines and provide a reliable basis for making optimal maintenance strategies.

3 Track stiffness measurement methods

3.1 Standstill measurement

The standstill measurement of overall track stiffness means that a measurement point is decided in advance, and then the overall track stiffness can be calculated by measuring displacement and vertical force exerted on this point.

The overall track stiffness, generally, can be measured based on four methods: (1) traditional hydraulic jack-loading method [1, 2], (2) impact hammer method [25, 26], (3) FWD method [9, 27], and (4) TLV method [2, 5, 28–31].

In the traditional hydraulic jack-loading method, a certain force is exerted on the rail and then rail deflection is measured with a displacement meter or a dial indicator, and thus the force–displacement curve can be obtained (see Fig. 2). The overall track stiffness can then be calculated according to stiffness definitions like secant stiffness or tangent stiffness.

The traditional hydraulic jack-loading method has been adopted since the beginning of the twentieth century. Figure 3 shows a flatbed trailer loaded with I-typed bars used by the Talbot Committee as a reaction wall to measure the track stiffness in 1918 [1].

In the impact hammer method, the track vibration is measured with acceleration transducers installed on rail or sleepers (track plates) after an impulse load is exerted on the track with an impact hammer. The hammer head is equipped with a force transducer to measure the impulse and thus the transfer function of the track can be obtained. At last, the track component stiffness and the overall track stiffness can be calculated through parameter identification (see Fig. 4). Typically, the impact hammer method can cover a frequency interval of 50–1,500 Hz, which depends on the material of the hammer head [32]. A soft-rubber

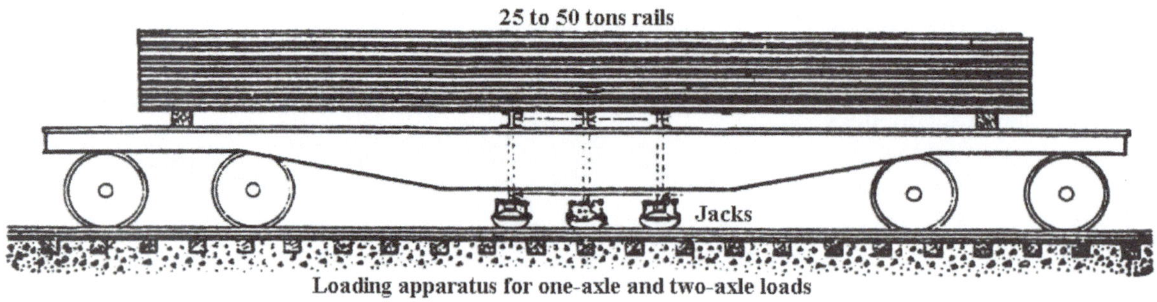

Fig. 3 Track stiffness measured by the Talbot committee of the USA (1918)

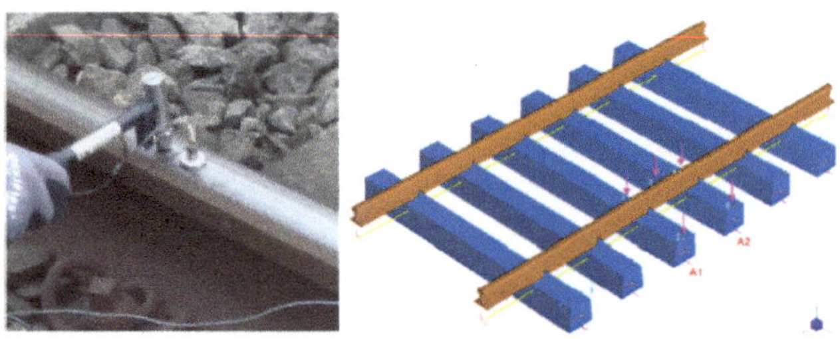

Fig. 4 Track stiffness measurement using the impact hammer method

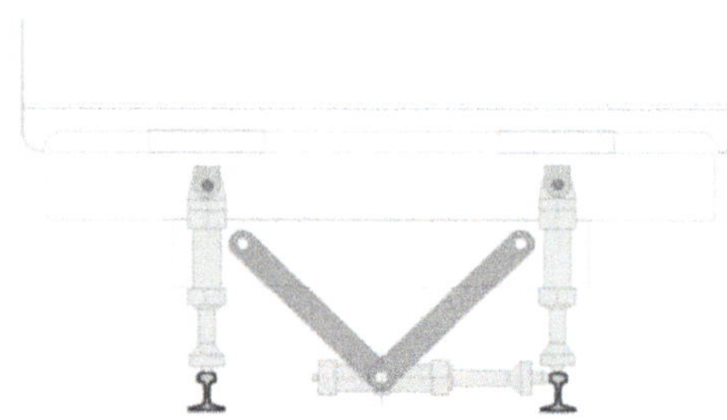

Fig. 5 Track loading vehicle

hammer head is suitable for lower frequencies than a hard-metal one. The impact hammer method is not reliable for the frequencies less than 50 Hz.

In the FWD (falling weight deflectometer) method, a mass impacts a track and the vibration response is measured at the same time. Afterwards, the track stiffness characteristics can be explored through transfer function calculation or other methods. The FWD method, to a certain extent, can reflect the impact effect on a track when a train is running at a high speed. The standard FWD method uses a 125 kN free-falling mass to impact the track. The vibration response of track is generally measured with velocity transducers or geophones.

The principle for the track loading vehicle (TLV) method is the same as that of the traditional hydraulic jack-loading method. However, this method is easier and can provide a larger vertical force (see Fig. 5). Organizations that have TLVs include Transportation Technology Center, Inc. (TTCI) and DECAROTOR of the USA, the South Africa BSSM, Delft of Holland, the railway department of Sweden, etc.

Queensland University of Technology in Australia improved a movable track stiffness measurement vehicle based on common TLV. This vehicle is composed of three cars. The front one is a six-axle towing locomotive (weight of 90 tons) to tow the vehicle to the specified measurement point. The middle one is a buffer car for buffering the influence of the front car on the rear one during the measurement. The rear car is equipped with measurement instruments and is the key part of the vehicle. The rear car weighs 57 tons totally (each part of 14.3 tons) and is equipped with hydraulic jacks and displacement transducers to exert vertical force and measure track deflection basin. When the vehicle reaches the specified measurement point, the hydraulic jacks extend from both sides of the rear car and act on the track separately. The deflection basin is recorded by 22 transducers to calculate the track modulus.

3.2 Continuous measurements

By the time of November 2015, there are 61 high-speed railway lines in China, and the total mileage has reached 12,000 km. The rapid development of high-speed railway lines has made the accurate and continuous measurement of track stiffness extremely important.

The above methods for standstill measurement of overall track stiffness are arduous, time consuming, and not suitable for a long-distance and multi-point measurement. The maintenance of railway lines requires continuous track stiffness measurement equipment. Therefore, many organizations have started to develop vehicles for continuous track stiffness measurement.

As early as in 1997, China Academy of Railway Sciences (CARS) put forward the conception of a vehicle for track elasticity measurement [33]. This vehicle comprises a heavy car in the front and a light car in the rear. The axle load of the heavy car can be varied by adjusting the number of concrete blocks in the range of 150–250 kN. Thus, the influence of different train axle loads on the measured results can be obtained. The light car weighs 40 kN and is used to eliminate the clearances between the rail and sleeper as well as between the sleeper and ballast bed. The measurement equipment on both heavy and light cars is the same except for the axle load.

This vehicle measures track geometric irregularities, which is similar to the chord measurement method. The main purpose of the vehicle is to measure the elastic deformation y_K of tracks, as shown in Fig. 6.

As shown in Fig. 6, the track stiffness can be expressed as

$$K = \frac{\Delta P}{\Delta y} = \frac{P_A - P_B}{y_{KH} - y_{KL}} = \frac{P_A - P_B}{y_H - y_L}. \tag{5}$$

Transportation Technology Center, Inc. in Pueblo, Colorado (USA), also developed a stiffness measurement vehicle, with similar measurement principle to CARS [34, 35]. As shown in Fig. 7, this vehicle comprises a heavy car, a light car, and a towing locomotive. The loading range of the heavy car is 1–55 kips (4–267 kN). The stiffness is measured under 10 kips (44 kN) or 40 kips (178 kN). The load for the light car is less than 3 kips, and the track geometric irregularity is measured under 2 kips (8.9 kN). This vehicle operates twice, under the static loads of 178 and 44 kN, respectively, and in two cases the static track

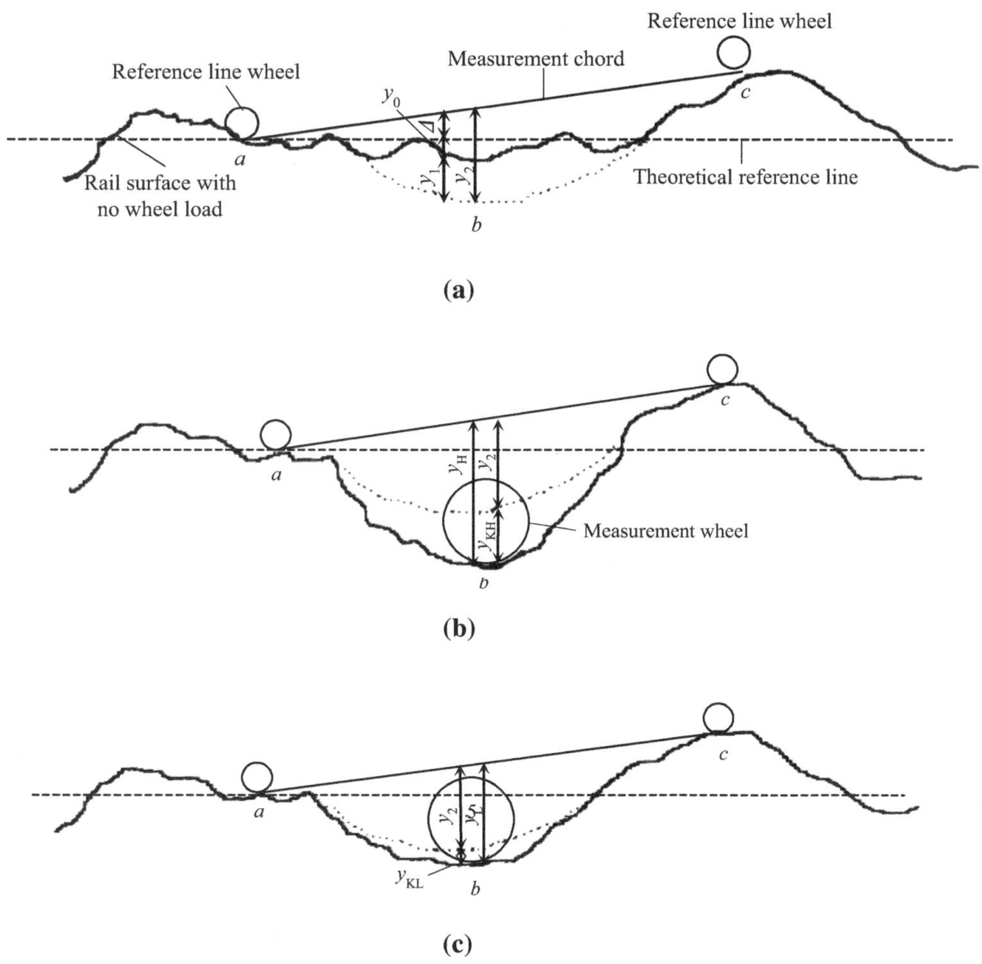

Fig. 6 Principle for track elasticity measurement of CARS-stiffness equipment. **a** Measurement with no wheel load on track. **b** Measurement with the wheel load of heavy car *a* on track. **c** Measurement with the wheel load of light car *b* on track. △ is reference line error. y_0 rail surface irregularities when there is no load on track, y_1 hidden gaps between rail, sleeper, and ballast, y_2 sum of *triangle*, y_0, and y_H chord measured value under the wheel load of heavy car *a*, y_L chord measured value under the wheel load of light car *b*, y_{KH} track elastic settlement under wheel load of heavy car *a*, y_{KL} track elastic settlement under wheel load of light car *b*

Fig. 7 Track stiffness measurement vehicle of TTCI

Fig. 8 Track stiffness measurement vehicle of SBB

irregularity must be excluded. In the dynamic measurement, it finds out areas with high or low stiffness (especially low-stiffness areas to identify ballast or subgrade diseases) and then sprays yellow coatings for performing standstill stiffness measurement. Under the load of 178 kN, the displacement measurement involves the whole track and subgrade, while under the load of 44 kN, it includes the rail, sleeper, and ballast. The measurement speed can reach 16 km/h.

In addition, the measurement principle for the stiffness measurement vehicle developed by Swiss Federal Railways (SBB 2007) is also similar to the above-mentioned ones [8]. This vehicle comprises a light car and a heavy car (see Fig. 8). The weight of the light car can be neglected. The weight of the heavy car is 20 tons. The vehicle speed is 10–15 km/h. Heidenhain LS 220 transducers are used for low-pass (cut-off wavelength of 10–20 m). The accuracy of displacement measurement can reach 0.2 mm.

At the beginning of the twenty-first century, the University of Nebraska (USA) commenced the research on track modulus measurement system [8, 36, 37]. This measurement system can also be used to explore the track modulus through measuring the vertical rail displacement according to the laser measurement method. The measurement vehicle is equipped with two laser sources, the vertical rail displacement can be obtained through measuring the distance between two laser lines, and then the track modulus could be calculated. The measurement system is shown in Figs. 9 and 10.

The measurement principle is shown in Figs. 11 and 12, where d is measured according to the camera view analysis and L_1, L_2, θ_1, θ_2, and H are all known values. After y_r is obtained, an equation is established based on the Winkler foundation model. Solving this equation, one can obtain the track modulus.

The ZOYON Technology Co., Ltd of Wuhan University (China), is now investigating a deformation rate-based track stiffness measurement method to be applied in the movable stiffness measurement vehicles [38]. This method was first introduced by scholars of Delft University of Technology in Holland; however, no prototype car has been made [2, 39–41].

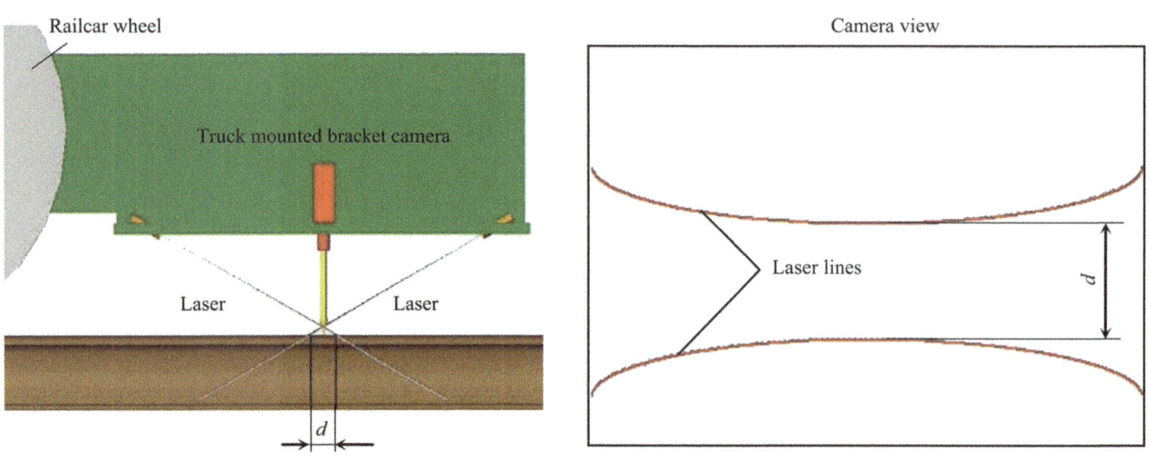

Fig. 9 Measurement approach of UNL-stiffness equipment

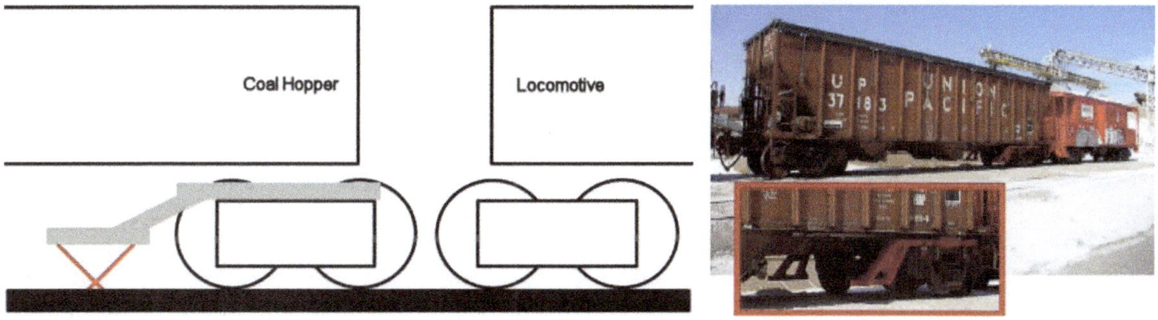

Fig. 10 Mounting position of the measurement system of UNL

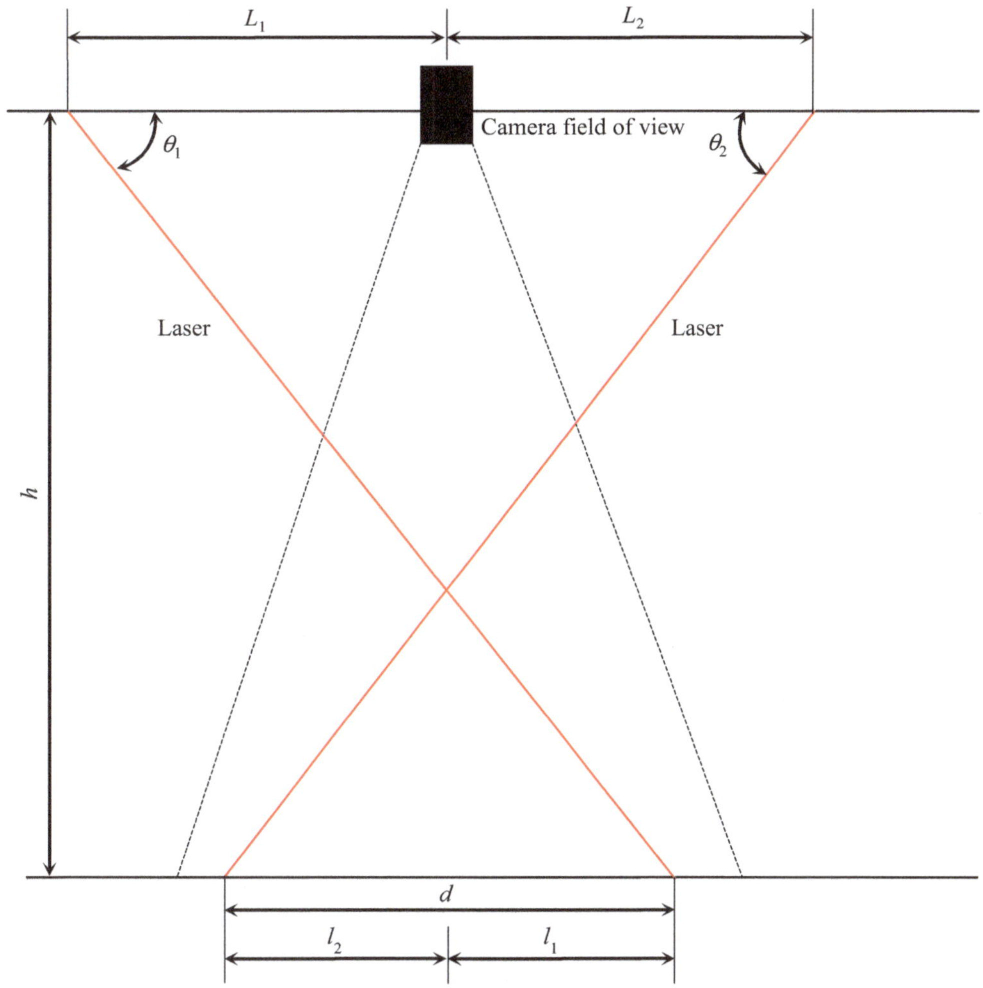

Fig. 11 Sensor geometry of UNL-stiffness equipment

According to this method, during load moving, the deflection basins at load-acting points are similar to each other. The ratio of the vertical track deflection rate at one point of a deflection basin to the load moving rate along the railway line is the slope of track deflection line at this point (see Fig. 13). Namely,

$$w'(x) = \text{Slope} = \frac{V_{\text{deflection}}}{V_{\text{ds}}}, \tag{6}$$

where $w'(x)$ refers to the vertical track deflection rate obtained based on the Winkler foundation beam model.

To measure the track deflection rate, the high-speed deflectograph system adopts laser Doppler sensors attached

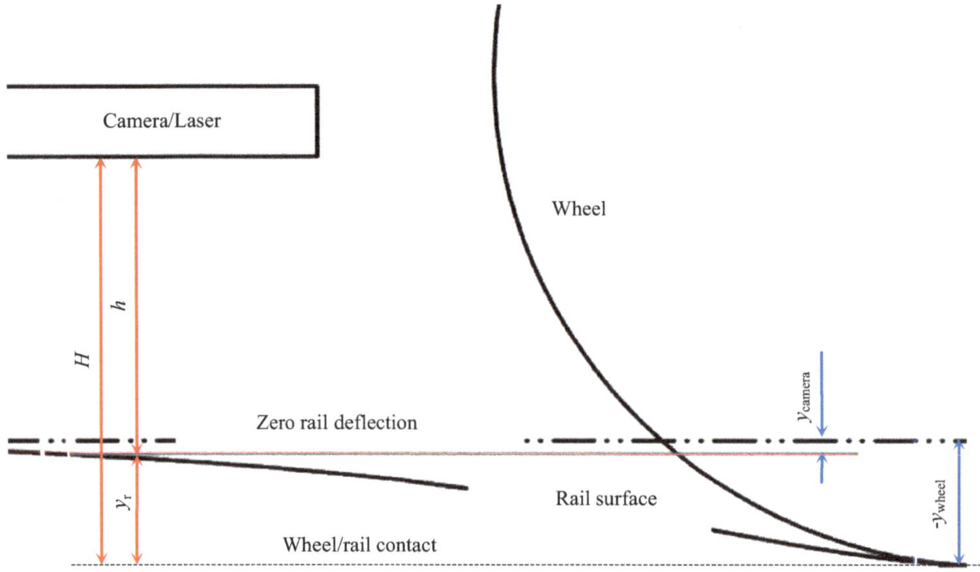

Fig. 12 Rail deflection and sensor measurement of UNL-stiffness equipment

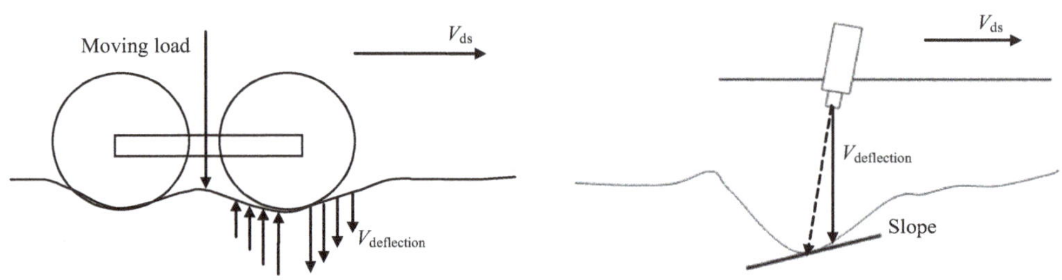

Fig. 13 Principle for movable track stiffness measurement of ZOYON

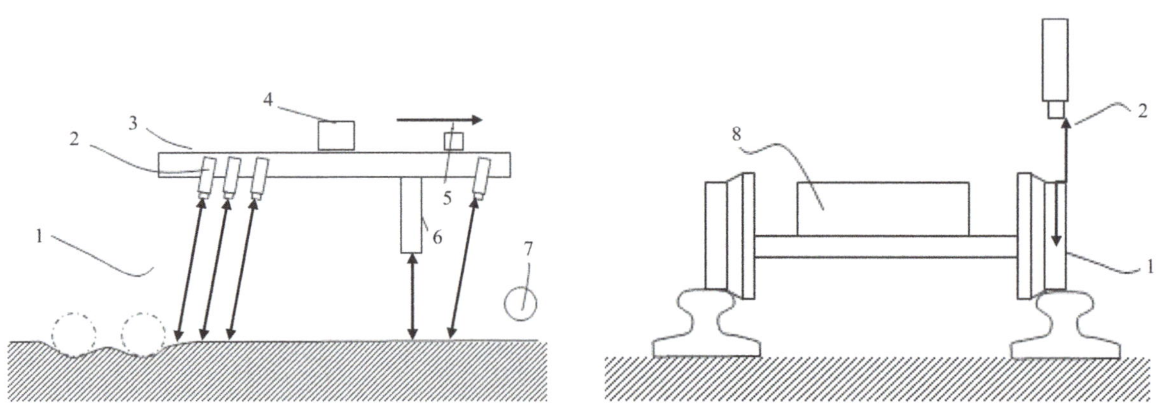

1. Loaded wheel 2. High-speed deflectograph 3. Measurement beam 4. Positioning and orientation unit
5. Temperature transducer 6. Laser distance meter 7. DMI 8. Balance

Fig. 14 Sensor layout and laser projection position of ZOYON

to a moving railway vehicle traveling at speeds of up to 130 km/h. The sensor layout and laser projection positions are shown in Fig. 14.

The European Research for an Optimised BALlasted Track (EUROBALT II) project indicates that track stiffness is another import parameter for optimal long-term maintenance strategies in addition to track geometric irregularity (Meissonnier 2000).

This project promotes Banverket (a railway department of Sweden) to develop a trolley for continuous

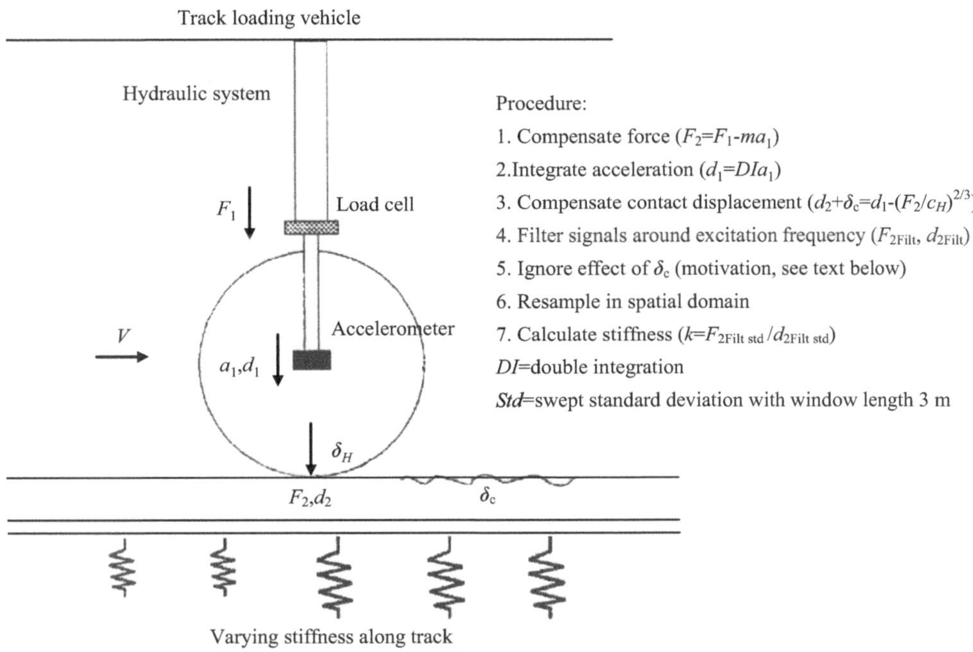

Fig. 15 Principle of Banverket's continuous track stiffness measurement method

Fig. 16 Measurement equipment of RSMV developed by KTH

measurement of vertical track stiffness, the measurement principle of which is shown in Fig. 15 [8, 36, 42–49].

The static load, dynamic load, and maximum speed of this trolley are 60 kN, 20 kN, and 30 km/h, respectively. This trolley can only run on straight sections and not on narrow curves or switches. This trolley can be excited with different frequencies but only one frequency for each run. This trolley has been used for many on-site measurements, which shows good repeatability and reproducibility.

Royal Institute of Technology (KTH) in Sweden built a new vehicle, called rolling stiffness measurement vehicle (RSMV), a rebuilt two-axle freight wagon. The RSMV, much more advanced than the prototype trolley, has one battery plate, one hydraulic system, and two oscillating mass bodies (see Fig. 16). The parameters of RSMV are listed: weight of each body of 4,000 kg, measurable axle weight of 180 kN or higher, max. oscillating amplitude for dynamic load of 60 kN, and measurement frequency of 50 Hz. The measurement speed can reach up to 50 km/h. Figure 17 shows the measurement principle of the RSMV, and its two sides are symmetrical. The measurement principle is similar to that of the trolley.

In order to compare the measurement results of the RSMV with that of the prototype trolley, a comparison test was performed, showing that, for those railway sections with the overall track stiffness less than 150 kN/mm, the results were almost the same, while for those hard sections, the results are quite different.

Sponsored by the Innotrack project (D2.1.9 INNOTRACK 2009), Portancemetre for measuring the overall track stiffness [8, 50, 51] was developed by the Centre d'Experimentation et de Rechrche (CETE–NC, Grand Quevilly, France) and Engineering Department (SNCF, Paris, France). This measurement vehicle comprises two parts, one is the core measurement system of demonstrator and the other is the technical carriage system. The technical carriage system carries the energy supply equipment, hydraulic system, and electronic devices. The core measurement system is installed with Type 417 single-axle wheel pairs. The technical carriage system is installed with Y25C bogie wheel pairs. The main transducers of the Portancemetre include unsprung mass accelerometer, chassis accelerometer, phase sensor (synchronous signal),

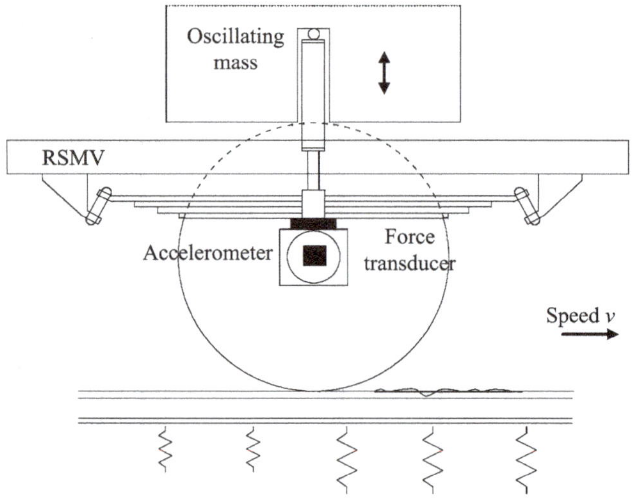

Fig. 17 Measurement principle of RSMV of KTH (single side, symmetrical two sides)

Fig. 18 On-site measurement with Portancemetre in Rouen, France

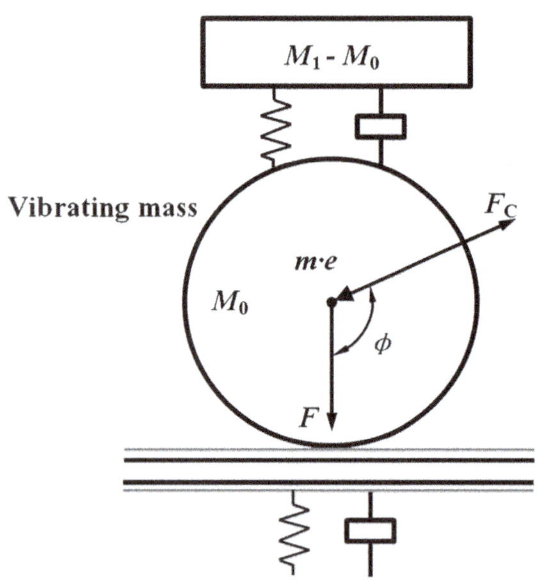

Fig. 19 Schematic of the mechanical vibrating wheel of Portancemetre

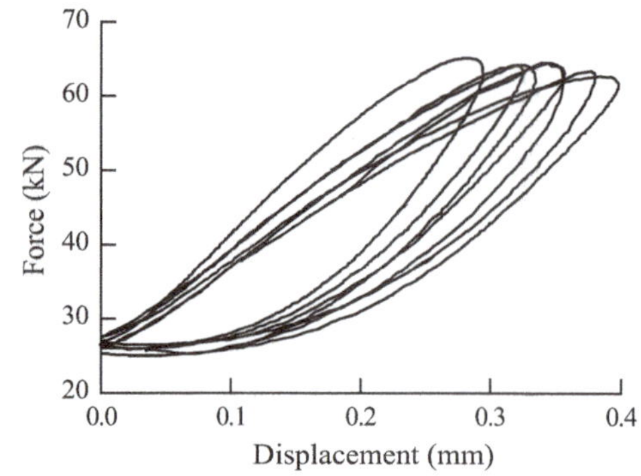

Fig. 20 Vertical force–displacement curve of "Des Jardins" track by Portancemetre

and incremental distance encoder. Figure 18 shows the on-site Portancemetre measurement in Rouen, France.

A camera with a linear CCD sensor is mounted under the technical carriage to record and monitor the measured track surface. All measurements are monitored by the portable computer placed in the locomotive.

Figure 19 shows the measurement principle for Portancemetre.

The force of the measurement system exerted on a rail can be calculated by

$$F = M_1 g = M_0 \Gamma_b + (M_1 - M_0)\Gamma_c + me\omega^2 \cos\phi, \tag{7}$$

where M_1 refers to the total mass, M_0 the unsprung mass, Γ_b the vertical acceleration of the vibrating wheel, Γ_c the vertical acceleration of the suspended mass, $m{\cdot}e$ the eccentric moment of the unbalanced system, ω the angular velocity of rotation, and φ the angle of rotation.

The vertical displacement z can be obtained by calculating the double integration of the vertical acceleration of the unsprung mass:

$$z(t) = \iint \Gamma_b(t)\mathrm{d}t\mathrm{d}t. \tag{8}$$

The overall track stiffness, over an average time period, can be obtained according to wheel–rail force and vertical displacement.

Figure 20 shows the typical force–displacement hysteresis curve. The static load and dynamic load per rail are about 50 and 20 kN, respectively.

4 Analysis and discussion of measurement methods

The standstill measurement is suitable for monitoring track stiffness and long-term stiffness of special sections; however, the development of continuous track stiffness measurement vehicles is still the main way for realizing long-

Table 1 Comparison of track stiffness measurement vehicles

No.	Measurement concept	Measurement accuracy	Measurement speed	Research progress
1	Standstill measurement, static stiffness	High accuracy for vertical measurement; unable to know the stiffness change along the line	Very slow	Mainly applied for heavy haul railway lines
2	Continuous measurement, static stiffness under dynamic load	Improved measurement accuracy due to enhancing displacement measurement	Adjustable, below 60 km/h usually	Mature
3	Continuous measurement, static stiffness under dynamic load, high-speed deflectograph	Unsure	130 km/h, even higher	Successfully applied to the road modulus measurement and the track modulus measurement is in progress
4	Continuous measurement, dynamic stiffness, eccentric wheel excitation, hysteresis curve measurement	High accuracy because of hysteresis curve fitting, stiffness and damping parameter identification	Slow, approximate 6 km/h	Data analysis and vehicle optimization
5	Continuous measurement, dynamic stiffness, eccentric wheel excitation	High accuracy and able to identify many railway line stiffness problems	50 km/h	Multi-line measurements and database construction

distance stiffness measurement of whole railway lines. Table 1 shows the performance parameter comparison of the above track stiffness measurement vehicles.

From Table 1, the existing continuous measurements of track stiffness have four problems.

4.1 Inconsistent measurement concepts

At present, the measurement methods target at different stiffness: static stiffness, dynamic stiffness, or the track component stiffness, which makes it rather difficult to compare the measurement results of these measurement methods.

For instance, the exciting frequency is unnecessary for static stiffness measurement, while for dynamic stiffness measurement, it is requisite. Likewise, when the deformation rate (TU Delft and ZONYON) is to be measured, the vehicle shall run as fast as possible; when the steady dynamic stiffness is to be measured (CETE-NC and SNCF), the vehicle speed cannot be too high. Hence, the measurement of track stiffness, firstly, requires a unanimous stiffness notion that can reflect the track stiffness characteristics completely.

Relatively speaking, the track stiffness notions of the RSMV (KTH, Sweden) and Portancemetre (CETE-NC and SNCF, France) are more advanced. The notions indicate that the track stiffness relates to the exciting frequency, which agrees with the conclusions of many literatures [52–54]. Other notions are based on the Winkler support model, which are not suitable for exploring deep information about track stiffness though they may solve some practical engineering problems.

4.2 Incapability for high-accuracy measurement

All the above-mentioned measurement methods are not suitable for high-accuracy measurement. The accuracy of the laboratory test results are acceptable; however, the accuracy of field test results is not satisfactory. The measurement accuracy mainly depends on the accuracy of force and displacement sensors or acceleration measurements. The force measurement includes strain measurement and pressure intensity measurement of hydraulic system. The strain measurement is more sensitive and accurate. Laser method is more suitable for displacement measurement, which provides high accuracy. Although acceleration transducers have high accuracy, in continuous track stiffness measurement, noise (especially low-frequency noise) can greatly affect the measurement. In addition, error cannot be eliminated when the displacement is calculated through the double integration. Hence, for the acceleration measurement method, appropriate data processing and signal analysis technologies are required to ensure high displacement accuracy. In fact, the continuous track stiffness measurement can be graded according to the desired accuracy (10, 1, 0.1 kN/mm and so on) to save cost and reasonably use the measurement sources.

4.3 Low measurement speed

Measurement speed is always very important for continuous track stiffness measurement. Theoretically, high measurement is likely to produce more accurate results. The track deformation rate measurement method (TU Delft and ZONYON) has the quickest measurement speed. However, it has two defects: the static stiffness notion takes no account of the influence of railway stiffness irregularity and geometric irregularity, and it is hard to deal with the dilemma between the measurement speed, transducer accuracy, and equipment operation reliability. These problems limit the application of this measurement method.

The measurement speeds of the methods designed by the University of Nebraska (USA) and CARS are moderate, which are limited by the accuracy of laser displacement transducers. The RSMV designed by the KTH (Sweden) has rather low measurement speed, which is limited by the reliability of the hydraulic system as well as the accuracy of transducers.

The measurement speed of Portancemetre (CETE-NC and SNCF, France) cannot be too high because the force–displacement hysteresis curve is to be measured.

4.4 Insufficient data analysis

From the perspective of the excitation-response measurement, the system transfer characteristics (the stiffness information) can be obtained regardless of the excitation type as long as there is sufficient response information. Hence, the track stiffness information, to a great extent, depends on the analysis technology of data.

KTH (Sweden) takes the leading position in the analysis technology of data, which has investigated the relation between track stiffness and time frequency/spatial frequency, the phase information of track stiffness, and amplitude information.

In fact, the laser measurement method (University of Nebraska, USA, and CARS, China) can also collect more data because the real-time track vibration can be reflected through the laser displacement data.

Up to now, no quantitative research on the relation between track stiffness and exciting frequency, exciting intensity as well as the running speed has emerged, which is the main reason for inadequate analysis of the measurement data of continuous track stiffness.

5 Conclusions and suggestions

This paper summarizes the significance of track stiffness measurement, takes a wide view of track stiffness measurement methods around the world, including the standstill measurement and the continuous track stiffness measurement, and performs a comparison analysis of all the measurement methods.

(1) Track stiffness is an important factor influencing the safety and reliability of train operation, the vibration and deformation of track structures, and the dynamic response of substructures like subgrades and bridges. The track stiffness measurement is of great theoretical and practical significance to the design of new railway lines and especially the railway maintenance works. The measurement of track stiffness for maintenance relates to four stiffness problems: low stiffness, variable stiffness, virtual stiffness, and assortative stiffness problems.

(2) The standstill measurement of overall track stiffness is time consuming and hard to realize a long-distance and multi-point measurement for whole railway lines. The existing methods for continuous track stiffness measurement have four problems: a. inconsistent measurement concepts. A unified stiffness notion has not been proposed to reflect the track stiffness characteristics completely; b. poor ability for high-accuracy measurement. The measurement accuracy for force, displacement, or acceleration is hard to ensure, especially for high-speed measurement; c. low measurement speed. It is difficult to solve the dilemma between the measurement speed, transducer accuracy, and equipment operation reliability; and d. insufficient data analysis. Much more stiffness information is expected to explore using the advanced analysis technology of the measurement data.

The following suggestions are proposed considering the imperfection of existing stiffness measurements:

(1) A definite stiffness notion should be established to completely reflect the track stiffness characteristics.

(2) A theoretical and systematic research should be performed on the quantitative relation between track stiffness and the factors like exciting frequency, running speed, and vehicle axle load, as well as the influences of track stiffness and track geometric irregularities on the vibration response of the vehicle-track-substructure coupled dynamics, respectively.

(3) In view of the accuracy requirement for stiffness measurement, the track stiffness measurement can be categorized based on three grades, namely low accuracy (10 kN/mm, substructure disease identification, for heavy haul railways mainly), medium accuracy (1 kN/mm, potential safety hazard checking), and high accuracy (0.1 kN/mm, evaluation of track regularity, for high-speed railways mainly). For example, for continuous measurement vehicle with low measurement accuracy, single-axle vehicle can be used. The axle load should be more than 150 kN to ensure the vertical measurement for subgrade, and one span length (sleeper pitch) along the railway line can be the measurement unit. For the continuous measurement with medium measurement accuracy, it is advised to use double-axle vehicle with axle loads

of 20 and 100 kN, the vertical measurement range is limited to the track structure, and the measurement unit ranges from 1 to 4 span lengths. For continuous measurement vehicle with high measurement accuracy, it is advised to use double-axle vehicle with axle loads of 10 and 50 kN, the vertical measurement range is limited to the rail and fasteners, and 4–10 span lengths can be adopted as the measurement unit.

For transition zones (railway–bridge, railway–tunnel, and bridge–tunnel areas), turnouts, and small-radius curve sections, the track stiffness characteristics are more complicated than those of straight lines [2, 49]. Thus, apart from the measurements, the stiffness of special track sections can be monitored.

In recent years, the features of wave propagation in track structures have attracted a wide attention, and the wave theory is applied to the high-frequency vibration of rails [55]. Rail support conditions affect the transfer laws of track vibration waves, and thus continuous track stiffness measurement equipment can be developed based on the wave transfer mechanism.

In addition, the vibrations of track components and vehicle components can be collected for power generation [56–58]. Different track stiffness values cause different vibration intensities, affecting the collection of electric energy. When the stiffness measurement accuracy is undesirable because of poor noise reduction in vibration displacement and vibration acceleration signals, the vibration energy storage ability or real-time electrical signals (voltage, power, etc.) might be a good index for continuous track stiffness measurement or long-term stiffness monitoring of special sections.

Acknowledgments This research was supported by the project (51425804) of the National Science Fund for Distinguished Young Scholars of China and the National Natural Science Foundation of China (NSFC) under grants U1234201, U1334203, and 51378439.

References

1. Kerr AD (2000) On the determination of the rail support modulus k. Int J Solids Struct 37:4335–4351
2. Esveld C (2001) Modern railway track (second edition). MRT-Productions, Delft, pp 71–90
3. Zhang GM (2002) Research on right level of track structure stiffness and track-part stiffness. China Railw Sci 23(1):51–57
4. Liu L, Wang W, Liu W (2000) Experimental study of the static stiffness of ballast. J Shanghai Tiedao Univ 21(4):1–6
5. Liu W, Wang W (2002) Study on dynamic stiffness and damp measurement of railway crushed stone ballast. J China Railw Soc 24(6):99–104
6. Ooi LE, Ripin ZM (2011) Dynamic stiffness and loss factor measurement of engine rubber mount by impact test. Mater Des 32:1880–1887
7. Sussmann TR, Ebersöhn W, Selig ET (2001) Fundamental nonlinear track load-deflection behavior for condition evaluation. Transp Res Rec 1742:01–2916
8. Berggren E (2009) Railway track stiffness—Dynamic measurements and evaluation for efficient maintenance. Ph.D. Thesis, Royal Institute of Technology (KTH), Stockholm
9. Wu TX, Thompson DJ (1999) The effects of local preload on the foundation stiffness and vertical vibration of railway track. J Sound Vib 215(5):881–904
10. Chen XP, Wang P (2006) Distribution regularity and homogenization of track rigidity for ballastless turnout. J Southwest Jiaotong Univ 41(4):447–451
11. Chen XP (2008) Study on theory and application of track stiffness in high-speed turnouts. Ph.D. Thesis, Southwest Jiaotong University, Chengdu
12. Liu XY (2004) Effect analysis of track stiffness on dynamic characteristics of wheel–rail system and its dynamic optimization. J Southwest Jiaotong Univ 39(1):1–5
13. Lopez Pita A, Teixeira PF, Robusté F (2004) High speed and track deterioration: the role of vertical stiffness of the track. J Rail Rapid Transit 218:31–40
14. Cai CB, Xu P (2010) Dynamic analysis of key design parameters for ballastless track of high-speed railway. J Southwest Jiaotong Univ 45(4):493–497
15. Li D, Selig ET (1995) Evaluation of railway subgrade problems, Transportation Research Record 1489, TRB. National Research Council, Washington DC
16. Selig ET, Li D (1994) Creating track modulus: its meaning and factors influencing it. In: Transportation Research Record: Journal of the Transportation Research Board, No. 1470, Transportation Research Board of the National Academies, Washington, DC, pp 47–54
17. Frohling RD, Scheffel H, Ebersohn W (1996) The vertical dynamic response of a rail vehicle caused by track stiffness variations along the track. Veh Syst Dyn 25:17–31
18. Zhai JQ (2007) Test and analysis of vibration propagation caused by the subway train. Dissertation, Tongji University
19. Wu JZ (2012) The impact of vibration on the building structure caused by metro and isolation research. Dissertation, South China University of Technology
20. Wang P, Xu JH, Wang L (2014) Effect of track stiffness on frequency response of vehicle-track coupling system. J Railw Eng Soc 9:46–52
21. Robinet A, Hosseingholian M, Quibel A, Froumentin M (2008) Track stiffness assessment. In: 8th World Congress on Railway Research, Seoul, Korea, pp 1–8
22. Zhai WM, Cai CB, Kai KY (2000) Effect of track stiffness on train running behavior. J China Railw Soc 22(6):80–83
23. Ebersöhn W (1995) Substructure influence on track maintenance requirements. Ph.D. Dissertation, Department of Civil and Environmental Engineering, University of Massachusetts at Amherst
24. Konstantinos G, Spyridon T (2012) Transition zone between ballastless and ballasted track: influence of changing stiffness on acting forces. Proc Soc Behav Sci 48:3548–3557
25. Kaewunruen S, Remennikov A (2005) Application of experimental modal testing for estimating dynamic properties of

structural components. In: Proceedings of Australian structural engineering conference 2005, Newcastle
26. Kaewunruen S, Remennikov A (2005) Monitoring structural degradation of rail bearing pads in laboratory using impact excitation technique. In: Proceedings of the 1st international conference on structural condition assessment, monitoring, and improvement, Perth, pp 389–394
27. Burrow MPN, Chan AHC, Shein A (2007) Falling weight deflectometer based inverse analysis of ballasted railway tracks. In: Geotechnical engineering, proceedings of the Institution of Civil Engineers, Issue GE3, pp 169–177
28. Kerr AD (1983) A method for determining the track modulus using a locomotive or car on multi-axle trucks. In: Proceedings AREA, vol 84
29. Kerr AD (1987) On the vertical modulus in the standard railway track analysis. Rail International, November Issue
30. Kerr AD On the reduced area method for calculating the vertical track modulus. In: Proceedings AREA, vol 86
31. Esveld C (1980) Track stiffness measurements using an adapted tamping machine. Rail International
32. D2.1.9 INNOTRACK (2009) Adapted "Portancemetre" for track structure stiffness measurement on existing tracks, Paris
33. Wangqing W, Geming Z, Kaiming Z, Lin L (1997) Development of inspection car for measuring railway track elasticity. In: Proceedings from 6th international heavy haul conference, Cape Town
34. Li D, Thompson R, Kalay S (2002) Development of continuous lateral and vertical track stiffness measurement techniques. In: Proceedings from railway engineering, London
35. Thompson R, Marquez D, Li D (2001) Track strength testing using TTCI's track loading vehicle. Railw Track Struct 97:15–17
36. McVey B, Norman C, Wood N, Farritor S, Arnold R, Fateh M, El-Sibaie M (2005) Track modulus measurement from a moving railcar. In: Proceedings of the AREMA annual conference, Chicago, September 25–27, 2005
37. Norman C, Farritor S, Arnold R, Elias SEG, Fateh M, Sibaie ME (2004) Design of a system to measure track modulus from a moving railcar. In: Proceedings from railway engineering, London
38. Information from http://old.zoyon.com.cn
39. Rasmussen S, de Man AP (2000) The high speed deflectograph and the hammer excitation test: two measurement techniques for determining track performance. Rail Eng Int 2000(1):13–16
40. Rasmussen S, Man AP (1999) Measurement techniques for track performance. In: Proceedings of conference on innovations in the design and assessment of railway track, Technical University of Delft, Delft
41. Rasmussen S, Krarup JA, Hildebrand G, Non-contact deflection measurement at high speed. In: 6th International conference on the bearing capacity of roads, railways and airfields, 24–26 June, 2002, Lisbon
42. Berggren E, Jahlénius Å, Bengtsson B-E (2002) Continuous track stiffness measurement—an effective method to investigate the structural conditions of the track. In: Proceedings of the conference railway engineering—2002, London, 3–4 July, 2002. ISBN: 0-947644-49-0
43. Berggren E, Jahlénius Å, Bengtsson B-E, Berg M (2005) Simulation, development and field testing of a track stiffness measurement vehicle. In: Proceedings of 8th international heavy haul conference, Rio de Janeiro, 13–16 June, 2005. ISBN: 0-646-33463-8
44. Berggren E, Kaynia A, Dehlbom B (2010) Identification of substructure properties of railway tracks by dynamic stiffness measurements and simulations. J Sound Vib 329:3999–4016
45. Berggren E (2006) Measurements of track stiffness and track irregularities to detect short waved support conditions. In: Proceedings of international conference on railway track foundations, Birmingham, 11–13 September, 2006. ISBN: 0-704426-00-5
46. Berggren E, Li M, Spännar J (2006) A new approach to the analysis and presentation of vertical track geometry quality and rail roughness. In: Proceedings of 7th international conference on contact mechanics and wear of rail/wheel systems (CM2006), Brisbane, 24–27 September, 2006. Also in journal of WEAR, vol 265, pp 1488–1496
47. Berggren E (2009) Efficient track maintenance—methodology for combined analysis of condition data, submitted to the conference IHHA 2009. Candidate paper to Special Issue of J Rail Rapid Transit
48. Li M, Berggren E (2009) A study of the effect of global track stiffness and its variations on track performance: simulation and measurement, submitted to the conference IHHA 2009. Candidate paper to Special Issue of J Rail Rapid Transit
49. Smekal A, Berggren E, Silvast M (2006) Monitoring and substructure condition assessment of existing railway lines for upgrading to higher axle loads and speeds. In: Proceedings of 7th World Congress on Railway Research, Montreal, 5–7 June, 2006
50. Alexander R, Sakdirat K Laboratory measurements of dynamic properties of rail pads under incremental preload. In: 19th Australasian conference on the mechanics of structures and materials, Christchurch, pp 319–324
51. Hosseingholian M, Froumentin M, Robinet A (2006) Feasibility of a continuous method to measure track stiffness. In: Proceedings from railway foundations conference, Birmingham
52. Fenander A (1997) Frequency dependent stiffness and damping of railpads. Proc Inst Mech Eng Part F 211:51–62
53. Fenander A (1998) A fractional derivative railpad model included in a railway track model. J Sound Vib 212(5):889–903
54. Frohling RD (1998) Low frequency dynamic vehicle/track interaction: modeling and simulation. Veh Syst Dyn 28:30–46
55. Ryue J, Thompson DJ, White PR, Thompson DR (2008) Investigation of propagating wave types in railway tracks at high frequencies. J Sound Vib 315:157–175
56. Pourghodrat A, Nelson CA, Hansen SE, Kamarajugadda V, Platt SP (2014) Power harvesting systems design for railroad safety. Proc IMechE Part F 228:504–521
57. Wang JJ, Penamalli GP, Zuo L (2012) Electromagnetic energy harvesting from train induced railway track vibrations. In: Proceedings of IEEE/ASME international conference on mechatronics and embedded systems and applications (MESA), pp 29–34
58. Gao MY (2015) Design and verification of a rail-borne electromagnetic transducer for railway energy harvesting. Research report, Key Laboratory of High-speed Railway Engineering, Ministry of Education, Southwest Jiaotong University, Chengdu

14

Micro-simulation model of two-lane freeway vehicles for obtaining traffic flow characteristics including safety condition

Yang Yue[1] · **Sida Luo**[2] · **Tianming Luo**[2]

Abstract Unidirectional two-lane freeway is a typical and the simplest form of freeway. The traffic flow characteristics including safety condition on two-lane freeway is of great significance in planning, design, and management of a freeway. Many previous traffic flow models are able to figure out flow characteristics such as speed, density, delay, and so forth. These models, however, have great difficulty in reflecting safety condition of vehicles. Besides, for the cellular automation, one of the most widely used microscopic traffic simulation models, its discreteness in both time and space can possibly cause inaccuracy or big errors in simulation results. In this paper, a micro-simulation model of two-lane freeway vehicles is proposed to evaluate characteristics of traffic flow, including safety condition. The model is also discrete in time but continuous in space, and it divides drivers into several groups on the basis of their preferences for overtaking, which makes the simulation more aligned with real situations. Partial test is conducted in this study and results of delay, speed, volume, and density indicate the preliminary validity of our model, based on which the proposed safety coefficient evaluates safety condition under different flow levels. It is found that the results of this evaluation coincide with daily experience of drivers, providing ground for effectiveness of the safety coefficient.

Keywords Micro-simulation · Two-lane freeway · Safety coefficient · Characteristic distance · Rules of state update

1 Introduction

Traffic flow models were developed to simulate and understand traffic operations. The models are mathematically theory-based or simulation-based [1]. Within these two categories, microscopic traffic flow models are extremely popular. It is believed that one of the earliest microscopic models is the car-following model proposed by Reuschel and Pipes [2]. This model regards vehicles as discrete particles and uses differential equation to capture the rule of motions of each vehicle under the situation that no overtaking behavior happens [4]. So the car-following model is basically a mathematical theory-based model. The original model was revised in the 90s by Bando [5] based on the dynamical non-linear effects proposed by Newell [6], after which a series of modifications were studied to optimize the model. More revised car-following models like generalized force model [7] were put forward and some revisions were made to take comfort factors during driving into account [8]. One revised psychological-and-physiological car-following model is now adopted by Vissim [9], a well-known microscopic traffic simulation software.

After the proposal of the car-following model, a cellular automation (CA) model was put forward by Cremer and Ludwig [10]. The model is discrete in both time and space, which is an effective simplification to approximate to the

✉ Sida Luo
whezlsd@126.com

Yang Yue
2473895863@qq.com

Tianming Luo
luotmseu@163.com

[1] Department of Road and Bridge Engineering, School of Transportation, Southeast University, Nanjing 210096, Jiangsu, China

[2] Department of Traffic Engineering, School of Transportation, Southeast University, Nanjing 210096, Jiangsu, China

solution of differential equations. Compared with many car-following models, CA model can well simulate overtaking behaviors in highway with unidirectional two lanes [11]. In general, CA model belongs to simulation-based models for its clear evolution rules of vehicle operations. It is easy to implement on computer for its discreteness. With the rapid development of modern computers, this model becomes increasingly popular in the 90s [12]. Its popularity gains partly due to its flexible framework to incorporate traffic control and many other interesting components. For example, Horni et al. (2013) used CA model to conduct parking simulation, and they incorporated agent-based techniques as well [13]. Chai et al. (2015) made use of CA model to simulate traffic streams at signalized intersections. The thing incorporated is fuzzy logic, and results show the model can well replicate decision-making processes [14]. CA model can even be used to evaluate vehicle load effect in bridge construction [15].

Mathematically, theory-based models are appealing since they can promote the understanding of the mechanism how traffic streams evolve over time and space. However, these models sometimes are hardly accessible, which makes it difficult to put into practice. As for simulation-based ones, CA model is widely used for microscopic simulation of vehicles on freeway though defects exist as well. Efficient as CA model is, its feature of temporal and spatial discreteness is not in coincidence with real situations, which probably leads to inaccuracy in simulation results. In addition, it is the traffic flow characteristics like speed, density, and delay that most traffic flow models lay emphasis on, including the CA model. But these models are unable to evaluate vehicle safety condition that ought to be seriously considered. Actually, safety is the most important factor concerning traffic operations on freeway where high speed may cause serious accidents [16].

Based on the analysis above, a micro-simulation traffic flow model is put forward in this paper. The model is aimed at two-lane freeway that is quite common. It is the simplest form of freeway, which acts as the foundation of potential models targeted at the freeway with more than three unidirectional lanes. The proposed model is discrete in time, but continuous in space. This is more close to real situations than that of CA model. Besides, the model takes into consideration different types of drivers in terms of their overtaking behaviors. Compared with incorporating human factors into classic car-following models [17], it is much easier for the proposed model to achieve, which is understandable and has no trouble in solving more complicated equations. Finally, the safety coefficient in our model can conveniently make an evaluation of safety condition.

The remainder of this paper is organized as follows. In Sect. 2, all aspects of the proposed model are overviewed. Section 3 provides a preliminary test of the model, and Sect. 4 shows how the model evaluates safety condition on freeway. Some discussions are presented in Sect. 5 which is followed by a conclusion section.

2 Model overview

2.1 Basic assumptions

(1) The freeway section is homogeneous.
(2) Vehicles travel on the right lane unless they are steered to pass the front car. The left lane gets used when vehicles are overtaking others, in which case they move one lane to the left, pass, and return to their former travel lane. This is what is called the keep-right rule, which is quite common in the United States.
(3) The arrival of vehicles conforms to Poisson distribution and we use the parameter of Poisson distribution to represent different traffic volume conditions.
(4) All vehicles are able to be classified into three types: small car, medium car, and large car. Each kind of vehicle has its own speed limits and vehicle length.
(5) Vehicles can be abstracted to a straight line from its tail end to head end. The crashes only happen at these two end points, or only rear-end crashes are considered.
(6) Traffic safety on the two-way freeway is assumed to only have something to do with lane-changing behavior while the effects of speed are ignored.

2.2 Explanation of terminology

Key parameters to the model are summarized as follows:

V_0: the speed of the vehicle backward in the left lane.

V_1: the speed of the vehicle backward in the right lane.

V_2: the speed of the vehicle ahead in the left lane.

V_3: the speed of the vehicle ahead in the right lane.

D_1: the distance of the two adjacent vehicles ahead in the left lane.

d_2: the distance of the two adjacent vehicles ahead in the right lane.

δ: the discount factor of safety coefficient.

$C(t)$: the safety coefficient at time t.

2.3 The micro-simulation model

Vehicle operation on two-lane freeway is determined by the driver behavior such as their driving skills, preferences, and so on. Besides, the type of vehicle, distance between two vehicles, speed limit, etc. also make a difference to traffic operations. Owing to these complicated factors, simplifications are needed to establish the micro-simulation model while it is indispensable to capture driver behaviors.

The simulation is set on a freeway section. Vehicles enter from one end of the section and exits from the other end. This model is discrete in time. In order to obtain accurate results from the simulation, the time interval cannot be too long. We set the time interval as 1 s. However, unlike cellular automata, our model is not spatially discrete. According to the basic assumptions, the arrival of vehicles conforms to Poisson distribution, so it is not difficult to figure out the probability of vehicle arrival in 1 s at the end where vehicles enter. We use a random number ranging from 0 to 1. When its value is bigger than this probability, there is a vehicle entering the section. The type of vehicle is determined by the percentage of various vehicle types. Thus, the actual situation of vehicle arrival is well simulated. The operations of vehicles will be updated every 1 s, including the instantaneous velocity of vehicles, the travel distance from the start, and the lane each vehicle occupies. Real-time traffic operations on the freeway section are reproduced in this way. Finally, a vehicle gets removed when its travel distance exceeds the length of the virtual section. The simulation ends when its operating time meets the requirement set by researchers.

The model has some application conditions. First, the freeway has just unidirectional two lanes. Second, the freeway section for simulation is supposed not to contain weaving areas or ramps. The proposed model can basically conduct simulation on any basic segment of two-way freeways. Anyway, basic segments constitute the most part of a freeway.

2.3.1 Explanation of parameters

a. The affected distance
The distance is defined according to which a driver decides whether to follow or overtake the front vehicle. The farthest distance that the driver can see under a certain speed is suggested to determine the affected distance in this paper.
b. The safety distance
The safety distance ensures there will approximately not be a crash even if drivers take extreme dangerous strategies to pass a car.
c. The feasible-passing distance
The feasible-passing distance determines whether or not a driver will brake when overtaking the front vehicle. If the actual distance between them is shorter than the feasible-passing distance, the driver should brake first and then speed up.
d. The critical distance
The critical distance is the minimum distance to pass a vehicle which ensures there will not be a crash if a driver takes safe strategies to make overtaking. It can be understood in another way that if the actual distance is shorter than the critical distance, no driver will choose to overtake.
e. The extreme distance
The extreme distance is the shortest distance that every two adjacent cars have to keep under any circumstance. The vehicle length is recommended to well characterize it.

Note that the distances defined above are all called "characteristic distance" in this study. They are some of the most important parameters to vehicle state update in the simulation.

f. The safety coefficient
We define the safety coefficient that ranges from 0 to 1. When the coefficient is close to 1, it means the vehicle operation is safe. The initial value of the coefficient is 1, and it will get lower and lower if vehicles meet safety problems. Mathematically, $C(0) = 1$, and the formula for updating the coefficient is

$$C(t+1) = C(t) \cdot \delta. \tag{1}$$

g. The discount factor of safety coefficient
δ is defined to reflect the degree of safety during lane-changing process based on assumption (5). Considering the safety is closely related to the distance between two adjacent vehicles while drivers change a lane [18], the method to figure out δ is determined as given below:

(1) Lane changing aimed at overtaking, then

$$\delta = k_1 \cdot k_2 \cdot k_3, \tag{2}$$

$k_1 = 0$ when the distance between two adjacent cars lies between the critical distance and the feasible-passing distance. Although drivers are able to overtake, it is very dangerous under this circumstance.
$k_1 = 0.5$ when the distance lies between the feasible-passing distance and safety distance. In this case, it is not safe enough to pass a car but there is little ground to blame drivers for this.
$k_1 = 1$ when the distance is longer than the safety distance, which means it is absolutely safe for drivers to make overtaking.
Similarly but slightly differently, k_2 has something to do with the front vehicle in the passing lane. Then the rule for determining k_2 is the same with that of k_1. In addition, the overtaking behavior will be affected by the vehicle behind in the passing lane. In this case, it should be more conservative of drivers who want to overtake to make judgments

because errors are bigger when figuring out the condition of vehicle operation behind. Meanwhile, the speed of the vehicle behind in the passing lane is probably not as low as that of the car which is to be overtaken. In summary, the passing behavior had better been finished within the feasible-passing distance. When the distance between two adjacent vehicles is shorter than the feasible-passing distance, $k_3 = 0$; when the distance lies between the feasible-passing distance and safety distance, $k_3 = 0.5$; otherwise, $k_3 = 1$. Accordingly, δ can be calculated in the end.

(2) Lane changing without overtaking
It is much safer when a vehicle changes back from the left lane to the right lane, which is only affected by the car in the front because it is much faster than the car behind. Similarly, when the distance lies between the critical distance and safety distance, $\delta = 0.5$, which means it is rather unsafe to change a lane; when the distance is longer than the safety distance, $\delta = 1$.

h. The average safety coefficient
The safety coefficient is used to evaluate the safety condition of a single vehicle. To make safety evaluation aimed at the whole traffic flow entering the section in simulation, the average safety coefficient is defined to be the arithmetic mean of the safety coefficients of all vehicles.

2.3.2 The rule of lane changing

a. Lane changing for overtaking
The lane-changing conditions are more complex when the aim is to overtake than that of non-overtaking lane changing. First, a driver needs to judge the distance and speed concerning the vehicle in front of him. Only if the distance is shorter than the affected distance and the speed is not lower than that of the front car, the driver will choose to overtake or follow the car. Once the driver chooses to make overtaking, he has to determine whether the conditions of passing are satisfied. The passing behavior is permitted only if the distance between his and the vehicle in front of him in both right and left lane is longer than the critical distance, and the distance between his and the car backward in the passing lane is longer than the feasible-passing distance. He, otherwise, can only choose to follow the car ahead, which suits the case in real life well.

b. Lane changing without overtaking
The normal situation is that vehicles drive on the right lane under the keep-right rule. As a result, the conditions for lane changing back to the right lane is given below: There is no vehicle ahead in the right lane, or the distance away from the vehicle ahead exceeds the affected distance, or this distance is within the affected distance but beyond the extreme distance while the speed of the vehicle ahead is higher than the instantaneous speed when the overtaking is just finished. A detailed explanation of the lane-changing rules will be seen in the flow chart below. Note that in the simulation process, the speed when a vehicle changes to the passing lane is a mathematical expectation. The expected value of speed is random that lies between the minimum and maximum speed limits of the passing lane.

2.3.3 The rule of speed update

In the driving process, it is impossible for any driver to maintain a certain speed, especially when following a vehicle in the front. Speed fluctuation is also influenced by the driving behavior of the front car. In order to characterize this speed fluctuation in the simulation, we formulate the following four rules of speed update: the rule of traveling freely, the rule of car-following with reference, the rule of normal car-following, and the rule of close car-following.

a. The rule of traveling freely
The rule is that there is no vehicle beyond the affected distance in front of a driver. In this case, the speed fluctuation only relates to the driver himself. According to the psychological characteristics of drivers, the variation of speed can be determined as positive and negative 3 km/h.

b. The rule of car-following with reference
The rule is that the distance away from the front vehicle is between the critical distance and the affected distance, and the speed is lower than that of the vehicle ahead. In this case, fluctuation characteristics of speed have something to do with the driver and the front vehicle. Less distance away from the front vehicle causes less speed variation. In addition, the driver will slightly accelerate with a tendency to reach the same speed as that of the vehicle ahead according to "reference dependence," a theory of behavior psychology. Consequently, the variation of speed is less than that of the rule of traveling freely, and the positive variation is dominant compared with negative variation.

c. The rule of normal car-following
The rule is that the distance away from the front vehicle is between the critical distance and the affected distance, and the speed is higher than that of the vehicle ahead. In this case, fluctuation characteristics of speed are related to the driver and the front vehicle as well. Therefore, the variation of speed is less than

that of the rule of traveling freely, but unlike the rule of car-following with reference, the positive and negative variations are of the same.

d. The rule of close car-following
The rule is that the distance away from the front vehicle is shorter than the critical distance and the speed is not lower than that of the vehicle ahead. In this case, the driver and the front vehicle also both have effects on the speed fluctuation. Moreover, the driver becomes the most cautious for the distance between the two vehicles is sufficiently small. Therefore the variation of speed is less than that of the rule of normal car-following.

Considering the precision of speed fluctuation does not make a great effect on simulating vehicle operations, the variation of speed can be produced using random number generation method while it coincides with those rules of speed update above.

2.3.4 The rule of distance update

Because a vehicle's trajectory is in two-dimensional plane, it is complex to determine the running distance of a vehicle directly. The traffic operations are approximate to be in one dimension. Therefore, the time interval is multiplied by the average speed at time t and $t + 1$ so as to obtain the running distance of a vehicle between the time t and $t + 1$.

The detailed framework of this simulation model is shown in the flow chart (Figs. 1 and 2). Through the simulation, traffic flow characteristics including speed, density, delay, and safety condition of different types of vehicles are obtained.

3 Partial test

The model is partially tested by comparing the simulation results with some qualitative analysis. On the basis of the analysis above, the extreme distance can be set as the maximum length of different types of small cars. Hence, the extreme distance is set to be 6 m in this test according to the official definition of small cars. Suppose that the critical distance, the feasible-passing distance, the safety distance, and the affected distance are 1.5, 2.0, 2.5, and 3.0 times as long as the extreme distance. The values of these distances are set solely to make a partial test to conduct a preliminary verification on the validity of our model. The annual average daily traffic volume of a two-way and four-lane freeway ranges from 25,000 to 55,000 veh/d [19], or 521 to 1,146 veh/h. Therefore, the traffic volume in this test is set to 1,050 veh/h, which is randomly picked out from the range [521, 1,146]. In this way, the saturation of traffic flow is not low without congestion occurring. We make the speed limit 80–100 km/h for small cars, and 60–80 km/h for medium and large cars for the right lane. With regard to the left lane, the speed limit is 100–120 km/h for small cars, and 80–100 km/h for medium and large cars. Meanwhile, the ratio of probability of the arrival for large cars, medium cars and small cars is 1:2:6.

After the simulation, the results of the delay are shown in Figs. 3, 4, and 5. From the three figures, it is found that the delay of small cars is the largest while that of large cars is the smallest. The reason is probably that the speed of small cars is affected by the medium and large cars ahead of them to a great extent. Chances are that it causes the difficulty in keeping normal speed for small cars based on the previous analysis that the saturation of traffic flow is not low. The number of points on these three graphs is different, which the ratio of arrival probability of all types of vehicles can account for. The number of points of small and large cars are the biggest and smallest, respectively.

The results of the speed can be obtained at the same time. Some statistics are calculated and listed in Table 1. It is obvious the average speed of the three types of cars is all within the range of speed limit, and the mean of small cars is larger than that of medium cars as with the case of medium cars and large cars. Furthermore, the variance of the speed for small and medium cars is larger than that of the large cars, meaning that the speed of large cars has the least fluctuation. It is known drivers of large cars have little tendency to make overtaking in general. Therefore, these results all have coincidence with common sense.

Last but not least, the validity of the proposed model can be tested using traffic fundamental diagram that captures the relationship between the density and flow (volume) of vehicles. Plenty of points can be determined by changing the traffic volume parameter in our model. Given the basic capacity of one vehicle lane that is about 2,000 veh/h according to the Ref. [20], the capacity of two-lane freeway is around 4,000 veh/h unidirectionally. Accordingly, the input of the mean demands (unit: veh/h) for each simulation are determined as is shown in the set S:

$$S = \{q | q = 50x + 2{,}050,\ x \in \{0, 1, 2, \ldots, 36\}\}. \quad (3)$$

In these 37 simulations, each time the traffic density can be figured out. Then the scatters are plotted in flow-density graph shown in Fig. 6. It can be concluded that these scatters generally accord with a triangular fundamental diagram. The critical density is about 44 veh/km and the maximum flow that just indicates the actual capacity of the freeway section is about 3,900 veh/h. Because 3,900 veh/h is smaller than 4,000 veh/h, this actual capacity seems fairly reasonable though hard to be proved. In general, the approximate reproduction of the fundamental diagram further verifies the validity of the proposed simulation model.

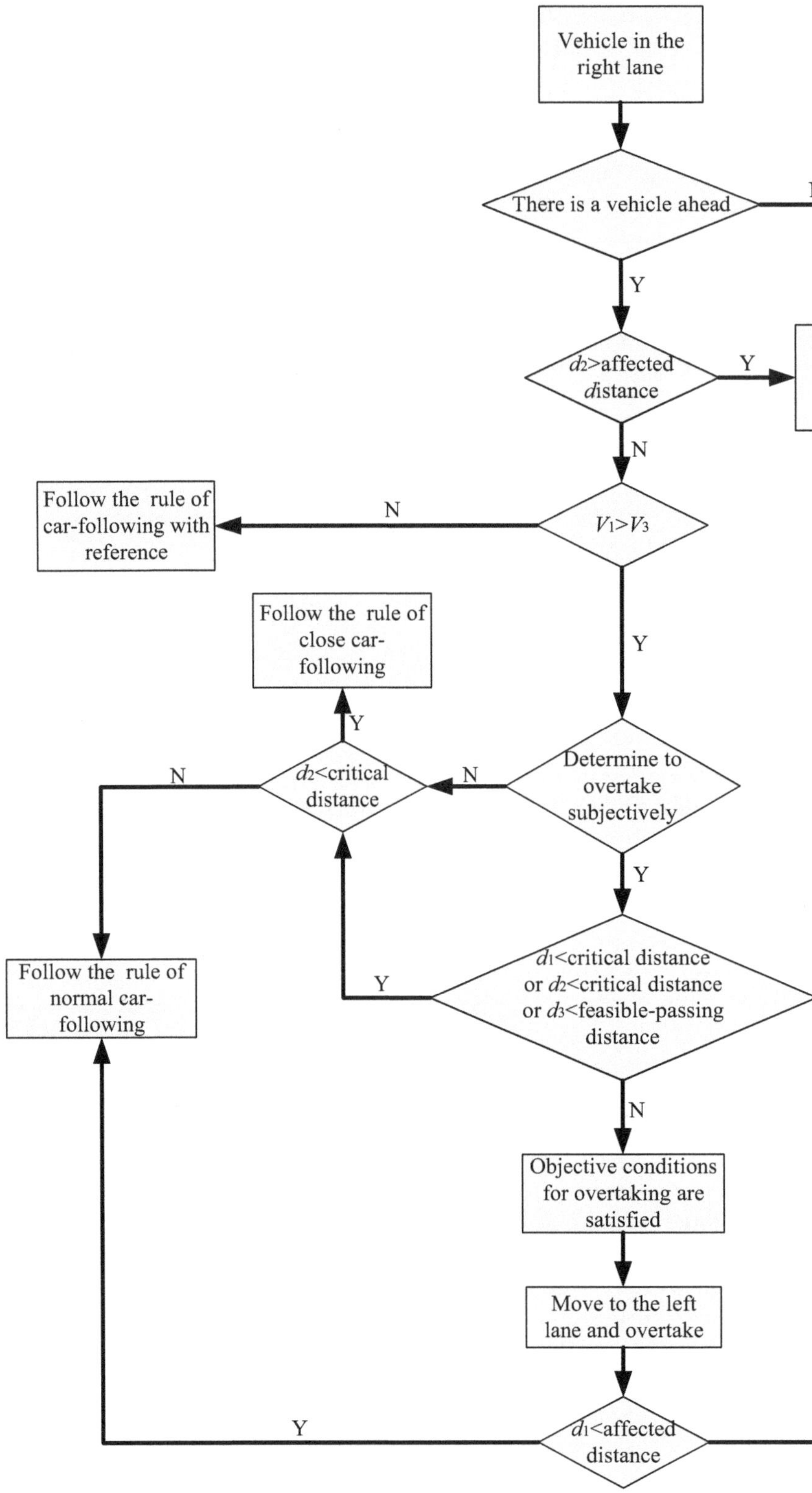

Fig. 1 The rule of vehicle state update in the *right lane*

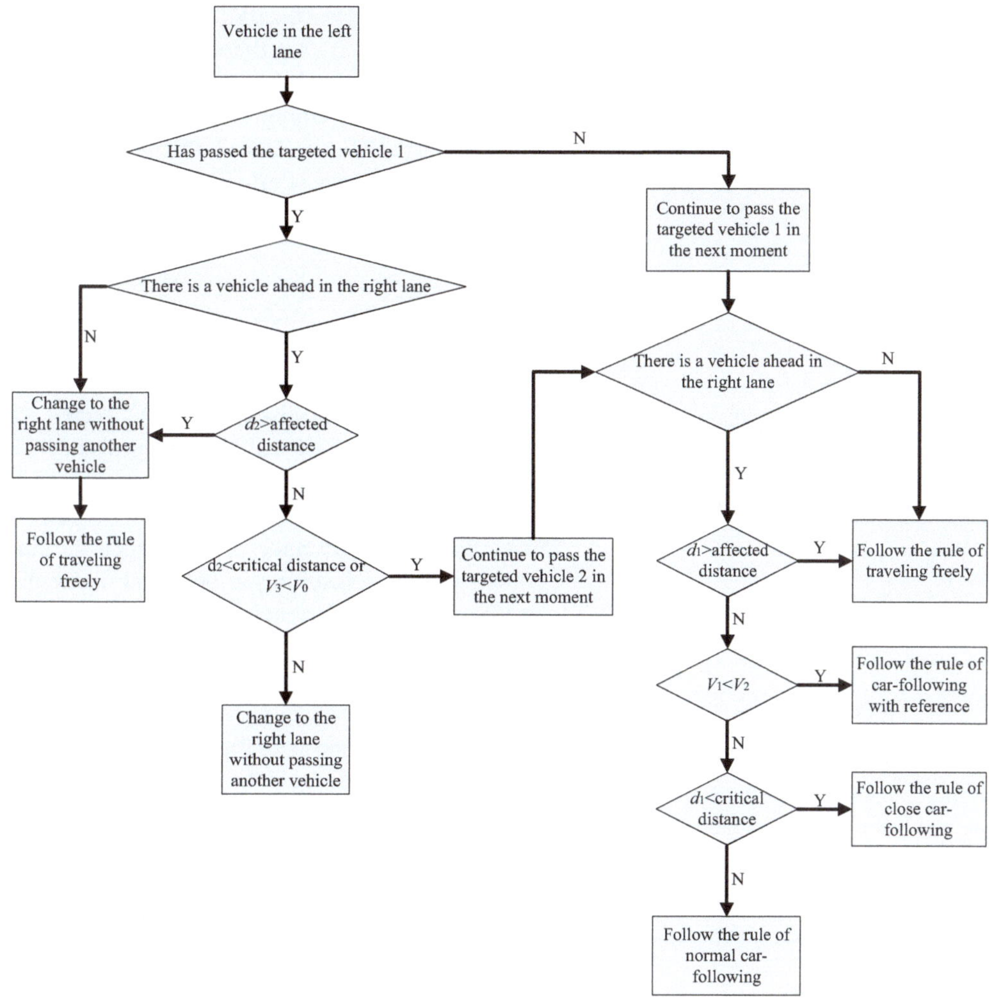

Fig. 2 The rule of vehicle state update in the *left lane*

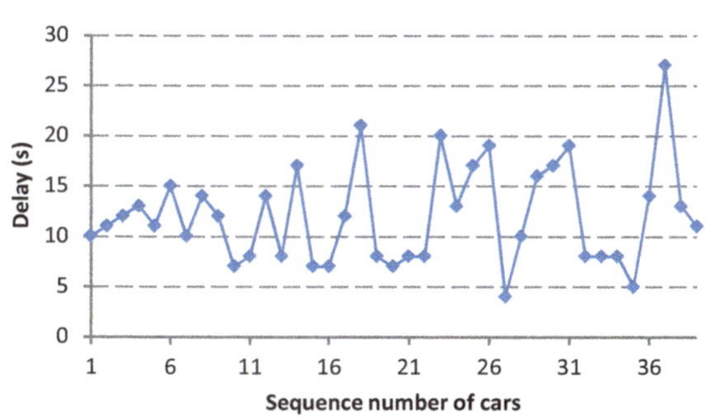

Fig. 3 The distribution of delay of small cars

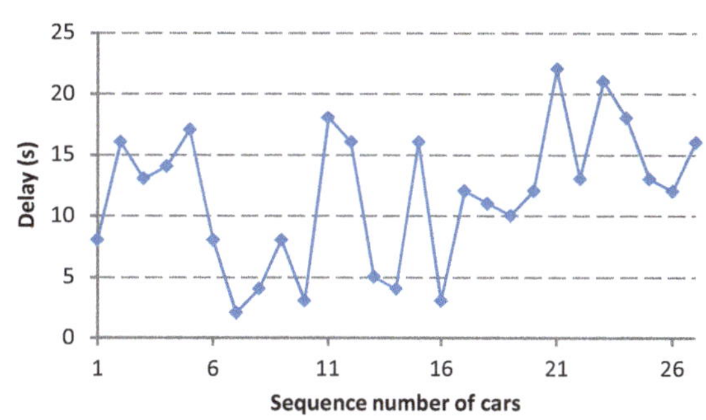

Fig. 4 The distribution of delay of medium cars

4 Safety evaluation

The safety coefficient defined in Sect. 2 is taken advantage of to make safety evaluation. The coefficient of each vehicle is recorded during the simulation, so the safety condition of different vehicles can be compared so that an evaluation is accessible. In addition, the average safety coefficient is able to reflect the safety of traffic flow. Some insights are expected to be made through having the traffic volume vary from light flow to heavy flow. The volume 1,050 veh/h discussed in Sect. 3 is continued to be used so as to represent a light flow whose V/C (the ratio of traffic volume to road capacity) is about 0.25 in this work. A medium flow and a heavy flow are also determined by

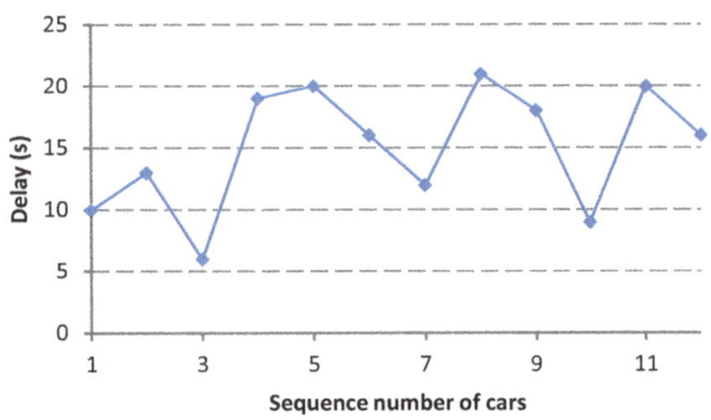

Fig. 5 The distribution of delay of large cars

Table 1 The mean and standard deviation of speed of different kinds of cars (unit: km/h)

Statistics	Small cars	Medium cars	Large cars
Mean	86.8	77.1	71.1
Standard deviation	9.0	17.0	6.8

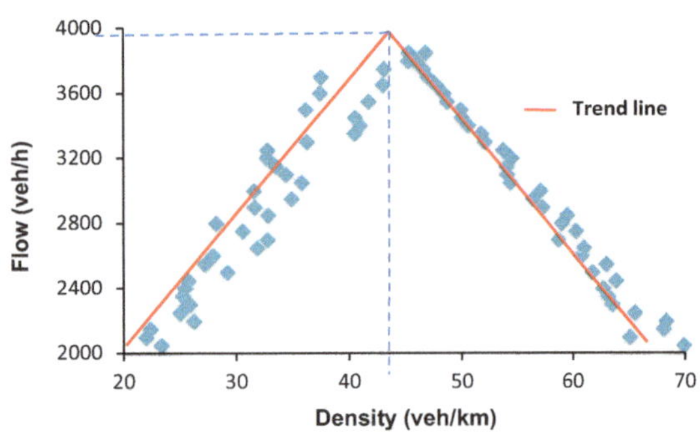

Fig. 6 The scatters obtained by micro-simulation

Table 2 The average safety coefficient under different flow levels

Traffic volume (veh/h)	Average safety coefficient
1,050 (light flow)	0.90
1,660 (medium flow)	0.84
2,300 (heavy flow)	0.88

random sampling like the way to determine 1,050 veh/h, and the results are shown in Table 2.

According to the results, the average safety coefficient is the highest when the traffic flow is light, and the most dangerous situation takes place when a medium traffic volume is observed. This summary is likely to be reasonable because when there is a little traffic, the safety condition should be good for a great distance between every two adjacent cars. The safety is not bad under the situation of heavy flow because drivers start to become cautious to avoid crashes and their speed is passively reduced for a fairly high vehicle density. The most dangerous condition on the freeway is that there exists a medium traffic flow while vehicle speed keeps at a quite high level. As a consequence, assuming the proposed model has been proved to be valid in Sect. 3, the average safety coefficient that is put forward is effective in evaluating traffic safety condition.

5 Discussions

The key parameters to the proposed simulation model is the four characteristic distance: the affected distance, the safety distance, the feasible-passing distance, and the critical distance when the extreme distance is determined to be the vehicle length. Although it is not involved how to determine these distances in this paper, the implications of them are clear, resulting in potential mathematical models to figure out the formulas for solving them.

The keep-right-except-to-pass rule is the premise of our work, but the proposed model is also applicable in countries like Britain where vehicles run on the left. In this case, the only alternation is to interchange the right lane and left lane. Besides, the rule is common in countries like the United States but not dominant in many other countries. In any case, the framework of our model is flexible and some rules for the evolution of vehicle operations can be modified to approximate to other similar driving rules.

It is the continuity in space that makes our model significantly differ from CA model, which is more coincident with real situations. Besides, the driver behavior used in this model is also close to real life. To be specific, drivers first decide whether to overtake a car according to the distance and speed difference. Then they judge whether the conditions for overtaking are all satisfied. Finally they take actions staying in the origin lane or changing a lane. Hereby our model is a typical simulation about decision-making process in lane changing. In addition, the proposed model can make a market segmentation on drivers. Different types of drivers behave distinctively in overtaking, which is significant in obtaining reasonable results from the simulation. Finally, the safety condition can be simply obtained in our simulation. These are the key advantages of the proposed model.

Weaknesses, however, exist in our model at the same time. First, there are many assumptions for the model. Tenable as many of these assumptions may seem, a few of them such as assumption (5) is not really grounded because speed difference has something to do with safety. And it is hard to assess the impacts they have on the accuracy and precision of the proposed model while the validity of model is solely preliminarily tested in this study. Second, the rules of state update for driving is quite complex.

Consequently, it is uneasy to achieve the simulation through programming, or extend the proposed model to freeway sections that have more than two lanes in one direction. Some simplifications could be made while valid model outputs are guaranteed.

6 Conclusions

In this paper, a micro-simulation model is put forward to make evaluation of traffic flow characteristics including safety condition. The model has some advantages over the classic CA model. A partial test has been conducted and the validity of our model is preliminarily verified. Future work can focus on the following aspects. First, some basic assumptions can be loosened. The probability distribution of the arrival of vehicles, for example, can be calibrated through real traffic flow data instead of Poisson distribution assumption. Second, methods to determine the model parameters can be focused on. Visibility is probably one of the most important factors that make a great difference to the proposed four characteristic distance. Third, the validity of our model requires further and more rigorous verifications. One possible method is to calculate the actual capacity of targeted freeway section on the basis of basic capacity, and comparison can be made between the calculation and the results obtained by our simulation. Another method is based on real flow data. Comparing, the data with the simulation results including speed, density, and delay, the validity of the model will be more strictly proved. Finally, multilane freeway traffic simulation can be achieved though the rules of vehicle state update in this study ought to be slightly revised. Taking three-lane freeway as an example, rules such that small cars never run in the right (nearside) lane while large cars never run in the left (fast) lane are necessary to make an extension of the proposed micro-simulation model.

Acknowledgments The authors thank the School of Transportation, Southeast University for providing the good environment to study on traffic operations.

References

1. Li L, Jiang R, Jia M, et al (2011) Modern traffic flow theory and its applications—volume I: freeway traffic flow. Tsinghua University Press, Beijing (**in Chinese**)
2. Reuschel A (1950) Vehicle movements in a platoon. Oesterreichisches Ingenieeur-Archir 4:193–215
3. Pipes LA (1953) An operational analysis of traffic dynamics. Appl Phys 24:274–287
4. Jiang R (2002) Research on the microscopic and macroscopic patterns of complex dynamical characteristics of traffic flow. Dissertation. University of Science and Technology of China (**in Chinese**)
5. Bando M, Nakayama A et al (1995) Dynamical model of traffic congestion and numerical simulation. Phys Rev E 51:1035–1042
6. Newell GF (1961) Nonlinear effects in the dynamics of car following. Op Res 9:209–229
7. Helbing D, Tilch B (1998) Generalized force model of traffic dynamics. Phys Rev E 58:133–138
8. Xu DJ, Mao BH, Chen SK, Bai Y (2015) Improved generalized force model considering the comfortable driving behavior. Math Probl Eng 19:1–9
9. PTV Group Germany (2013) PTV Vissim 6 user manual
10. Cremer M, Ludwig J (1986) A fast simulation model for traffic flow on the basis of Boolean operations. J Math Comput Simul 28:297–303
11. Hua XD, Wang W, Wang H (2011) A two-lane cellular automaton traffic flow model with the influence of driving psychology. Acta Phys Sin 60(8):084502 (**in Chinese**)
12. Jia B (2003) The cellular automata simulation for complex dynamical characteristics of traffic flow at bottlenecks. Dissertation. University of Science and Technology of China (**in Chinese**)
13. Horni A, Montini L, Waraich RA, Axhausen KW (2013) An agent-based cellular automaton cruising-for parking simulation. Transp Lett 5(4):167–174
14. Chai C, Wong YD (2015) Fuzzy cellular automata model for signalized intersections. Comput Aided Civ Infrastruct Eng 30(2015):951–954
15. Wu J, Yang F, Han WS et al (2015) Vehicle load effect of long-span bridges assessment with cellular automation traffic model. Transp Res Rec 2481:132–139
16. Lslam MS, Lvan JN, Lownes NE et al (2014) Developing safety performance function for freeways by considering interactions between speed limit and geometric variables. Transp Res Rec 2435:72–81
17. Saifuzzaman M, Zheng ZD, Haque MM, Washington S (2015) Revisiting the task-capability interface model for incorporating human factors into car-following models. Transp Res Part B 82(2015):1–19
18. Hassan Y, Easa SM, Abdelhalim AO (1997) Automation of determining passing and no-passing zones on two-lane highways. Can J Civ Eng 24(2):263–275
19. Yang SW et al (2009) Survey and design of roads, 3rd edn. China Communications Press, Beijing (**in Chinese**)
20. Chen KM et al (2003) Highway capacity analysis, 2nd edn. China Communications Press, Beijing (**in Chinese**)

15

Comparative study of China and USA public private partnerships in public transportation

Zhijie Dong[1] · Meicheng Wang[2] · Xianfeng Yang[3]

Abstract As an effective alternative approach to provide goods and services for public infrastructure, the Public Private Partnership (PPP) has been studied extensively over the past few decades. On a global scale, China and the United States have developed cooperation on PPP projects in various areas. To perform a comparative study, this paper analyzes how PPP projects work in both countries for public transportation. The basic features, types, and phases of PPPs in public transportation are introduced first, followed by a thorough discussion on their advantages and challenges. This paper adopts a case study method to analyze the achievements and problems of PPP projects in both countries and then proposes important findings and suggestions for future research.

Keywords Public private partnership · Public transportation · China · USA

✉ Zhijie Dong
zd57@cornell.edu

Meicheng Wang
mw833@cornell.edu

Xianfeng Yang
xyang@mail.sdsu.edu

1 School of Civil and Environmental Engineering, Cornell University, Ithaca, NY 14853, USA

2 Department of City and Regional Planning, Cornell University, Ithaca, NY 14853, USA

3 Department of Civil, Construction, & Environmental Engineering, San Diego State University, San Diego, CA 92182, USA

1 Introduction

Governments and the private sector deliver goods and services with different methods, where they share responsibilities and claim rights in various degrees. Samuelson [1] noted that public service provision does not necessarily imply that the government is also the producer of the services. In fact, the private sector has been involved in the service delivery for decades. In the early 1990s, Public Private Partnerships (PPP) projects were first launched in the U.K. on toll-road concessions. After these first experiments in the U.K., PPPs have been growing in popularity as an effective alternative instrument for the provision of goods and services by governments, especially in the infrastructure industry for both developed and developing countries, such as Greece [2], Poland [3], Indian [4], and Ghana [5].

In China, with markets opening up for public facilities and services at the end of 1980s, growth in the number of PPPs in infrastructure has increased fast. For example, since late 1990s, the number of expressways under PPP contracts, programmed in the Chinese National Expressway Network Plan (The 7918 Plan), has increased from 0 to 122 PPP contracts at present [6]. Such a rapid growth was the result of the gap between public funding and actual project costs. Governments were hoping to seek more funding from private sector and meanwhile reallocate responsibilities of design, construction, operation, and maintenance by means of PPPs. At the same time, the trend towards PPPs in the provision and maintenance of infrastructure projects in the United States has been increasing mildly due to the strengths of financial and institutional arrangements that support the traditional procurement across the country. However, as the U.S. infrastructure system matured, the needs for repairing and expanding urban networks of roads, bridges, and tunnels have escalated beyond the fiscal and managerial capabilities of traditional procurement. This has led to an increasing willingness by infrastructure agencies at each government level to consider and in some cases apply alternative structures of public service delivery, including financing, contract delivery, and life-cycle preservation methods to leverage the scarce public resources.

As two of the world's largest economics, China and the United States have developed cooperation on many PPP projects in various industrial and technology fields, such as electricity production and transmission [7], and the project of building high-speed railway between China and US is currently being considered [8]. Although there is much literature on PPP practices in China (e.g., [9, 10]), and more researchers in the United State are paying attention to this delivery method, less work has been done to investigate how PPP works differently in these two countries. Since public transportation is a major contributor to the economy in both countries, a comparative study on relevant PPP projects will lead to potential benefits. This paper focuses to fill such a research gap.

2 Features of PPPs

2.1 Attributes of PPPs compared with public provision and privatization

Definitions of PPPs vary broadly in the literature depending on the specific backgrounds of countries and regions as well as different research interests. Most definitions highlight the unique features of PPPs that distinguish it from public provision, privatization, or both (e.g., [11–13]). As defined by Federal Highway Administration [14],

> A public–private partnership is a contractual agreement formed between public and private sector partners, which allows more private sector participation than is traditional. The agreements usually involve a government agency contracting with a private company to renovate, construct, operate, maintain, and/or manage a facility or system. While the public sector usually retains ownership in the facility or system, the private party will be given additional decision rights in determining how the project or task will be completed.

This definition emphasizes that in a PPP structure the public and private sectors share responsibility for the delivery of the project and/or its life-cycle services.

PPPs usually incorporate a long-term contractual relationship between public and private sectors, as against to the short-term procurement in traditional public provision. Another essence of PPPs is the substantial role of government by either the eventual ownership or the life-cycle supervision, as against to the result of privatization. The purpose of introducing PPPs in between public provision and privatization is to deliver public services with more competent resources aggregation and risk allocation.

Traditional public provision separates the construction phase from the sequent operation and maintenance (O&M) phase by only purchasing the construction service from the private firms. This procurement structure often leads to insufficient O&M and service deterioration in the long term. In addition, political pressure incentivizes governments to prefer introducing new projects to the O&M phase of existing projects. For these reasons, PPPs offer a viable alternative, in that the private sector is responsible for not only asset delivery but also overall project management and successful operation over the contract life [15]. Since returns of investment for private sector depend on long-term project performance than bundling the construction and O&M phases, service providers will seek efficient resource allocation to reduce life-cycle costs. Figure 1 summarizes key differences between PPPs and traditional procurement methods.

Privatization, referring to the shift of government functions from a public sector to the private agency, is also an alternative of service delivery [16]. In most cases privatization and contracting out are often used synonymously [17], but "contracting out" tends to indicate some parts of services being transferred to a number of private firms. Privatization has the potential to take more advantage of market competition. However, when certain services demonstrate increasing return and natural monopoly, or when they are not excludable and the society insists not to charge users, privatization does not work well for achieving users' welfare [13]. Transportation infrastructure projects demonstrate such features, thus require strong network planning and substantial roles government during the entire process. Meanwhile, the quality of transportation infrastructure is more contractible than that of education or health care, making it qualified for the application of PPPs.

2.2 Various types of PPPs

PPP projects can be structured either "vertical" or "horizontal" in nature [18], which are also referred to "contractual" and "institutional," respectively. In a vertical partnership, a concession agreement or PPP contract assigns the responsibility of service delivery to the private sector over its entire life cycle (i.e., "design, build, finance and operate"), and transfers the service to public control at the end. In a horizontal partnership, both public and private sectors are responsible for providing infrastructure services as shareholders in a special purpose vehicle (SPV), in which public and private shares are jointly ventured for the project purpose. In contrast to the temporary and limited nature of vertical partnership, the horizontal partnership transfers ownership and operating function to the joint venture

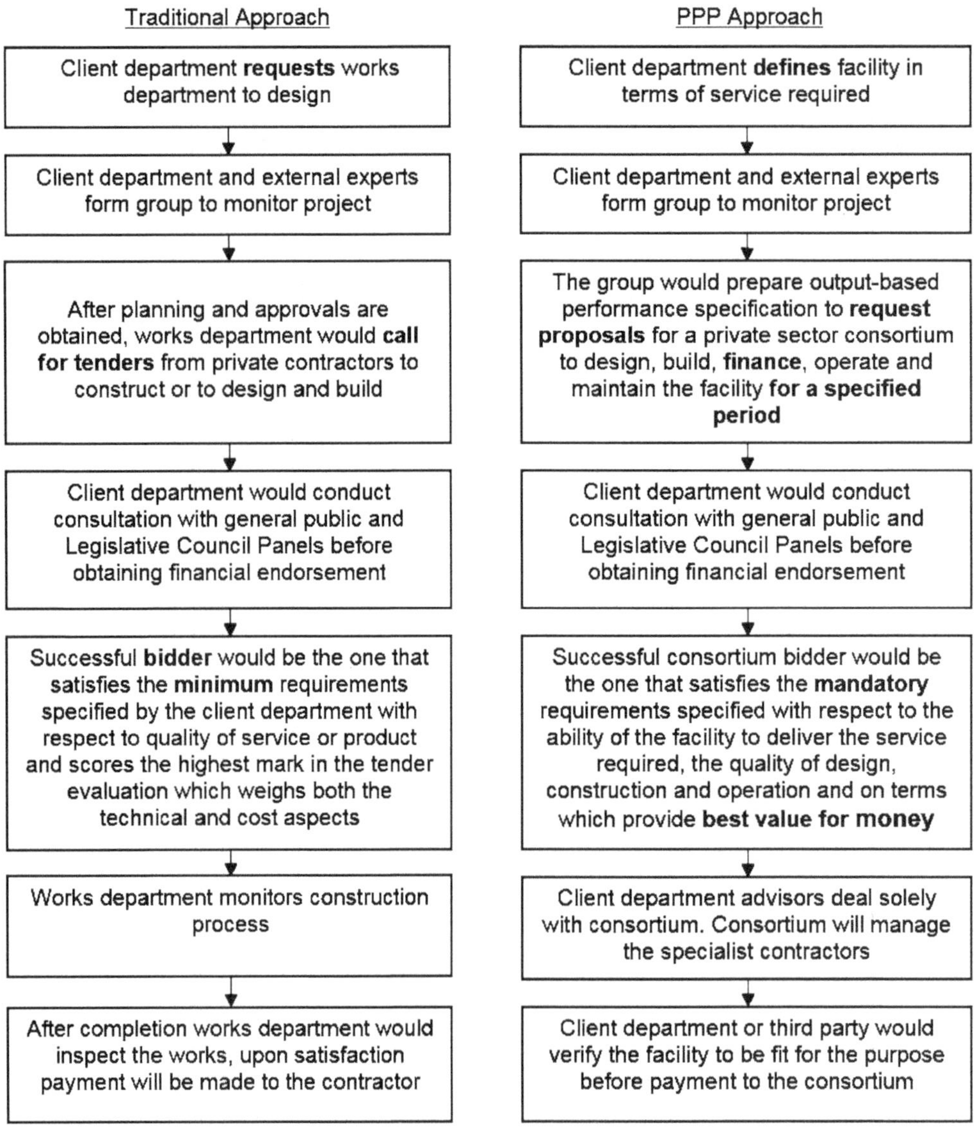

Fig. 1 Differences between PPP approaches and the traditional practice

permanently. In extreme cases, government may hold no equity in the joint venture, but participate in the PPPs as a bond issuer or supervisor. Figure 2 demonstrates the different structures of vertical and horizontal partnerships.

In practice, PPPs take on many different forms that are typically known by acronyms. Most vertical PPPs are regarded as the variants of Build-Operate-Transfer (BOT), indicating that the private sector is responsible for building and operating specific facilities or services during a relatively long term, and eventually transfers the facility to the public sector. BOTs allow sufficient time for the private sector to internalize life-cycle cost and get returns on investment from fees charged to the government [19]. To mitigate the risk of overestimated demand, BOTs usually ask limited access to private capital, which means governments hold more financial responsibility and bears the equity risk.

A major variation of the BOT is build-own-operate-transfer (BOOT), also known as design-build-finance-operate (DBFO), involving significant finance from the private sector. In this structure, private firms maintain ownership until transfer it to the government, during which they are able to charge user fees and assume most equity risks. BOOTs are most suitable when the public sector faces a large financial gap or is not able to bear the financial risks [20].

3 Promises and challenges of PPPs in public infrastructure

The PPP gains the popularity in the infrastructure industry over recent years for its promises to allocate resources more efficiently and mitigate risk more effectively. The

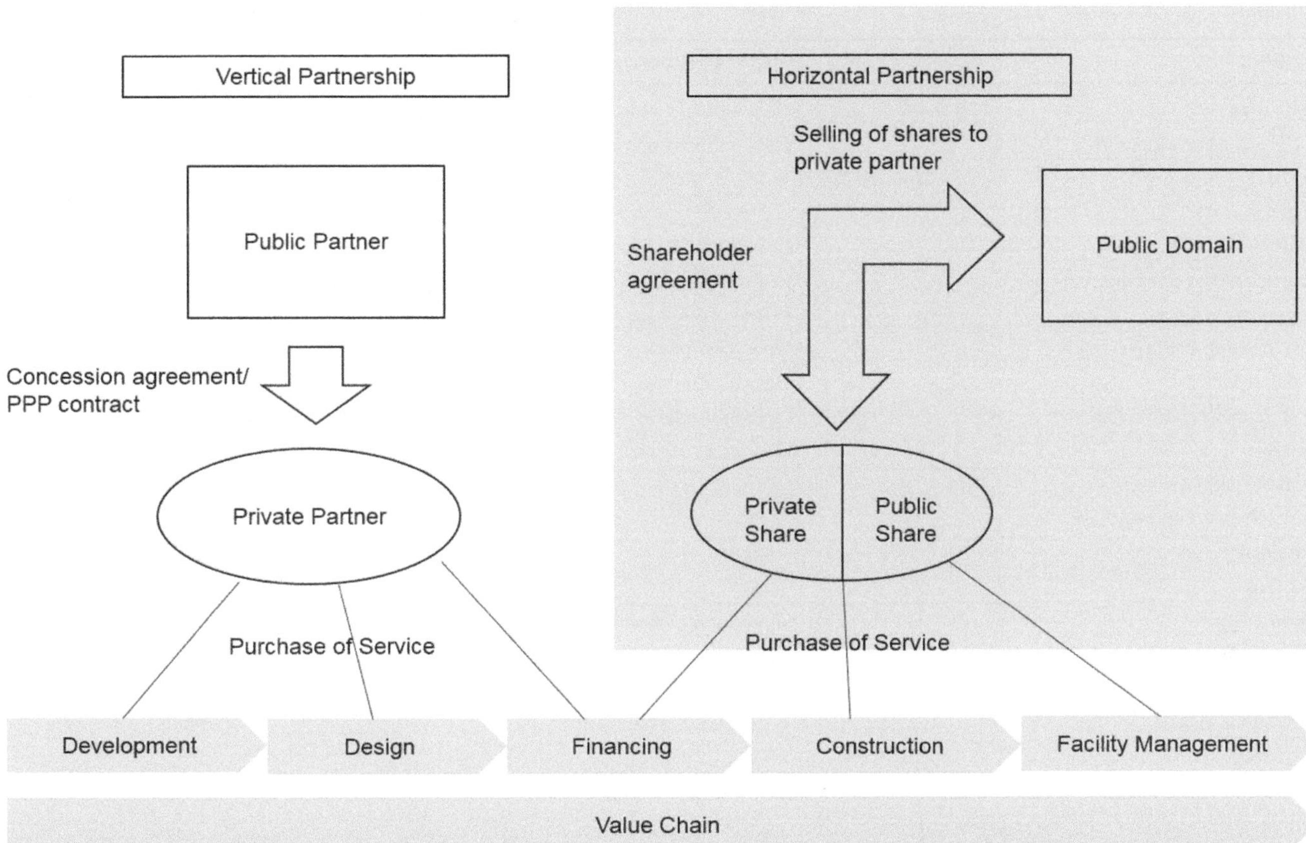

Fig. 2 Structures of horizontal & vertical partnerships

most highlighted advantage is the enhancement of "value for money" (VfM), meaning delivering specific facility or service with the same quality at a cost lower than conventional procurement or spending the same amount of money while gaining a better service quality [21]. Walker and Smith [22] argued that the private sector possesses better mobility than the public sector and is more capable to raise massive funds for the large-scale infrastructure projects. This is doubtful because the involvement of private firms does not intrinsically generate revenue or lower the cost of capital. The economic gains from PPP structure derive largely from the bundling of construction and O&M, minimizing the life-cycle cost and preventing government from obtrusively starting new projects rather than maintaining existing ones. Secondly, competition for the contract will lead to the pricing more economically viable than the inefficient user fees in public provision under political pressure. In addition, the private sector knows how to charge the premium that matches the risks they bear. Other reasons supporting PPPs include the innovation brought by private firms and the on-time service delivery incentivized by economic gains.

Ideally, most advantages mentioned above can also be realized in a competitive market, but considering the increasing return to scale and thus the natural monopoly in most transportation infrastructures, a long-term public role to conduct network planning and regulation is indispensable. The urban railway network that will be discussed in the following section is one of the most salient examples. Since the quality of transportation service is relatively contractible, the public sector could partner with the private sector for enhancing completion but meanwhile retain long-term regulation and eventual ownership to avoid the loss of users' welfare. For these reasons, whether the cost is covered by user fees or government transferring payment, PPPs in the arena of transportation win both complete privatization and traditional procurement.

Besides the promise of integrating resources more efficiently, the PPP becomes preferable because of its vision of allocating risks to the parties that are best able to manage them. Risk management is a dynamic process over the life cycle of the project, which can be broken down into the identification, assessment, and mitigation of risks through optimal allocation between public and private sectors. Comparing to traditional procurement, one of the major risks of public infrastructure, the unavailability risk, is borne by private firms instead of public users under PPPs. On the other hand, each sector will assume risks with

Table 1 Typical categories of risk management

Risk categories	Stages of the project	Allocation of risk
Land acquisition	Development phase	Public/private party
Delays in project Development	Development phase	Private party
Design risk		
Planning risk		
Project completion risk	Construction phase	Private party
Project cost risk/cost overruns	Construction phase	Private party
Technology risk	Construction/operation Phase	Private party
Regulatory & administrative risk	Operation phase	Public/private party
Commercial risk	Operation phase	Private party
Operations & maintenance risk	Operation phase	Private party
Financial risk	Operation period	Private party
Interest rate risk		
Foreign exchange exposure risk		
Tax rate change risk		
Inflation risk		
Termination risk	Operations phase	Private party
Insolvency and outside editor risk	Throughout project life cycle	Private party
Environmental risk	Construction/operations Phase	Public/private party
Political & social risk	Throughout project life cycle	Public/private party
Events of wars		
Nationalization or revocation		
Social Risk		

corresponding benefit claim. For example, lenders will ask for higher risk premiums if the private sector need to bears demand risk (i.e., collecting availability payment rather than user fee). Table 1 lists typical categories of risk management.

It is worth noting that the advantages of PPP structure do not guarantee the promises of PPP projects. Instead, there are certain preconditions to apply PPPs and a number of shortcomings that often impede the success of PPPs. Given that competitive auction can only be achieved when real competition exists for the contract [23], any institutional or financial barriers that inhibit competitions could become challenging for PPPs, including but not limited to weak protection of property rights, incapable budgetary control and incompetent financial market.

The more intrinsic shortcoming of PPPs is the high transaction cost during long-term contractual relationship. A lengthy bidding process in PPPs could impose high contracting cost at the beginning, which is the reason why many countries impose a threshold or value to select projects qualified for PPPs. Moreover, since the contract is not likely to be perfectly detailed, there is room for opportunistic renegotiation, which could enable the public and private sectors to take advantages of each other's efforts. For example, Gomez-Ibanez and Meyer [24] noticed that governments in the transportation sector often bail out concessionaires. Table 2 depicts typical steps of a vertical PPP project. If each sector is not bear the full responsibility of its own actions, such renegotiation is likely to cause inefficient lobbying and even corruption.

In sum, if the cost of renegotiation negates the efficient gain, or if the speculation by any partner could escape the penalty under an unsupportive institutional and finical environment, PPPs may not be the best option for transportation infrastructure projects.

4 PPP applications of publication transportation in China and the USA

Growing demand for mobility has created serious traffic congestion and deteriorating air quality in urban areas. This has encouraged huge investments in the viable public transport infrastructure, such as subway systems, as alternatives to car travel. Many large transport infrastructure projects used to be financed by the government directly, but the financial burden far exceeds the availability of public funds. This section focuses on analyzing PPP public transportation projects in between both China and the United States using the method of case study.

Table 2 Typical steps of a vertical PPP project

Phase I: Project Identification & Early Consideration	
⇩	• Assessment of need, economic & financial feasibility • PPP test
Phase II: Preparation & Project Approval	
⇩	• Development of PPP delivery method • Development of traditional delivery method • Efficiency comparison test (Value for Money test)
Phase III: Project Delivery & Contract Award	
⇩	• Preparation and prequalification • Negotiation process • Selection based on criteria • Contract award and closing the deal
Phase IV: Project Implementation & Contract Management	
⇩	• Construction and Operation • Performance control by the government
Phase V: Contract Termination	
	• Transfer

4.1 Beijing No. 4 subway line

Beijing No. 4 subway line is a 17.4-mile project, which runs through the north and south of the city passing by many central business districts and universities [25]. This is the first project adopting PPP delivery approach in Chinese metropolitan railway system. This line is regarded as the "Golden Line" by Beijing municipal government, and it was partially opened in time for the 2008 Beijing Olympics.

In 2004 after the proposal of allowing investments from outside Chinese mainland was approved by the State Development of Reform Commission, the Beijing municipal government opened up participation in the building and operation of Beijing No. 4 subway line to domestic and foreign companies. Three private players won the contract after a competitive process, in which only a small number of prequalified bidders were invited to take part. These three companies are Hong Kong Mass Transit Railway Corporation (MTR), Beijing Capital Group (BCG) and Beijing Infrastructure Investment Corporation Ltd (BIIC). The former one is a private company from Hong Kong and the others are state-owned.

The signed contract covers a period of 30 years, with the initial 4 years reserved for construction and the rest for operation. The project is divided into two parts (part A and B) according to the characteristics of construction missions. Part A involves civil engineering and physical infrastructure of the project (e.g., earthwork, tunnels and stations), with an investment about USD 1.5 billion (70 % of the total expenses) paid by the Beijing municipal government (through the No. 4 Beijing Subway Line Investment Company). Part B covers the operational aspects (e.g., vehicles, ticket machines, signaling systems, air-conditioning, fire protection, escalators, elevators, control devices, and power supply facilities), and represents the 30 % of the total expenses [26]. It is privately funded and managed by a SPV, in which the shares of MTR, BCG, and BIIC are 49 %, 49 %, 2 %, respectively, functioning as partners in a regular BOT project [27, 28]. The No. 4 subway line will be transferred back to the Beijing government by 2039. Figure 3 illustrates the joint venture structure and investment agreement of the project.

There was also an asset lease agreement between the No. 4 Beijing Subway Line Investment Company and MTR-BCG allowing the latter to use the infrastructure. During the concession period, MTR-BCG would obtain revenues from ticket fares and the commercial operation of the subway stations. Besides monitoring assets, quality, and safety management, BIIC also plays a role in guaranteeing a certain level of profits for MTR-BCG. Besides subsidy on ticket fares, BIIC promised to make up the difference if the annual revenue is less than 70 % of the projected level, while if it is higher, the exceed part will be absorbed by the equity investors National Development and Reform Commission [26]. In reality, annual ridership of 4th subway has constantly exceeded the predicted level by more than 10 %, generating considerable profit for investors. Since there is no upper limit of revenue gains in the contract, 4th subway faces critics of leaving too much profit to the private sector.

There are many factors driving the operation of this project. First, it is a product under the investment system reform led by the central government. Second, it helps the Beijing municipal government to avoid raising large amount upfront cost in a short term. Last but not least, the intense interaction between different partners has resulted in an environment where mutual learning on technical, management, and economic aspects can flourish, leaving experiences on how to share risks and revenues among public and private stakeholders. In order to govern the PPP activity of Beijing's 4th subway line, the municipal government had to issue several supplementary documents, including "Regulations on Developing Urban Infrastructure with PPPs Arrangement," proving that certain institutional environment is necessary to promote PPPs. On the other hand, such amount of efforts led to 18-month negotiation preceding the final contract, which inevitably caused the high transaction costs. Although this cost of time is common in well-designed PPP projects, some government officials may not choose PPPs if they have to delivery public services faster.

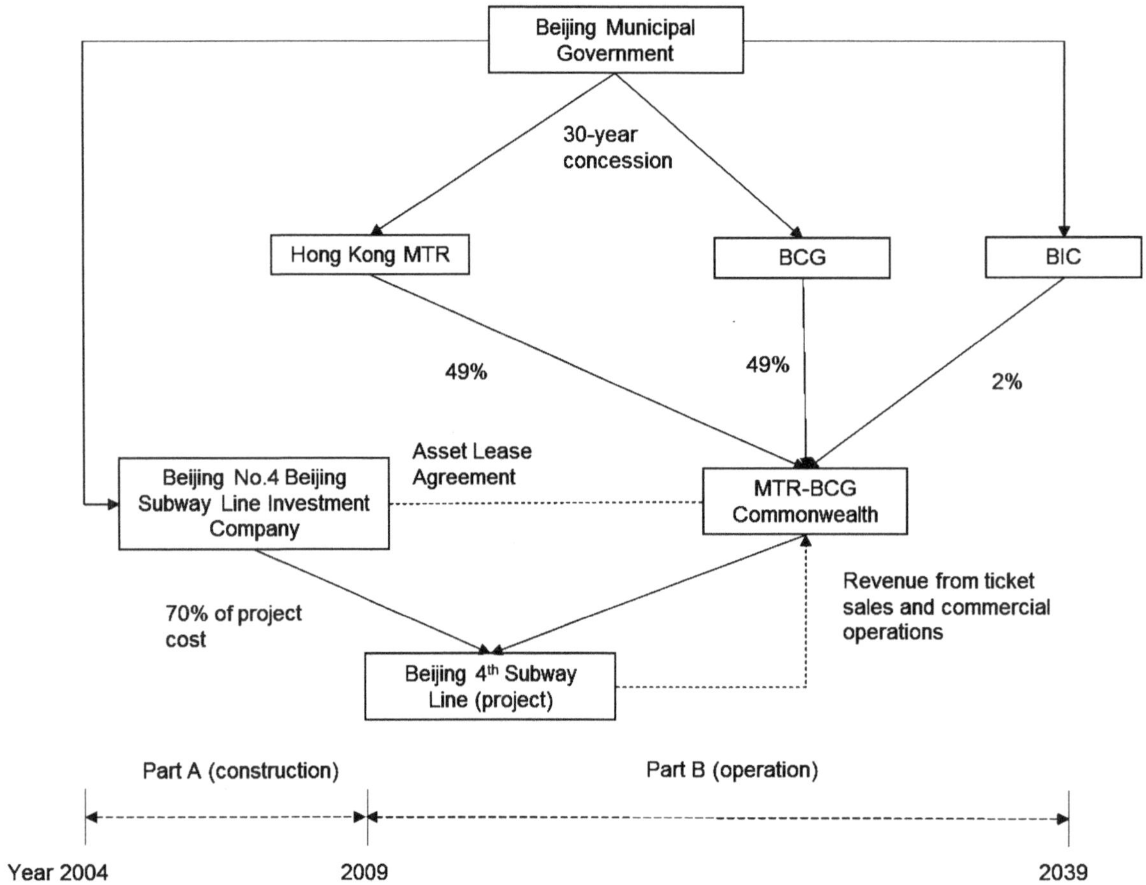

Fig. 3 Joint venture structure in Beijing No. 4 subway line

4.2 Las Vegas Monorail

The Las Vegas Monorail is a 3.9-mile rail transit system located in Clark County, Nevada, connecting major hotels and casinos along the world famous Las Vegas Strip. It was built on an existing free monorail jointly owned by MGM Grand and Bally's Monorail. LLC. The tourist growth stimulated the expansion from the original 1-mile system to the current Las Vegas Monorail to satisfy the increasing transportation demand.

In 1997 the State of Nevada passed legislation for the expansion plan to allow a private company to own, operate, and charge a fare as a public monorail system. Later the non-profit corporation, Las Vegas Monorail Company (LVMC), was formed in 2000. LVMC gained a 50-year contract with the original developer, and signed an agreement with Transit Systems Management LLC (BOO partner), for the construction, operation, and management of the project. Different from most public transportation PPPs, the government does not own a share in the venture, nor does the private sector transfer the project to the government. Instead, the public sector participating in this unconventional PPP by providing policy supports.

The primary funding resource of Las Vegas Monorail project is the tax-exempt revenue bonds issued by the State, with further user fares and advertisement to fulfill the debt service. Salomon Smith Barney and the Nevada Department of Business and Industry issued revenue bonds, and LVMC would pay the expenses of operations and maintenance after receiving the bonds. It is worth noting that the State of Nevada was only responsible for bond issuing, and most construction costs were funded by the private sector without government investment, guarantee, or subsidy [29]. As pointed out by the Federal Highway Administration [30], one major innovation in financing arrangement of this project is that the LVMC is the first and only privately owned public transportation systems in the U.S. and operates with no public subsidies. This innovative funding solution using PPP also promoted the development of monorail technology. For example, Las Vegas Monorail is the nation's first fully automated, line haul electric transit system with zero emissions. In 2009 only, tourists who rode the Monorail helped reduce emissions by more than 48 tons of carbon monoxide, volatile organic compounds and nitrogen oxides over the course of the year by saving about 2.7 million vehicle miles [31].

When this PPP project was being implemented, many difficulties have been overcome such as overruns, schedule delays, mechanical failures, and legislation. However, due to inaccurate ridership prediction, LVMC filed for bankruptcy protection in January 2010 when it became unable to pay debt service on the bonds and its other expenses [32]. Following by the recovery of the entire project, the ticket fee of Las Vegas Monorail is now at least double than most metropolitan cities in the U.S.

5 Discussion

Although public private partnerships have different development paths in China and the United States, valuable lessons learnt from these two typical cases in public transportation infrastructure will help understanding the contexts of conducting PPPs in these two countries.

First of all, how the government plays its role is crucial for implementing the PPPs. On the one hand, local governments should fully respect the agreements and give considerable freedom to private partners. One counterexample is that the second-largest shareholder of Beijing No. 4 subway line, Zhiqiang Ren, has recently complained about the government's arbitrary decision makings including pricing, investment, and line extension, which led his investment to be a "generous donation" [33]. On the other hand, the lack of the ability for the local government to conduct proper risk analysis and negotiate with private parities inhibits the development of PPPs. Las Vegas Monorail's filing bankruptcy demonstrates the challenges too limited involvement of the government (i.e., the State of Nevada) on the PPP project. The role of the public sector is typically defined by the PPP agreement; thus the core task is to keep the resource allocation and risk taking balanced so that both sectors are able to maximize their capacities [34]. From a financial point of view, before an attempt secure equity financing or debt, it is important for PPPs to have an equitable risk-sharing system.

Second, efficiency and innovation should be sought before funding PPPs. Up to now, most contributions from the private sector have been financed by short-term bank loans, often with the backup from the local governments. However, a bottleneck is caused by the lack of financing sources in future PPP infrastructure developments with the large scale of investment required and the limited guaranteeing capacity of local governments under new regulations. One important alternative is infrastructure bonds, which have been used for years in China and the United States. For example, Beijing recently raised $1 billion bonds for the development of Beijing metro system in three-year and five-year notes [35]. Another possible alternative to enlarge the investing base is to create opportunities for pension funds and insurance companies to invest in infrastructure funds on a long-term basis. The project of Las Vegas Monorail has made some innovative achievements by introducing the entirely private-owned public transportation systems. Besides bringing in innovation in project finance, patterning with the private party under a PPP agreement will also attract advanced technology, which make the services sustainable for public purposes [36].

Third, an effective evaluation and renegotiation system shall be developed. If a PPP is considered as an approach of accessing financing only, the chances of failure will be increased because of a high risk of choosing inappropriate projects (also known as "white elephants"). The core criterion while selecting and designing a project is whether it will bring value for money, which is termed VfM analysis. In the context of PPPs, such analysis is conducted to decide whether it would be eventually more beneficial to the public users [30, 37]. If a PPP project does not enhance social benefit, the public sector needs to consider renegotiation the contract or choose other delivery methods.

6 Conclusions

The public private partnership (PPP) model offers significant advantages over traditional public procurement in terms of efficiency, service quality, and value for money. The main factor that drives PPPs' prosperity in many counties is the long-term capacity of project finance brought by the private sector, which works along with government supervision and network planning.

Government agencies and companies in the U.S. and China have been collaborating on PPP projects in the area of energy and climate; however, few research studies have been conducted to analyze the difference of PPPs applications in both countries. Since transportation is the major application of PPP all over the world, China and the United States have huge potentials for exploring cooperation opportunities on transportation infrastructure projects.

Among the discussions about promises and challenges of PPPs, public transportation system, especially urban rail system, differs from other infrastructure such as bridges or highway in that: (1) it is more dependent on network planning and coordination at municipal level; (2) the upfront cost such projects takes a larger part of life-cycle cost; (3) to reach a socially optimal scale of service, government has to restrict the level of the user fee, which usually requires government support to sustain the cash flow.

Two typical PPP cases in China and the United States, the Beijing No. 4 subway line and Las Vegas Monorail respectively, are selected for discussion. With analysis of their different PPP structures, valuable suggestions are proposed from the case study that more attention should be given to the roles of governments during the long-term partnership. Governments need to take discreet supporting and supervising steps of promoting diversity of finance, encouraging innovation of skills and technologies, and creating responsible evaluations. Future research could focus on more recent projects using PPP approach in China and the United State within the field of public transportation, comparing their financing mechanism and institutional systems.

References

1. Samuelson P (1954) The pure theory of public expenditure. Rev Econ Stat 36(4):337–389
2. Roumboutsos A, Anagnostopoulos KP (2008) Public–private partnership projects in Greece: risk ranking and preferred risk allocation. Constr Manag Econ 26(7):751–763
3. Staebner A, Marchand JP (2012) Public-private partnership for motorway in Poland: research at the national and European levels. Eur Roads Rev 20:61–67
4. Narayan V (2013) Public private partnership for urban infrastructure and service in India. In J Logist Supply Chain Manag Perspect 2(3):405–410
5. Robert OK, Dansoh A, Ofori-Kuragu JK (2014) Reasons for adopting public-private partnership (PPP) for construction projects in Ghana. Int J Constr Manag 14(4):227–238
6. China Audit Office (2008) Audit report on Chinese toll roads. China Audit Office, Beijing
7. U.S. Department of the Treasury (2008) The five areas are: clean, efficient, and secure electricity production and transmission; clean water; clean air; clean and efficient transportation; and conservation of forest and wetland ecosystems. Joint U.S.–China fact sheet: U.S.–China ten year energy and environment cooperation framework, Washington, DC
8. Tharoor I (2014) China may build an undersea train to America. Washington post. https://www.washingtonpost.com/blogs/worldviews/wp/2014/05/09/china-may-build-an-undersea-train-to-america/
9. Cheung E, Chan APC, Lam PTI, Chan DWM, ke Y (2012) A comparative study of critical success factors for public private partnerships (PPP) between Mainland China and the Hong Kong Special Administrative Region. Facil Manag Dev 30:647–666
10. Ke YJ, Jefferies M, Shrestha A, Jin XH (2014) Public private partnership in China: where to from here, organization, technology and management in construction. Int J 6(3):1156–1162
11. European Investment Bank (2004) The EIB's role in public-private partnerships (PPPs). European Investment Bank, Luxembourg
12. International Monetary Fund (2006) Public Private Partnerships, Government Guarantees, and fiscal risk. Fiscal Affairs Department, International Monetary Fund, Washington, DC
13. Engel E, Fischer RD, Galetovic PA (2014) The economics of public-private partnerships: a basic guide. Cambridge University Press, New York
14. Federal Highway Administration (2004) Report to congress on public-private partnerships. U.S. Department of Transportation, Washington, DC
15. Martimort D, Pouyet J (2008) To build or not to build: normative and positive theories of public–private partnerships. Int J Ind Organ 26(2):393–411
16. Savas ES (2000) Privatization and the new public management. Fordham Urban Law J 28(5):1731–1737
17. Starr P (1988) The meaning of privatization. Yale Law Policy Rev 6(1):6–41
18. Grabow B, Reidenbach M, Rottmann M, Seidel-Schulze A (2006) Public private partnership projects in Germany. A survey of current projects at federal, land and municipal level. German Institute of Urban Affairs, Berlin
19. World Bank Group (2014) Concessions, build-operate-transfer (BOT) and design-build-operate (DBO) projects. http://ppp.worldbank.org/public-private-partnership/agreements/concessions-bots-dbos
20. World Bank Group (2009) PPP Modalities. Toolkit for public-private partnerships in roads & highways. https://www.ppiaf.org/sites/ppiaf.org/files/documents/toolkits/highwaystoolkit/6/pdf-version/5-36.pdf
21. McCann S, Aranda-Mena G, Edwards PJ (2014) Delivering value-for-money in the operating phase of public private partnership: interview findings. Int Public Manag Rev 15(2):91–110
22. Walker C, Smith AJ (1995) Privatized infrastructure- the BOT approach. Thomas Telford Inc, New York
23. Demsetz H (1968) Why regulate utilities? J Law Econ 11(1):55–66
24. Gomez-Ibanez JA, Meyer JR (1993) Going private: the international experience with transport privatization. The Brookings Institution, Washington, DC
25. de Jong M, Rui M, Stead D, Ma Y, Xi B (2009) Introducing public–private partnerships for metropolitan subways in China: what is the evidence? J Transp Geogr 18(2):301–313
26. National Development and Reform Commission (National DRC) (2015) Beijing No. 4 subway line project. http://tzs.ndrc.gov.cn/zttp/PPPxmk/pppxmal/201507/P020150721602804672682.pdf
27. Wang H (2005) Chengshi Guidao Jiaotong Tourongzi Wenti Yanjiu (studies on urban mass transit issues). Chinese Finance Press, Beijing
28. Wang H (2009) Chengshi Guidao Jaotong Rongzi MoshiYanjiu (studies on urban mass transit finance). Chinese Architecture Industry Press, Beijing
29. Kirk RS, Mallet WJ (2013) Funding and financing highways and public transportation. Congressional Research Service, Washington, DC
30. Federal Highway Administration (2012) Value for money assessment for public-private partnerships: a primer. U.S. Department of Transportation, Washington, DC
31. Las Vegas Monorail (2015). http://www.lvmonorail.com/
32. Planetizen (2014) Las Vegas Monorail bondholder sues for fraud. http://www.planetizen.com/node/53621
33. Zhang XM (2014) Why public-private partnership doesn't work in China. EJ Insight. http://www.ejinsight.com/20141201-why-public-private-partnership-doesnt-work-inchina/
34. Zhang S, Gao Y, Feng Z, Sun W (2014) PPP application in infrastructure development in China: institutional analysis and implications. Int J Proj Manag 33(3):497–509
35. Wong CM (2014) Beijing infrastructure prices $1b bond. Finance Asia. http://www.financeasia.com/News/392094,beijing-infrastructure-prices-1b-bond.aspx
36. Asian Development Bank (2014) Public-private partnerships in urbanization in the People's Republic of China. Asian Development Bank, Mandaluyong City
37. Martin H (2013) Value-for-money analysis—practices and challenges: how governments choose when to use PPP to deliver public infrastructure and services. The World Bank, Washington, DC

16

Improvement of railway performance: a study of Swedish railway infrastructure

Yamur K. Al-Douri[1] · Phillip Tretten[1] · Ramin Karim[1]

Abstract The volume of rail traffic was increased by 5 % from 2006 to 2010, in Sweden, due to increased goods and passenger traffic. This increased traffic, in turn, has led to a more rapid degradation of the railway track, which has resulted in higher maintenance costs. In general, degradation affects comfort, safety, and track quality, as well as, reliability, availability, speed, and overall railway performance. This case study investigated the needs of railway stakeholders responsible for analysing the track state and what information is necessary to make good maintenance decisions. The goal is to improve the railway track performance by ensuring increased availability, reliability, and safety, along with a decreased maintenance cost. Interviews of eight experts were undertaken to learn of general areas in need of improvement, and a quantitative analysis of condition monitoring data was conducted to find more specific information. The results show that by implementing a long-term maintenance strategy and by conducting preventive maintenance actions maintenance costs would be reduced. In addition to that, problems with measured data, missing data, and incorrect location data resulted in increased and unnecessary maintenance tasks. The conclusions show that proactive solutions are needed to reach the desired goals of improved safety, improved availability, and improved reliability. This also includes the development of a visualisation tool and a life cycle cost model for maintenance strategies.

✉ Yamur K. Al-Douri
yamur.aldouri@ltu.se

Phillip Tretten
phillip.tretten@ltu.se

Ramin Karim
ramin.karim@ltu.se

1 Division of Operation and Maintenance Engineering, Luleå University of Technology, Luleå, Sweden

Keywords Condition-based maintenance · Condition monitoring · Maintenance · eMaintenance · Railway track

1 Introduction

Railway infrastructure covers a large area and is often difficult to maintain given its complexity, number of stakeholders, weather variations, and physical components designed and developed for previous traffic and track conditions. This infrastructure is divided into several components, including administration, maintenance, traffic, and new investments. Of this system, the track is the central part. The whole railway infrastructure is focused on the track; therefore, good maintenance of the track is essential. Track maintenance must consider availability, reliability, and safety, and also be cost effective [1–3]. The purpose of the track is to provide a stable, safe platform for the train to operate at various speeds. The main issues in track deterioration are deformation of the rail head, dangerous rail cracks, and damage to the sleepers, caused by speed and load of the rolling stock. The problems increase as the train speed increases, resulting in considerable negative effects on the track components' life cycle. When the track breaks down, the loss of service, property, and even of life may occur, resulting in high maintenance costs, decreased overall track performance, unsatisfied passengers, and damaged goods [3–5].

Railway maintenance is crucial for increasing safety and to reduce costs [6]. Maintenance conducted by the Swedish Transportation Authority (TrV) is necessary to support railway resources and facilities [3, 7]. After railway stakeholders consider the data and the maintenance requirements, they can make policy decisions assisting Trafikverket. For this reason, support planning is used to identify the necessary resources for the desired maintenance tasks. The goal is to increase availability and reliability, while decreasing the risk of failure [8, 9]. The type of maintenance strategy to be implemented is based on the railway track components under specific conditions [10, 11]. Preventive maintenance is one of these strategies. It is used to define maintenance over predetermined time intervals or according to specific monitoring condition-based maintenance [10]. Condition-based maintenance (CBM) is another strategy for sufficient maintenance before an impending failure; it triggers maintenance when degradation occurs in the track. To realize this, Condition Monitoring (CM) is needed. CM means that the status of the railway track condition is monitored, recorded, and reported so that maintenance actions can be conducted before a problem occurs, thus reducing breakdowns which improves the overall track performance and reduces costs [11, 12].

Presently, Trafikverket is responsible for gathering track condition data. Schunnesson et al. [13] described tamping, a maintenance strategy at Trafikverket, and evaluated the measurement effectiveness. Tamping aims to improve the track quality by optimising the track geometry maintenance strategy. Bergquist and Söderholm [14] proposed an approach to be used for condition assessment of tracks. This approach ensures earlier fault warnings while the traditional approach uses a control chart to assess the track twist condition. Schunnesson's strategy relies on the real track geometry data, while the approach in Ref. [14]. relies on visual inspection of the track condition. Sadeghi and Askarinejad [15] developed a new track degradation model in two formats. This model can provide better understanding of the long-term behaviour of railway track system. Vale and Ribeiro [16] proposed a condition-based maintenance model formulated as a nonlinear program that characterised the deterioration rate of the standard deviation. They simulated the deterioration rate using Monte Carlo techniques on the real data, with the goal of optimising tamping and predicting its need in ballasted tracks.

A vast amount of information generated over several parameters indicates the condition of the railway track. Controlling the information is essential to improve the quality of the information and allow stakeholders to make better decisions [17–19]. In addition, information exchanges within maintenance should be enhanced using today's digital technologies. Maintenance draws on new technologies that take advantage of the enormous data available for the railway track. It facilitates the collection, forwarding, and processing of the information for the relevant maintenance activities and stakeholders [19, 20]. Many stakeholders have a share in the system [21, 22]. The maintenance stakeholders involved in analysis and decision making are particularly interested in condition-based maintenance. They are responsible for investigating, designing, and developing the railway maintenance strategy; to make the best decisions, certain needs must be met.

The purpose of this study is to investigate the needs of maintenance stakeholders. As noted previously, they are responsible for analysis of track irregularities and maintenance decision support. The aim of this study is to find out if the stakeholders have access to robust track information and is the information acceptable for track maintenance. Additionally, potential problems in data by analysis are investigated. To fulfil this goal, the study aims to find out what data are necessary and how they are to be collected so that track performance could be improved through proactive railway track maintenance.

2 Railway infrastructure

Railway infrastructure is divided into technical subsystems, including substructure, track, electrical system, signalling system, and telecom system [23]. The rolling stock is treated as a different technical subsystem, depending on what is being transported, passengers or goods [1, 2]. For trains to run on time the condition of the infrastructure must be sufficient [3]. The purpose of a railway track is to provide a stable, safe, and efficiently guided platform for the train to run at various speeds with different axle loadings [24]. There are two types of railway tracks: conventional ballasted tracks and slab tracks [5]. Current studies indicate slab tracks may be more cost effective than conventional ballasted tracks, considering life cycle, maintenance cost, and traffic disruption during maintenance [25]. The slab track incurs higher initial construction costs and repair costs, but structural damage takes longer to occur [5, 25]. Most railway tracks in Sweden are of the traditional ballasted type.

Figure 1 shows the traditional ballasted rail track. The rails are made of I-shaped bars on which the rail vehicle wheels run. Smooth and comfortable rails should have lengths up to 400 m, which are soldered together. Sleepers transfer the load from the vehicles into the ballast and are constructed of concrete. The ballast bed anchors the track, creates bearing capacity for the track, and prevents lateral and longitudinal movement [27]. The sub-ballast is the prime track foundation; it prevents vertical movements based on the thickness of the subgrade [26]. The most physical failures and/or degradation over time occur in the track. The track affects the performance of the whole system and can lead to huge failures. Failure during operation can be costly, including costs linked to loss of

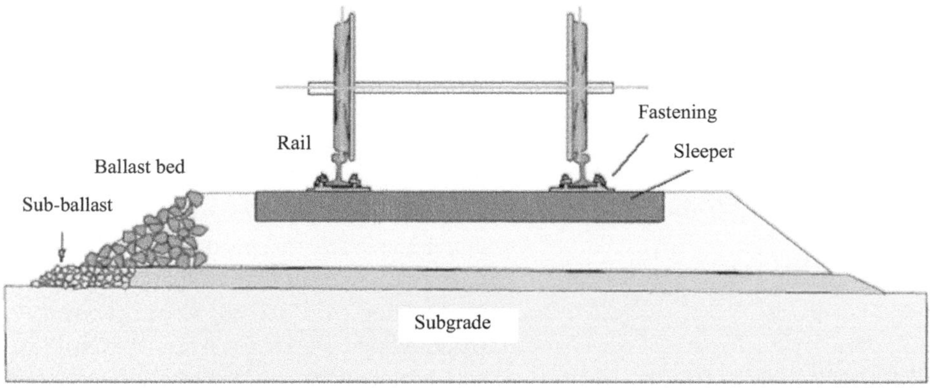

Fig. 1 Example of ballasted rail track [26]

service, property, or human life. Good track quality means lower cost and longer service life. Often, the quality is mixed because elements with different ages and statuses have to work together. Moreover, the track is not perfectly levelled and aligned; track irregularities cause the train to vibrate. Passengers are less comfortable, and goods may be damaged. This explains the need for maintenance [3, 5].

2.1 Maintenance of railway infrastructure

Proper maintenance of railway is necessary for rail traffic. The Swedish standard SS-EN 13306 p. 5 [28] defines maintenance as "the combination of all technical and administrative actions with the aim of retaining or restoring an object in such a condition as to perform its proposed function". On a basic level, maintenance is conducted on the infrastructure as reactive or proactive. Reactive maintenance refers to actions after failure has occurred, while proactive maintenance takes place at regular intervals or following certain criteria to restore the desired functionality. The maintenance objective is to minimise safety, health, environment, and financial risks [9].

Overall, maintenance plays an important role in the railway industry, supporting both safety and profit-making [6, 9]. Maintenance support includes such resources as documentation, information, information systems, personnel, support equipment, materials, spare parts, and facilities [29]. The trend in the railway has been towards increasing automation and increasing productivity. As a result, it has become larger, more expensive, and more technically complicated than ever. The track infrastructure is a significant part of the railway system, and its maintenance plays a vital financial role in technical, administrative, and managerial decisions [3, 5, 7].

Monitoring the condition of the railway track is part of the maintenance strategy [3, 4] and it is described as an "arrangement of the 'type' of maintenance, which is to be done to the track and its components, at a specific rate or frequency, under specific operating conditions" [3]. This can include the total life cycle, that is, costs associated with the lifetime of the track such as operating costs, maintenance costs, energy costs, and taxes. Also are included costs associated with decisions related to maintenance actions and probabilistic assessment estimations of train delay, derailment, and traffic disruption. All these consider the operational and technical aspects of the track, as well as the maintenance actions [30]. For example, by adding a life cycle costing (LCC) model for the track the total cost over its total lifespan will be calculated. It would be used to systematically process and evaluate the cost impacts of the whole life cycle of the infrastructure [3]. A LCC model is a simplified representation of a real situation in that it extracts the product cost and estimates the relationships of the products to each other and their costs in a model. Each model has its own application of the contents based on the amount of information needed and identified with the results [31].

Presently, Trafikverket is developing a LCC model of the railway infrastructure in the Mainline project [32]. This model incorporates mechanisms to estimate track deterioration with the financial and operational impacts. In this project, a tool is being developed, although not to be a fully functioning piece of software, consisting of recommendations and algorithms to be used in developing a commercial product. This Life Cycle Analysis Tool (LCAT) provides elements of Life Cycle Analysis (LCA) to assist work in railway track to justify decisions based on evidence, develop a budget for more efficient maintenance, define levels of safety and service, and predict timing for future expenditures [32].

2.2 Condition-based maintenance for railway infrastructure

A particular maintenance strategy should be determined for railway infrastructure components for specific operating

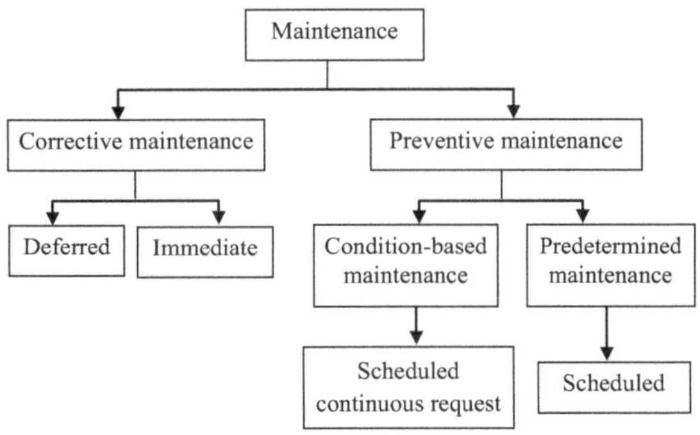

Fig. 2 Description of the maintenance strategy [10]

conditions [3]. Generally speaking, railway maintenance strategy can be one of the two types, corrective maintenance or preventive maintenance (see Fig. 2 [9]). Corrective maintenance, immediate or deferred, is carried out after fault recognition [10]. Immediate maintenance is done as soon as the fault is detected, while deferred maintenance is delayed according to given maintenance rules. Preventive maintenance takes place at predetermined intervals or according to specific criteria to reduce the probability of failure or the degradation in the functionality of the equipment. Preventive maintenance can be predetermined maintenance following established intervals of time or number of used units or condition-based maintenance [10].

Condition-based Maintenance (CBM) for the railway is a maintenance strategy that can provide sufficient warning of an impending failure. It will allow the equipment to be maintained only when there is objective evidence of failure. The methods used to monitor tracks have been improved over recent decades. Sensor fusion techniques are now commonly in use [6]. Moreover, a variety of techniques in vibration, temperature, acoustic emissions, ultrasonic, lubricant condition, and time/stress analysis have received considerable attention. The components are usually replaced or repaired as soon as the monitoring level value exceeds the normal value [6, 29]. Track condition monitoring is an important part of CBM. Condition monitoring (CM) monitors the current railway track condition and aims to predict the future condition of the operation [33]. CM can decrease the operational risk, improve the performance, and reduce the operation cost. Being able to monitor and detect the degradation of the equipment improves the ability to prepare the necessary maintenance. The required maintenance will be performed efficiently, reducing downtime and associated costs [12, 34].

The critical part of the railway infrastructure is the track; track failures potentially affect safety and may cause delays in the speed [35]. Thus, the main objective is to ensure the performance, safety, reliability, and cost-effectiveness of the track. The inspection methods used by Trafikverket to control track irregularities and note any displacement from the design are conducted by inspection trains. The STRIX/IMV100 monitors the track using a contactless measurement system, based on an inertial measurement system and an optical system. The measurement is based on the standard EN 13848-2. Trafikverket track condition data are collected by the measurement wagon. The wagon goes along the track at speeds up to 200 km/h and measures each track section five to six times per year, depending on the section. Up to 35 track geometry parameters are obtained and stored for every 25 cm [35]. Different measurements and ways of measuring are available to assess the railway track components in direct physical contact with each other. For instance, the longitudinal level measurement is a vertical position of the right and left rails measured by a position sensor and accelerometer. The acceleration measure should be integrated vertically twice over time to determine the position of the wagon. The result is summed to identify the vertical position of each rail [11, 36].

To measure track condition, Trafikverket uses the Q value. This value is calculated based on the standard deviation. The Q value includes the contract between the rail administrator and contractor. These values show how the maintenance action is related to the maintenance strategy, i.e. corrective or preventive maintenance [11, 36]. The measurement parameters give early warning of deterioration. The available information from the monitoring and control systems includes performance data from sensors [10, 29]. The information generated in railway track monitoring is used to support different stakeholders. Information exchange and knowledge management affect the decisions made and the system performance [19]. Controlling the information assures better decision-making quality through information verification and improved processes [37].

In addition, ICT (Information and Communication Technology) is used to increase accessibility and usability. ICT provides several technologies to enhance information exchange over the maintenance processes. This, in turn, suggests the need to define and develop new technologies able to manage and use this information [17]. An example of ICT is the use of eMaintenance to manage railway tasks and information using digital technologies [19]. eMaintenance provides diverse benefits, for example, increased value to customers, increased availability, and reduced life cycle cost of railway track. However, developing eMaintenance requires the transformation of traditional "fail and fix" maintenance practices to a "predict and prevent" methodology [38, 39]. It can integrate production systems, operations systems, and remote customer sites, like those found in railway industry [38]. In this context, eMaintenance provides dynamic information to the customers and to the entire railway track system. Developing and implementing this type of system allows information to be processed in real-time. In this way, maintenance information can be used throughout the whole system life cycle [40].

2.3 Stakeholders in condition-based maintenance

ISO/IEC 15288 (2002) defines a stakeholder as "an interested party having a right, share or claim in the system or in its possession of characteristics that meet that party's needs and/or expectations". In this definition, the railway stakeholder is seen as a party with needs requiring appropriate actions [21]. Particular stakeholders in the railway include those involved in operation and maintenance decisions and activities [6]. CBM processes have a number of stakeholders. They may be active within this process internally or externally. Internal stakeholders perform technical and administrative actions on an item. External stakeholders are concerned with the required function of the item. Internal stakeholders may be determined by applying inductive reasoning to the condition-based maintenance process, using empirical data to develop a theory. External stakeholders may be identified by applying deductive logic to the condition-based maintenance process, using general theory to explain a specific case [21, 41]. As an example, stakeholders in condition-based maintenance are categorised in the following groups based on the Swedish railway system:

- *Measurement stakeholders* They are responsible for measuring the railway infrastructure using the measurement wagon. They are responsible for collecting data, filtering the measured data at different levels, and assuring the data accuracy. Their goal is to record the railway infrastructure state.
- *Data quality stakeholders* They are responsible to assess the data fitness to serve a specific purpose. The assessment uses statistical methods, comparing new data with previous data records. They report the data quality to the measurement stakeholders to solve problems.
- *Analysis stakeholders* They are responsible for analysing the incoming data using different factors. This analysis will help to identify the railway infrastructure state and decide when, where, and how maintenance should be done. A maintenance decision is made by analysing a track section. Most of the decisions are corrective maintenance.
- *Decision-making stakeholders* They have the responsibility for designing and developing the railway infrastructure maintenance strategy. They also define the cost of and the budget for maintenance. For instance, they plan tamping and determine the cost.
- *Maintenance stakeholders* They are responsible for performing the required maintenance on the infrastructure. The required action is based on the decisions of the decision-making stakeholders.

The stakeholders in condition-based maintenance are responsible for analysing the railway infrastructure state and setup and developing the maintenance strategy. Obviously, stakeholders who make decisions have certain needs; if their needs are met, in terms of the correct data gathered by selecting the correct measures, their analysis will be better, they will make better decisions, and the performance of the railway system will improve.

3 Case study design and participants

A case study is generally used to test assumptions and to answer specific research questions. The design of case study established the chain of evidences to create a draft of information, in order to capture the possible information. Through our case study, we seek to improve the railway infrastructure performance [41].

In what follows, first, we do a review of the literature on maintenance, condition monitoring, and condition-based maintenance. Next, we conduct a case study using mixed methods research. We use material from interviews to prepare the questions. The results help us to recognise the data, parameters, identify the problems, and determine the corresponding needs to correct the problems and improve track performance. Finally, the study analyses track condition data with respect to the identified problems/needs. The mixed study combines qualitative and quantitative approaches, with two types of data sources.

3.1 Interviews

Individuals interviewed held positions where they were responsible for data quality, data analysis, and maintenance decision making for Trafikverket. Eight interviewees work fulltime analysing the railway track performance, and one spends five percent of his time on analysis. Table 1 shows the participants, their position, roles, and the years of experience. The interviews were open-ended lasting from two to four hours per individual. The interviews were either face-to-face or via videoconference so as to ensure better understanding. This method also yielded a high volume of information. Some interviewees were followed up by telephone to complement missing information. Questions used served as guidelines to gain information for four different areas; data, parameters, problems, and needs. The questions are found in Appendix.

The interviews were structured around what was required by decision-making stakeholders and the problems

Table 1 Interviewees' position, role in the organisation, and experience

No. of interviewees	Position	Roles	Experience
Interviewee 1	Track engineer	Developing Life Cycle Model for Railway Superstructure and introducing "Standard" element which is a calculation block within LCC depending on certain parameters	10 years
Interviewee 2	Researcher	Research on railway maintenance optimisation by applying maintenance decision support tools (RAMS & LCC)	7 years
Interviewee 3	Track engineer	Specialist in track analysis	13 years
Interviewee 4	Track maintenance engineer	Maintenance planning, Planning of major replacements of track components, Technical support to local track managers	5 years
Interviewee 5	Maintenance engineer	Long-term maintenance planning and technical support	20 years
Interviewee 6	Track engineer	Management leader Optram	6 years
Interviewee 7	Project Manager	Project Manager of the contract for Track measurements	9 years
Interviewee 8	Maintenance engineer	Presenting long-term maintenance plans	10 years

they faced. The interviewer wanted to get to the point without missing important issues. The interviewees were experienced and understood the environment. All interviews were tape recorded and transcribed to make data analysis easier and capture the relevant information. The interviews results considered the expert personnel answers. The analysis of the interview data was used to make empirically based conclusions on the case study data. Specifically, the case study used a pattern matching analytic technique, comparing an empirically based pattern with a predicted one or with several alternative predictions. The pattern matching technique finds gross matches or mismatching, so interpretation is less likely to be challenged.

3.2 Data collection and treatment

The case study is based on the Trafikverket database Optram (optimised track management system). These data collected with a measurement vehicle, operated by a contractor, were used to observe more than 30 parameters of the track condition. The parameters are measured up to six times per year. The measurement depends on the nature of the section, such as curves or stations, and its relation to GPS positioning and kilometre markers on the track. The resulting data on track quality are divided into five classes, depending on the load and the speed [11, 12]. Optram data consist of track geometry measurements, overhead wire measurements, real profile measurements, corrugation measurements, ballast profile measurements, ground penetrating radar, and stiffness measurements. These data are used in planning the maintenance strategy. The data have been chosen for the lines section BDL111 and BDL119 between 2007 and 2014.

The Optram track geometry parameters are longitudinal level (short, middle, and long wavelength), alignment (short, middle, and long wavelength), track gauge, cant, twist (3 m and 6 m), and the standard deviation for longitudinal level and cooperation. Each of these parameters is calculated on the EN 13848 standard. Also, the indicator identifies the track state is Q value. Q value is the sum of standard deviations of the cant error and the standard deviation of the average longitudinal level on the comfort limits for the acceptable standard deviation.

Track section 111 is a 130-km section of single track from Kiruna to Riksgränsen. It is mainly used for iron ore transport, which means that this line demands much maintenance at a high cost. The train quality class is 3, 80–120 km/h, the maximum axle load is 32 t, and the annual accumulated tonnage is over 22 MGT. However, the track structure on this line section was renewed with ballast between 2006 and 2009. In addition, it should also be noted that this line is operated in extreme climate conditions, which can influence the track performance. The performance has effects on the availability and safety aspects of the track. The annual climate varies from warm summers to heavy snow, and the temperatures vary between +27 °C and −34 °C.

The track section 119 is a 34-km section of single track consisting of six meeting stations from Boden to Luleå. Two of these six meetings allow for passenger exchange. The traffic consists of both passenger and goods traffic, and has an annual volume corresponding to approximately 23 MGT. On average about 50 trains pass the section daily, including ore trains with an axle load of 30 tonnes, freight trains with axle loads of 22.5–25 tonnes, and passenger trains. The speed varies from 50 km/h for loaded ore trains up to 140 km/h for the passenger trains. Figure 3 shows the track sections 111 and 119 on the Sweden map.

The data analysis was conducted using Excel based upon measured and filtered track quality values; longitudinal level, alignment, track gauge, cant, twist, STD for

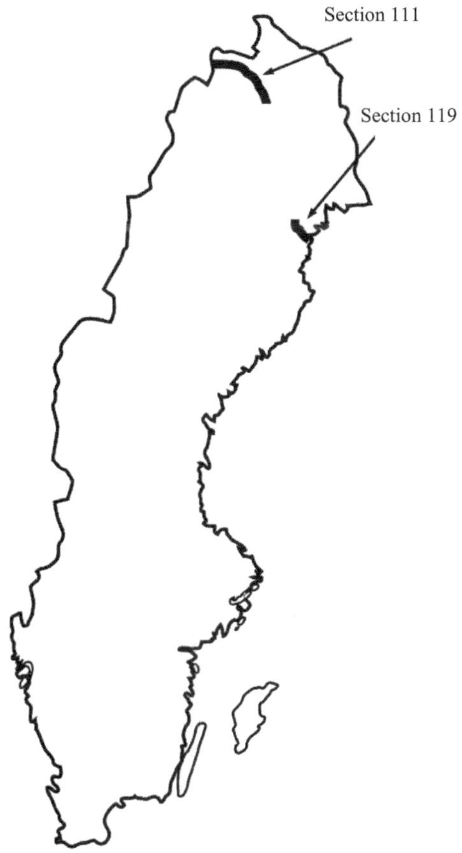

Fig. 3 Sweden map track sections 111 and 119

longitudinal level, and the relationship of track distance (km) or over time. This study analyses longitudinal level, alignment, track gauge, cant, and twist over the kilometre. The distance was analysed for 200 m of each track section. This was done to gain an understanding of the track behaviour and quality for facilitating maintenance decision making. In addition, STD for longitudinal level and cooperation analyses over time was used since it aims to understand the track deviation over the years and recognise the problems from measurement or the data quality.

4 Results and discussion

We selected a single case study to describe the needs of stakeholders, along with the problems encountered, with the goal of improving the performance of the railway infrastructure. Ultimately, the findings should help companies make better maintenance decisions about track and asset. They will be able to define what work needs to be done, and when and where it should take place. To this end, the study correlates track data from many resources and provides features for analysing railway data. The interviews provided main four points: data, parameters, problems and analysis, and the needs of analysis and decision-making stakeholders.

4.1 Data

Trafikverket collects data by monitoring the infrastructure. These data are collected by many systems: assets BIS, failure system 0felia, inspection system Bessy, track recording measuring data, and wayside monitoring systems for wheel flats and hot bearings. For the track condition data, the data quality stakeholders agree that the accuracy of data measured or calculated is good. They agree that the collaboration between different types of data measurements works well and the content accuracy is acceptable. Some stakeholders at Trafikverket are working on assuring data quality by comparing many measurements at the same time and in the same place. This will help them understand the overall situation, for example, by comparing the data of two different car measurements IMV100/EM80 and IMV200/STRIX over the same distance. Figure 4 shows the comparison of these measurements. This comparison was done using Optram data. IMV100/EM80 was measured on 12 September 2014. IMV200/STRIX was measured on 3 October 2014.

Figure 4 shows the difference between two measurements through the longitudinal level (mm). Longitudinal level (mm) is a parameter measured by Trafikverket to express a mean vertical position covering the wavelength ranges stipulated, and is then calculated through successive measurements. Figure 3a shows the IMV200/STRIX car has a smooth trend over the distance, with only about 2 mm difference in some places at the same position. Figure 3b shows IMV100/EM80 has a harsh trend, with around 2 m shifting over the same distance measurement in some places. Car inspection uses more or less the same equipment but at different speeds or with different collaborations between car equipment. The car runs once per year. During this time, if the car inspection finds a problem, an alarm will be recorded in Optram. However, the car inspection cannot measure the data accuracy. Additionally, it recognises neither empty values nor zero values in the data; this affects processing. For this reason, measurement is uncertain, for example, when the track is covered by snow.

4.2 Parameters

Parameters are measurable factors setting the conditions of track operations. Trafikverket measures more than 30 parameters used in variant analysis and prediction. The analysis of parameters shows the quality of the track and indicates failures requiring maintenance. Trafikverket contractors are responsible for doing maintenance when the

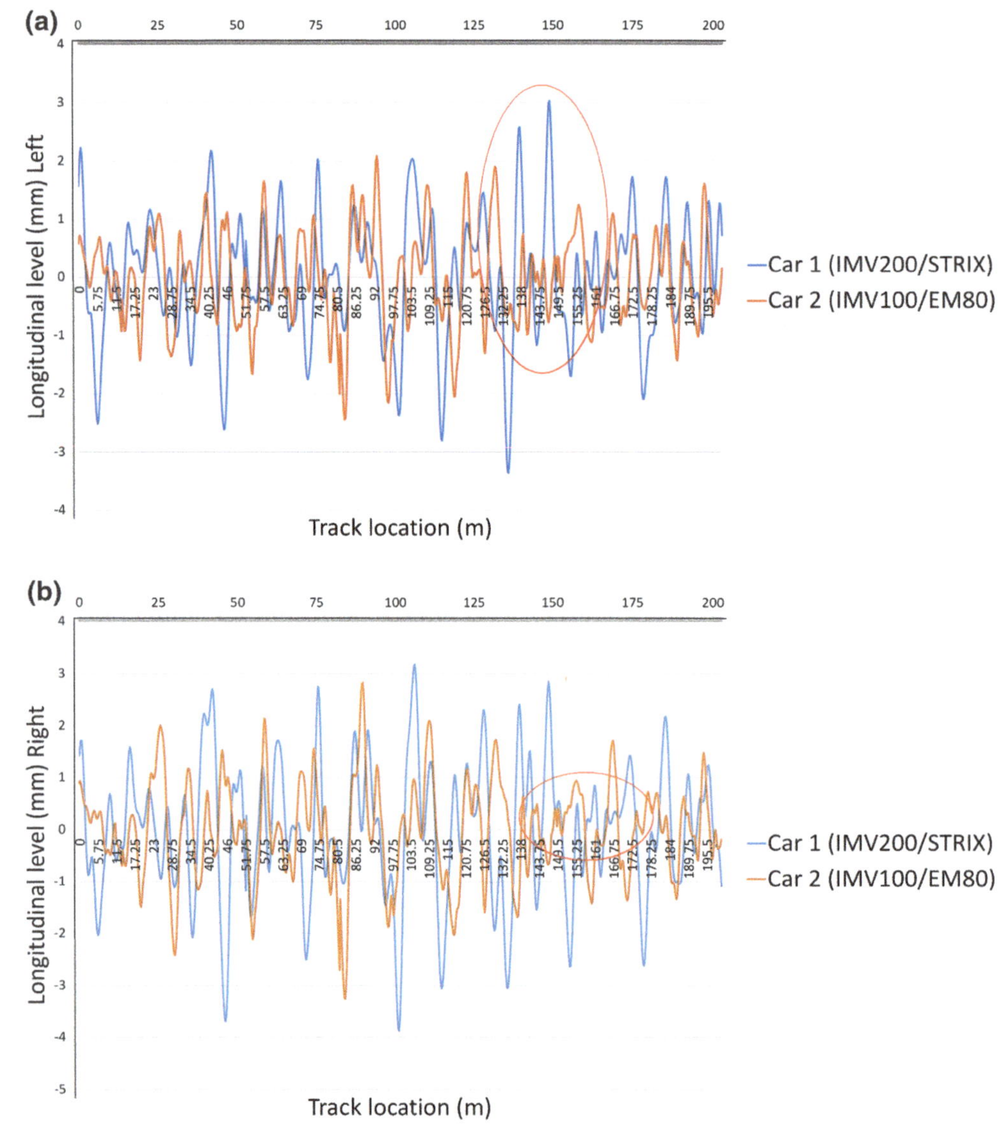

Fig. 4 Measurement of track line BDL111 at kilometre marker 1446 **a** from vehicle IMV100/EM80 and **b** from vehicle IMV200/STRIX

parameters' threshold is exceeded. Although it is difficult to know what other contractors did or how they delivered the data, the main parameters are track geometry. The degradation of these parameters is extremely dangerous; degradation also has an effect on passenger comfort. Track geometry parameters are track gauge, alignment, longitudinal level, cant (cross level), and twist. Figure 5 shows four of these parameters within a spot failure on track number 1464, measured on 29 September 2014.

Figure 5 presents the analysis of four parameters, longitudinal level, track gauge, twist, and cant of the failure spot, measuring a 200-m distance. These figures show the degradation of track performance between 13 m and 30 m. Obviously, the degradation over this distance appears with the analysis of the four parameters. However, these parameters have different measurements and meanings.

- *Track gauge* is the distance between the gauge faces of the two adjacent running rails.
- *Alignment* is the mean horizontal position covering the wavelength ranges stipulated, then calculated from successive measurements.
- *Longitudinal level* is the mean vertical position covering the wavelength ranges stipulated, then calculated from successive measurements.
- *Cant (cross level)* is the height of the vertical side of the right-angled triangle related to nominal track gauge plus the width of the rail head.

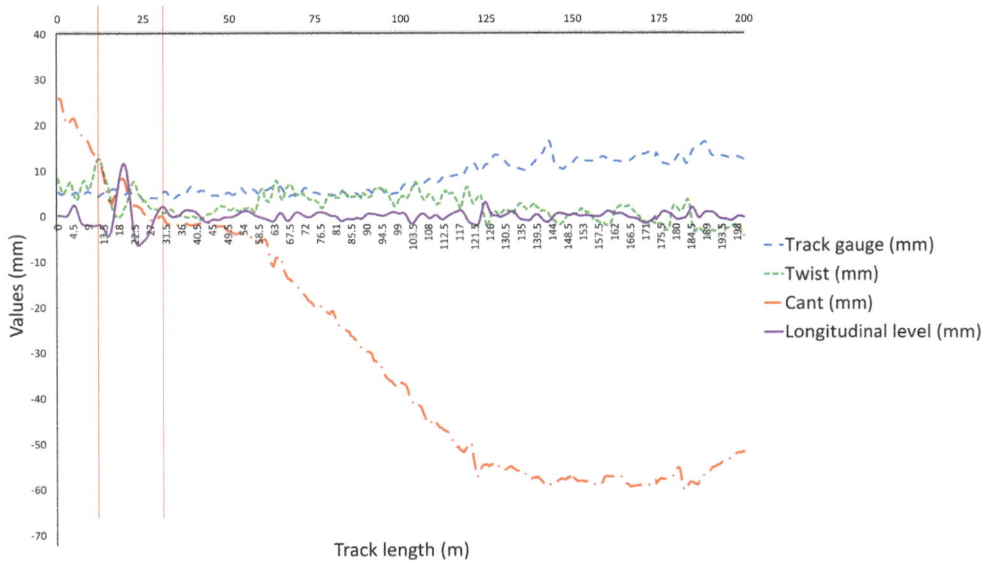

Fig. 5 Track geometry parameters for track line BDL111 km 1464 and the measurements within *red* column show spot failure

- *Twist* is the gradient between the two points of measurement.

4.3 Problems

Trafikverket stakeholders recognised problems after performing data analysis and comparing the historical data. Different types of analysis show the problems and give a better understanding of the nature of the problem.

4.3.1 Positioning problems

Positioning means finding the exact position on the track that needs maintenance. Two types of maintenance actions are considered, tamping or manual action. The positioning data are integrated from BIS (Ban Information System). BIS describes the railway in the form of nodes and links. Then, it classifies the railway into systems, areas, and distance, based on the map of Sweden. Positioning problems occur because kilometres (km) given in the BIS system are different from reality. The positioning within kilometres (km) is correlated using different track areas and stations. There might be additional uncertainty in measurements of alignment when the track is covered by snow. Trafikverket analysis stakeholders observed the difficulties in determining the exact position which posed a problem for doing tamping. The contractor with Trafikverket should deliver positioning of less than 5 m, but the track positioning is currently ±20 m. Therefore, the measurement data are neither good nor accurate. This obviously appears in comparisons using other historical data for the same distance, such as longitudinal level measuring for a 200-m distance. Figure 6 shows the positioning problem by comparing the data of 30 January 2014 and 28 May 2014 over the same 200-m distance.

Figure 6 presents the difference between these two measurements for the same track number and the same distance, around 20 m. This difference will increase difficulties in figuring out the exact failure point; this, in turn, will increase maintenance cost. Most of the interviewees agreed that this problem has a negative effect on maintenance decision making. It might lead to wrong decisions. At the same time, uncertainty about the position might have an effect on the selection of a maintenance strategy. Other interviewees disagreed, saying that positioning is not a huge problem and does not greatly affect the maintenance decision. However, no statistical study shows the number of positioning problems over the whole data. There is some discussion about integrating a GPS system for more accurate position, but Trafikverket contractors have no desire to integrate it.

4.3.2 Limitations of mean Q value

Calculating the mean Q value could be useful to estimate the track quality, and then compare the tracks to get an idea about the budget. Figure 7 shows a Q value curve over 200 m, calculated on 29 May 2014 with the measured track length of 1464 kilometre.

Figure 7 shows the quality of the track. Over this distance, the quality is high. The quality is decided based on SS-EN13848 standard. The standard categorises the track classes; however, the Q value limitation is based on an arbitrary judgment of the historical mean values. It is more

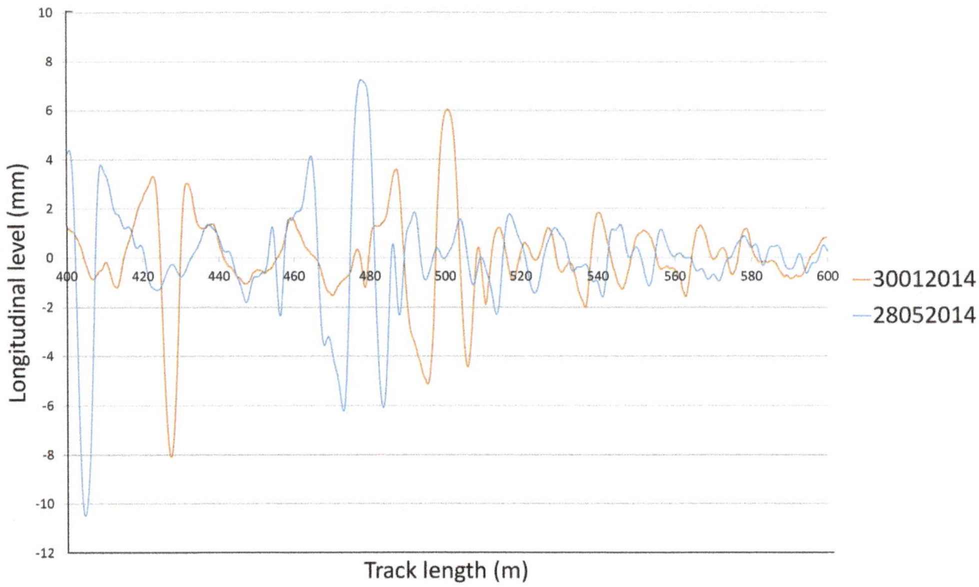

Fig. 6 Measurements of location show a variation of circa 20 m with the same measurement vehicle on track line BDL119 at kilometre marker 1174 in Jan and May 2014

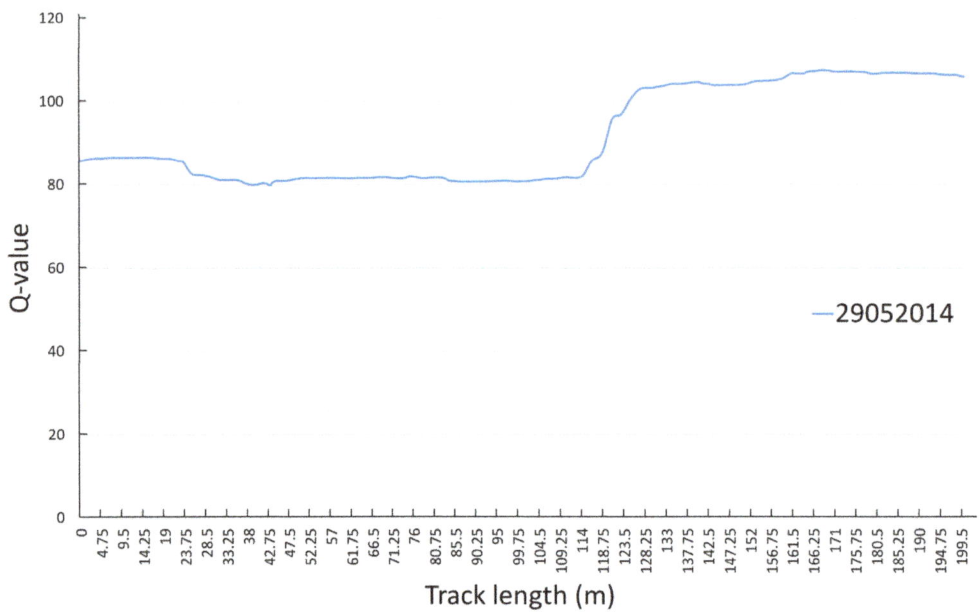

Fig. 7 Track quality Q value of track line BDL111 at kilometre marker 1464

common to divide the track into small sections to get more control and detect the track quality better, but it is critical for track quality to set up an appropriate Q value limit that should be maintained. The formula calculating the Q value has no accurate value limits on track quality and age. Additionally, it shows the need to have more knowledge about the degradation of track quality to set up the proper limitations. The interviewees mentioned that the limit for the Q value is not high. This means the performance-based contracts need to ensure a better standard and quality. Also, it is not a sure-fire method for keeping high track quality in the long run with low maintenance cost.

4.3.3 *Difficulties of analysis*

The problem is that the systems use different condition monitoring data for the railway infrastructure. Trafikverket has many systems responsible for recording all the

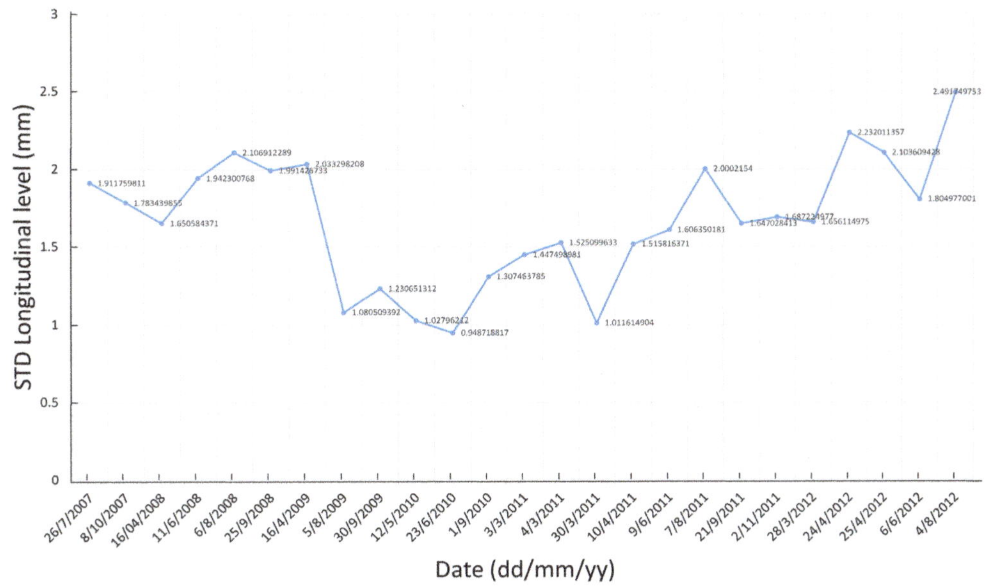

Fig. 8 Values from same track section showing variations in STD of longitudinal level from 2007 to 2012

infrastructure state. Trafikverket's software tools for analysis do not provide all the required functionalities; therefore, it is difficult to do additional analysis or calculations to better understand track degradation. Hence, different systems must be integrated. For instance, Optram can offer analysis over time or distance as shown in Fig. 8. This analysis shows the need for doing tamping based on the STD values, although specific problem and problem area are not available. Figure 8 shows the results of data from a 100-m section that was recorded between 2007 and 2012.

It is not possible to plot specific positions due to the nature of the system and the type of data available in the system. Presently, personnel export data to other tools, such as Excel, for further data analysis. However, it is still difficult to accurately analyse different track sections simultaneously. Track quality is an important analysis and for this to be done tracks must be compared.

4.3.4 Short-term strategy

Trafikverket has a short-term strategy whereby their contractors fulfil the demands of the Q value, using corrective tamping to decrease track failure. If irregularities in ballast occur too often, Trafikverket will lose control of track quality since there is no specific long-term strategy. The cost of using the tamping machine is uncertain. Figure 9 shows the short-term strategy based on the analysis of longitudinal level (mm) on 29 May 2014. The figure suggests the need to do tamping between 18 and 25 m. Over this distance, we see very high peaks and very low peaks. Note that longitudinal level is one of the parameters used to calculate the Q value. Therefore, this variant will affect the Q value.

However, throughout a year, tamping is conducted based on the track quality analysis. Yet, the results show that the analysis is inaccurate and ineffective. Therefore, discussions and on-going projects are conducted to identify future tamping plans. An example of this is that the Q value has no specific limit and the measured parameters are not reliable enough to make tamping decisions. It is proposed that other measured parameters should be used to predict track degradation and determine tamping actions.

4.3.5 Incomplete track measurement

Incomplete track measurement has been recognised in data analysis. In such cases, the measurement wagon was supposed to measure the whole track, but was suddenly stopped or changed direction. The measurement stakeholders were supposed to ensure the continuity of the measurement wagon but sometimes they did not realize a distance was not measured. Trafikverket measures 5–6 times per year. Software tools have indicated some places are not measured. Figure 10 shows the STD of longitudinal level of 100 m of track. The distance between 1164,661 and 1165,561 km has not been measured over several measuring times: 1 January 2010, 3 March 2011, and 4 March 2011.

The measurement of the track has been disrupted, with negative effects on decision making. Some parts have not been measured, so there is no accurate assessment of track situation.

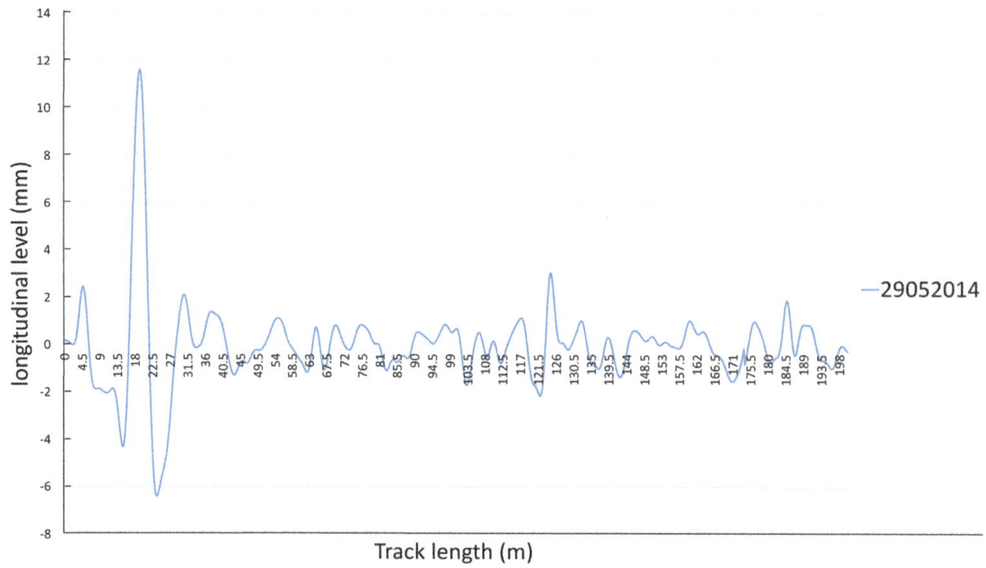

Fig. 9 The high longitudinal level variation above shows that tamping actions must be taken

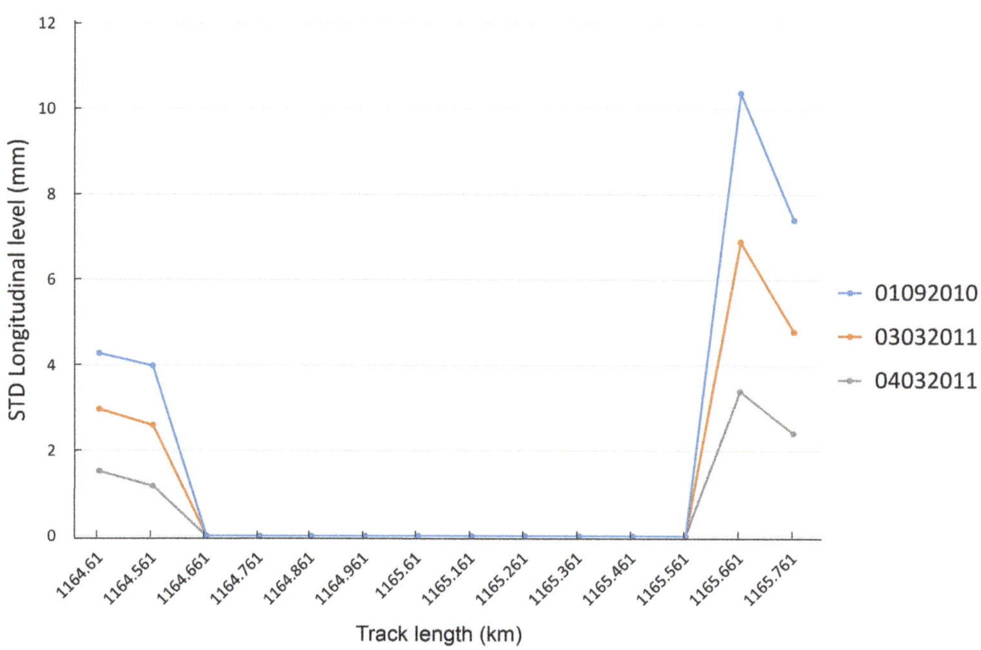

Fig. 10 STD of longitudinal level of 100 m of track

4.4 Needs

Trafikverket stakeholders that are responsible of analysis and maintenance decision making want to improve maintenance performance and decrease costs by improving analysis and prediction. If done correctly they will increase railway availability, improve safety, increase the use of resources, enhance reliability, and overall performance. As a result of this, track degradation and maintenance costs will decrease. To meet these aims, the following must happen.

4.4.1 Analysis enhancement

Enhance analysis by integrating data resources. Trafikverket has several data resources such as assets BIS, failure system 0felia, inspection system Bessy, track recording measuring data, and wayside monitoring systems for wheel flats and hot bearings; these should be integrated into one system. This will facilitate the analysis of axle weight, speed, or total load instead of just time analysis, as shown in Fig. 8. The trends found in the analysis suggest that specific parts of the track need maintenance and give a better understanding of the situation. For instance, the analysis presents the standard deviation of alignment over time. Other factors must be considered for analysis as well, such as axle weight or speed, as these are partially to blame for failure. At the same time, a general framework capable of integrating all these systems is required. Most of the interviewees have used other software tools in Trafikverket to integrate systems and enhance analysis to assist them in maintenance decisions and estimation of maintenance costs.

4.4.2 *Further analysis*

After analysis, store the highest and lowest degradation values in the data source for each track section. Facilitate analysis based on the calculated values, to ensure better decision making and to predict track degradation. Storing the values for each track will facilitate the comparison of railway tracks. Figure 8 shows the importance of this for understanding degradation. A comparison is used to estimate the failure spot. It aims to estimate the existing problem and understand the whole performance of the railway infrastructure. One of the interviewees said he stores these values manually in an individual file for further calculation. They might help understand and minimise track degradation. As the analysis in Fig. 5 indicates, considering the parameters of two tracks can better explain the behaviour by highlighting the track situation and allowing stakeholders to compare the highest and lowest values in each track to estimate the current problem. It will also allow the stakeholders to determine the highest values on track quality and performance.

4.4.3 *Track maintenance strategy determining*

Develop and determine a maintenance strategy for each track section. Today the railway track has two maintenance strategies, either, reactive or proactive. Many software tools in Trafikverket are supposed to do proactive maintenance, but Trafikverket uses them for reactive maintenance because there is no clear limitation for maintenance cost. There is a need to shift the culture of Trafikverket, specifically among the analysis and decision-making stakeholders, from a reactive to a proactive approach. The maintenance of the railway tracks needs to integrate proactive solutions into its plan to reach the desired goal of improving safety, security, and reliability. Figure 9 shows the preferred case for deciding maintenance. Most of the interviewees said defining the strategy depends on the track degradation type. They agreed that a proactive strategy can decrease the maintenance cost and improve performance. They also said that they have a limitation for each track parameter, which has two levels UH1 and UH2 defined by the SS-EN13848 standard. The maintenance action is taken based on degradation reaching the UH1 as reactive maintenance. The UH1 values are lower than UH2. There is no reason for contractors to work on a proactive strategy when the degradation reaches UH2.

4.4.4 *Track maintenance planning*

Plan the maintenance for each track. Maintenance planning is done by the Trafikverket contractor. The car inspection records the faults, as shown in Fig. 4, and equates risk level to required maintenance. Corrective tamping maintenance is governed by regulations in the SS-EN13848 standard and preventive tamping should be planned in advance. There are two main aspects to consider in maintenance planning: applying for track possession time and decreasing the maintenance cost by predicting the required tamping. For instance, predicting tamping for nine months is enough time to consider the maintenance plan and budget. However, the costs depend on the distance of the track needing maintenance and the location needing maintenance. Another person claimed setting up a maintenance plan for a certain part of the track will help estimate the cost required to order the tamping machine. At the same time, it will decrease the cost and improve the track quality and performance.

4.4.5 *Time plan for maintenance action*

Create a time slot for maintenance. It is important to determine the time required for maintenance and estimate the cost required for each part of the maintenance plan. Estimating the extra time required to finalise the task is a learning process for other maintenance actions as well. Setting a time slot is crucial in estimating the time required for the maintenance plan, the extra time needed to complete the maintenance and the required cost. Determining the extra time required will help to set up a better maintenance plan.

4.4.6 *Track visualisation*

Visualise the railway track data geographically. This can be done through GPS coordinates that measures Trafikverket during the measuring track quality. The visualisation will simplify managing the railway. The mapping can show the track parts and give information on the geographic track location. This method allows

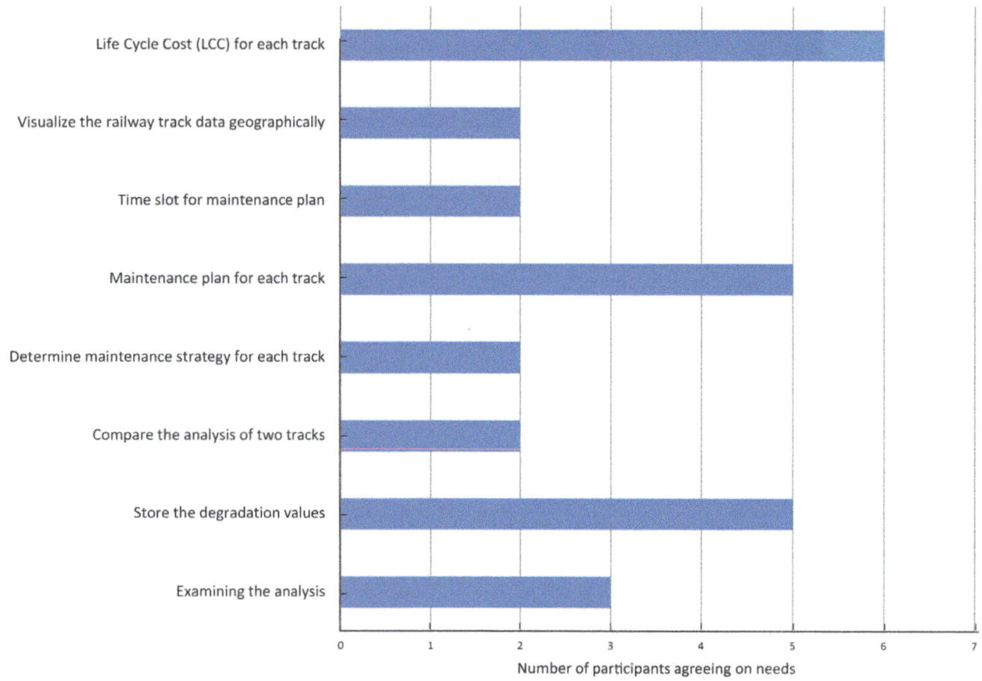

Fig. 11 Condition-based maintenance stakeholders' agreement on needs

stakeholders to connect the results of the analysis with the track infrastructure. However, visualising the track can help analysis and decision-making stakeholders to determine the more seriously affected areas by offering a clear picture and facilitating comparisons with other track sections at the same time. A clear picture of the track will show the influence of various factors on the track status. Also, it makes easier to understand how they influence one another. This technology has the ability to examine individual aspects of the tracks and would provide evidence to justify decisions to funders and policy makers.

4.4.7 Track life cycle cost (LCC)

Determine life cycle cost (LCC) for each track. This requires a comparison of different maintenance, intervention, and renewal strategies for track sections based upon the costs and risks over a period of time. Track deterioration modelling techniques are used to evaluate the deterioration processes for sleepers, transport loadings, track alignments, lengths of radius, the operational speed, sublayer conditions to renew the track structure. However, the track LCC model inputs are initial track condition data, intervention triggers, uplift effects, and the cost of interventions and environmental impacts. LCC's effective decision supports system for maintenance of the railway track. At the same time, the output cost takes into account all factors associated with the lifetime of the system, such as operating costs, maintenance costs, and energy costs. The maintenance cost plays an important role in track infrastructure. The cost of maintenance of LCC track infrastructure has two different levels of uncertainty. Costs are due either to such factors as train delay, traffic disruption or derailment, or to the reliability and maintainability of the track. The main issue for decision-making stakeholders is determining the tamping cost. Hence, it is necessary to order tamping and set up the cost for about a year. Therefore, Trafikverket wants to connect the life cycle costs within certain areas needing maintenance, e.g. tamping. Ultimately, determining the life cycle cost for the track will be based on the equipment needing maintenance.

Figure 11 shows the interviewees' agreement on needs. Many stakeholders want LCC, a maintenance plan for each track and the ability to store values after the analysis. However, two of Trafikverket's condition-based maintenance stakeholders want a maintenance strategy for each track; they want to compare the analysis of two tracks at the same time, clearly visualise the track on a geographical map, and ascertain a time slot for each maintenance plan. Achieving these needs will improve railway track performance, facilitate the analysis of the track, and help predict the future state of the track.

5 Conclusion

The purpose of this study was to investigate the needs of maintenance stakeholders and to find out if the stakeholders' knowledge needs are satisfactory. The results showed that problems with data measurement and analysis should be dealt with specific needs for more consistent data and information that can assist them in maintenance decisions. These needs are:

- Positioning irregularities of measured data made it difficult to determine where maintenance activities were to be conducted.
- Over 30 railway track parameters are measured but only five show to be significant for track irregularities. Degradation of these affects safety.
- Long-term maintenance strategy is lacking, therefore, only arbitrary judgement and historical data are used to plan maintenance actions.
- Limitations in analysis of data since the software and data sources are not compatible with each other.
- Missing data, even when measurements are conducted regularly on the track sections. It is not known why the information is lost or missing.

In addition to the problems, areas of improvement were found. These are:

- The maintenance of the railway tracks needs to integrate proactive solutions in order to reach the desired goals of improved safety, improved availability, and improved reliability. This includes analyses based on the calculated values to understand the process of track degradation and what needs to be done to improve maintenance decisions.
- Develop a visualisation tool that shows the railway track geographically, using both mile markers and GPS coordinates, for information concerning track quality. This mapping should show the track parts clearly and be used to facilitate the connection between the results and the track infrastructure. This is to also include maintenance windows (time slots), which are crucial in assisting the estimation of maintenance actions.
- Develop Life cycle cost (LCC) model for track degradation where maintenance interventions can be compared with different interventions and maintenance strategies.

These findings are applicable for the Swedish railway network in that the areas with specific problem and the areas chosen for improvement are critical for maintenance decisions. For good decision support, it is necessary to use accurate condition monitoring data, from all sources, to gain a complete understanding of the railway network. And then a more long-term strategy can be set up which can then support a LCC model for railway infrastructure. For that to occur, a more specific development, in the areas of condition monitoring data collection and in the development of analysis models, is needed. When these are complete, track degradation models necessary for a LCC model, with clear input definitions for data, triggers, uplift effect, cost, and environmental impacts can be developed. In conclusion, by applying the knowledge gained in this study, railway track availability can be increased, as well as, operation and maintenance costs can be reduced.

Limitations of this study are that there were a limited number of interviews conducted, although key people were chosen, and that only Optram was studied. It is true that there are other systems used in the railway network. Even if some do report more accurate data they do not specifically deal with track condition data. Additionally, by analysing several systems together, a better understanding of the whole rail network could be possible; however, this study does show where improvement work should begin.

Appendix

The questions focus on two main areas:

(a) Identify the problem.
(b) Define the stakeholders needs.

Data

(1) What do you think of the data accuracy? Why?
(2) Is the contractor responsible for assuring the accuracy or only for data gathering? How?
(3) How can the contractor assure the quality of data? In each point give me an example.

Parameter

(4) What are the parameters that are more sufficient and important? Why?

Problems

(5) What are the problems recognised after analysis?
(6) What are the problems that exist in different areas such as functionality and data? (As an end-user)

Needs

(7) What analysis is needed within and does it not exist now? Why? and How?
8) What is needed to improve the prediction? Why? and How?

References

1. Nissen A (2009) Development of life cycle cost model and analyses for railway switches and crossings. Lulea University of Technology
2. Svensson N, Eklund M (2007) Screening of environmental pressure from products in the Swedish railway infrastructure: implications for strategic environmental management. Resour Conserv Recycling 52(2):248–265
3. Tzanakakis K (2013) The railway track and its long term behaviour: a handbook for a railway track of high quality. Springer Science & Business Media, Berlin
4. Esveld C (2001) Modern railway track
5. Indraratna B, Salim W, Rujikiatkamjorn C (2011) Advanced rail geotechnology–ballasted track. CRC Press, Balkema
6. Murthy DP, Kobbacy KAH (2008) Complex system maintenance handbook. Springer, New York
7. Karim R (2008) A service-oriented approach to e-maintenance of complex technical systems. Luleå University of Technology, Division of Operation and Maintenance Engineering
8. IEC (2004) 60300 (3-14): Dependability management—Part 3–14: Application guide Maintenance and maintenance support
9. Kumar U, Nissen A, Schunnesson H et al (2010) Operational Safety and Maintenance. LTU, Luleå
10. Crespo Márquez A (2007) The maintenance management framework: models and methods for complex systems maintenance. Springer Verlag, London
11. Iman Ak (2013) Cost-effective maintenance of railway track geometry: a shift from safety limits to maintenance limits, Luleå tekniska universitet
12. Asplund M (2014) Wayside Condition Monitoring Technologies for Railway Systems, Luleå tekniska universitet
13. Schunnesson H, Juntti U, Nissen A et al (2013) Evaluation of track geometry maintenance for heavy haul railroad in Sweden-a case study. Proc Inst Mech Eng Pt F: J Rail Rapid Transit:0954409713482239
14. Bergquist B, Söderholm P (2014) Data analysis for condition-based railway infrastructure maintenance. Qual Reliabil Eng Int
15. Sadeghi J, Askarinejad H (2010) Development of improved railway track degradation models. Struct Infrastruct Eng 6(6):675–688
16. Vale C, Ribeiro IM (2014) Railway condition-based maintenance model with stochastic deterioration. J Civ Eng Manag 20(5):686–692
17. Karim R, Candell O, Söderholm P (2009) E-maintenance and information logistics: aspects of content format. J Qual Maint Eng 15(3):308–324
18. Haseloff S (2005) OPUS—Context Awareness in Information Logistics
19. Candell O, Karim R, Söderholm P (2009) eMaintenance—information logistics for maintenance support. Robot Comput Integr Manuf 25(6):937–944
20. Gutsche K, Böhm T (2011) e-Maintenance of railway assets based on a reliable condition prediction. Int J Perform Eng 7(6):573
21. Söderholm P, Holmgren M, Klefsjö B (2007) A process view of maintenance and its stakeholders. J Qual Maint Eng 13(1):19–32
22. S\oderholm P (2005) Maintenance and continuous improvement of complex systems: linking stakeholders requirements to the use of built-in test systems
23. Espling U, Kumar U (2008) Benchmarking of the maintenance process at Banverket (the Swedish National Rail Administration). In: Anonymous Complex System Maintenance Handbook. Springer, pp 559–583
24. Chiacchiari L, Loprencipe G Measurement methods and analysis tools for rail irregularities: a case study for urban tram track. J Mod Transport:1–11
25. Pingrui Z, Xueyi L, Guan L (2014) Experimental study of temperature gradient in track slab under outdoor conditions in Chengdu area. J Mod Transport 22(3):148–155
26. Lander S, Petersson J (2012) Tamping planning in railway maintenance: improvement potential for optram as decision support
27. Chiacchiari L, Loprencipe G Measurement methods and analysis tools for rail irregularities: a case study for urban tram track. J Mod Transport:1–11
28. Swedish Standards Institute (2001) Maintenance Terminology. Research Report SS-EN 13306
29. Holmberg K, Adgar A, Arnaiz A et al (2010) E-maintenance. Springer Publishing Company Incorporated, New York
30. Patra A, Söderholm P, Kumar U (2009) Uncertainty estimation in railway track life-cycle cost: a case study from Swedish National Rail Administration. Proc Inst Mech Eng Pt F: J Rail Rapid Transit 223(3):285–293
31. International Electrotechnical Commission (2005). Dependability management. Part 3-3: Application guide–life cycle costing
32. MAINLINE (2014) MAINtenance, renewal and Improvement of rail transport iNfrastructure to reduce Economic and environmental impacts SST.2011.5.2-6. http://www.mainline-project.eu/
33. Randall RB (2011) Vibration-based condition monitoring: industrial, aerospace and automotive applications. John Wiley & Sons, Chichester
34. Utne I, Brurok T, Rødseth H (2012) A structured approach to improved condition monitoring. J Loss Prev Process Ind 25(3):478–488
35. Bergquist B, Söderholm P (2014) Control Charts supporting Condition-Based Maintenance of Linear Railway Infrastructure Assets. In: Anonymous Proceedings of the 3rd international workshop and congress on eMaintenance: June 17-18 Luleå, Sweden: eMaintenance, Trends in technologies & methodologies, challenges, possibilites and applications. Luleå: Luleå tekniska universitet
36. Granstr\om R (2005) Maintenance for improved punctuality: a study of condition monitoring technology for the Swedish railway sector
37. Jagersma PK, Jagersma PK (2011) Competitive information logistics. Business Strategy Series 12(3):136–145
38. Kajko-Mattsson M, Karim R, Mirjamsdotter A (2010) Fundamentals of the eMaintenance Concept. In: Anonymous 1st international workshop and congress on eMainteance, p 22
39. Lee J, Ni J, Djurdjanovic D et al (2006) Intelligent prognostics tools and e-maintenance. Comput Ind 57(6):476–489
40. Candell O, universitet Lt (2009) Development of Information Support Solutions for Complex Technical Systems Using EMaintenance. Division of Operation and Maintenance Engineering, Luleå University of Technology
41. Yin RK (2014) Case study research: design and methods. Sage publications, Thousand Oaks

17

Safety evaluation for railway vehicles using an improved indirect measurement method of wheel–rail forces

Jing Zeng[1] · Lai Wei[1] · Pingbo Wu[1]

Abstract The wheel–rail force measurement is of great importance to the condition monitoring and safety evaluation of railway vehicles. In this paper, an improved indirect method for wheel–rail force measurement is proposed to evaluate the running safety of railway vehicles. In this method, the equilibrium equations of a suspended wheelset are derived and the wheel–rail forces are then be obtained from measured suspension and inertia forces. This indirect method avoids structural modifications to the wheelset and is applicable to the long-term operation of railway vehicles. As the wheel–rail lateral forces at two sides of the wheelset are difficult to separate, a new derailment criterion by combined use of wheelset derailment coefficient and wheel unloading ratio is proposed. To illustrate its effectiveness, the indirect method is applied to safety evaluation of railway vehicles in different scenarios, such as the cross wind safety of a high-speed train and the safety of a metro vehicle with hunting motions. Then, the feasibility of using this method to identify wheel–rail forces for low-floor light rail vehicles with resilient wheels is discussed. The values identified by this method is compared with that by Simpack simulation for the same low-floor vehicle, which shows a good coincidence between them in the time domain of the wheelset lateral force and the wheel–rail vertical force. In addition, use of the method to determine the high-frequency wheel–rail interaction forces reveals that it is possible to identify the high-frequency wheel–rail forces through the accelerations on the axle box.

✉ Jing Zeng
zeng@home.swjtu.edu.cn

1 State Key Laboratory of Traction Power, Southwest Jiaotong University, Chengdu 610031, China

Keywords Wheel–rail force · Safety evaluation · Indirect method · Union safety domain · Wheelset derailment coefficient · Hunting motions · Cross wind · Low-floor vehicle

1 Introduction

Nowadays, a great deal of attention to railway industry is focused on the condition monitoring of railway vehicles. The importance of this issue is highlighted by occurrences of problems caused by poor vehicle dynamic conditions or operational environment. For instance, the hunting motions of railway vehicles caused by the wear of wheel and rail profiles may lead to safety problems; a high-speed train subjected to strong cross winds has the potential risk of overturning. For a new opened railway line, the track inspection train is commonly used to evaluate the quality of the tracks. For a new designed vehicle, however, the derailment safety and running behavior must be evaluated by on-track tests. When using dynamic tests to measure the dynamic indices of a track inspection vehicle and a new designed vehicle, we need instrumented wheelset and some accelerometers installed on the tested vehicles. However, the instrumented wheelset is known as an expensive device and involves a complicated calibration technology [1–7]. In addition, some structural modifications to the wheelset, e.g., drilling holes on the wheel hub, are necessary for the signal transmission and power supply, which means that this device is not suitable for long-term commercial operation of the train. Therefore, it has been a great concern to find some alternatives to the traditional method for wheel–rail force measurement.

The inverse identification method is a new approach to determine the wheel–rail interaction forces. In the Chinese

Standard GB5599-85 [8], the derailment safety of railway vehicles are evaluated by measurement of the bogie frame force and the wheelset derailment coefficient. In this method, strain gages are attached on the bogie frame and a calibration test needs to be conducted to obtain the relationship between the forces and structure deformations. In literatures [9, 10], inverse dynamic models combined with force identification methods were commonly used to measure the forces on other structures as the direct measurement requires either expensive device or complicated instrumentation. The methods were also applied to the railway industry. For instance, Xia et al. [11, 12] developed an inverse wagon model to estimate the wheel–rail interaction forces by measuring only the wagon body responses as inputs. In their work, the vehicle system was modeled with rigid bodies and the partial modal matrix method was adopted to predict the input loads. Uhl [13] put forward an inverse method for the identification of contact forces, using accelerometers on the axle box and the transfer function between response points and wheel–rail contact points. Zhu et al. [14] established an inverse dynamic model to identify the dynamic forces in railway vehicles, in which the validation tests were conducted on a full-scale roller rig. Mehrpouya et al. [15] used a finite element model for a railway freight vehicle to identify the forces applied to the wheelset. Tanaka et al. [16] proposed a method for estimating the extraordinary values of wheel load and lateral force using the easily measured axle box accelerations. Gao et al. [17] and Wang et al. [18] utilized specially calibrated force measuring devices to obtain the forces exerted on the axle box spring and the swing arm.

However, the inverse methods for determining the wheel–rail interaction forces described above are usually derived from vehicle system models. Consequently, these methods cannot avoid the complexity of wheel–rail creep force and normal contact force and are not applicable to nonlinear vehicle systems. To identify the nonlinear wheel–rail contact forces, solutions are obtained by combination of instrumented wheelset technology and advanced sensors in Japan. That is, Matsumoto et al. [19] developed a new method to measure the wheel–rail forces without sticking the strain gage or installing the slip ring on the wheelset. In their work, the lateral wheel–rail force was obtained from the wheel deformation by several non-contact gap sensors, while the wheel–rail vertical and longitudinal forces were obtained from the deformation of axle box spring and the strain of swing arm.

An indirect method for wheel–rail force measurement has been proposed by the authors in Ref. [20]. In the previous work, we compared the forces applied to the instrumented wheelset with those obtained using the indirect method. This indirect method was validated on the curved track in the loop test line, from which it was concluded that the identified forces had good coincidence in the time domain with the instrumented wheelset [21]. It was also validated by a high-speed passenger vehicle subjected to cross winds, which also showed that the identified results agreed well with that from instrumented wheelset [22]. In general, the relative error of the peak values was lower than 15 %.

In this paper, an improved indirect method for wheel–rail force measurement is applied to evaluate the running safety of railway vehicles. The wheelset is treated as a rigid body, which means that although it is possible to measure the total lateral force of the wheelset, it is impossible to separate the lateral forces between the two wheels. To deal with this problem, a derailment safety domain by a combination of the wheelset derailment coefficient and the wheel unloading ratio is developed based on the classic Nadal's derailment criterion. For demonstration of its effectiveness, this indirect method are applied to the safety evaluation of railway vehicles in different scenarios, for instance, the cross wind safety of a high-speed train, and the safety of a metro vehicle with hunting instability. In addition, the feasibility of applying this method to identify the wheel–rail forces for low-floor light rail vehicles with resilient wheels, as well as high-frequency wheel–rail interaction forces, is also investigated.

2 Derailment evaluation by indirect method for wheel–rail forces measurement

The mechanism of the indirect method for wheel–rail force measurement is to establish the force and moment equilibrium equations by analyzing the forces exerted on the wheelset, as shown in Fig. 1. The unknown wheelset lateral force and wheel–rail contact forces can then be calculated from the inertial forces and suspension forces, which are obtained by measuring the accelerations on the axle box and the relative displacements of the primary suspension. For the vehicle with primary suspension, the wheelset or wheel axle lateral force, i.e., the so-called H force, is given in Ref. [20] as follows:

$$H = -m_{\mathrm{w}} a_y - F_{\mathrm{s1}} - F_{\mathrm{s2}}, \tag{1}$$

where m_{w} is the wheelset mass; a_y is the lateral acceleration of the axle box; F_{s1} and F_{s2} are the lateral forces of the left and right primary suspensions.

Ignoring the shift of wheel–rail contact points and the roll acceleration of wheelset, the wheel–rail vertical contact force sat the two sides of the wheelset, i.e., Q_{L} and Q_{R}, are calculated as

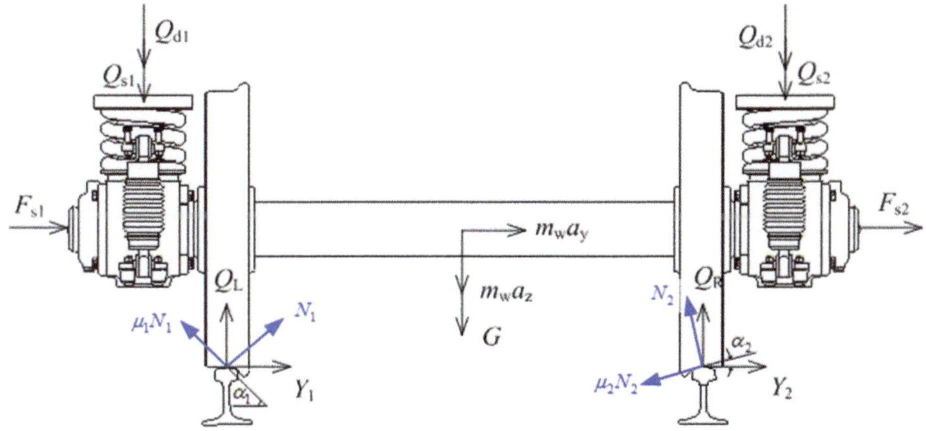

Fig. 1 Analysis of forces exerted on a suspended wheelset

$$\begin{aligned} Q_L &= Q_{s1} \cdot (l_s/2 + l_c/2)/l_c + Q_{d1} \cdot (l_d/2 + l_c/2)/l_c, \\ &- Q_{s2} \cdot (l_s/2 - l_c/2)/l_c - Q_{d2} \cdot (l_d/2 - l_c/2)/l_c, \\ &+ G/2 + m_w a_z/2 + H \cdot r_0/l_c, \end{aligned} \quad (2)$$

and

$$\begin{aligned} Q_R &= Q_{s2} \cdot (l_s/2 + l_c/2)/l_c + Q_{d2} \cdot (l_d/2 + l_c/2)/l_c \\ &- Q_{s1} \cdot (l_s/2 - l_c/2)/l_c - Q_{d1} \cdot (l_d/2 - l_c/2)/l_c, \\ &+ G/2 + m_w a_z/2 - H \cdot r_0/l_c, \end{aligned} \quad (3)$$

where Q_{s1} and Q_{s2} are the vertical forces of the primary springs; Q_{d1} and Q_{d2} are the vertical forces of the primary dampers; a_z is the vertical acceleration of the axle box; l_c is the lateral distance between the wheel nominal running cycles; l_s is the lateral distance between the primary springs; l_d is the lateral distance between the primary vertical dampers; G is the gravitational force of the wheelset; and, r_0 is the wheel radius.

The commonly used wheel derailment coefficient *Y/Q* is defined as the ratio of the wheel–rail lateral force to the wheel–rail vertical force. According to the critical force state of wheel climbing derailment (see Fig. 1), the single-wheel derailment coefficient at the left and right wheel–rail contact point can be derived [23]:

$$\frac{Y_L}{Q_L} = \frac{\tan \delta_L - \mu_L}{1 + \mu_L \tan \delta_L}, \quad (4)$$

$$\frac{Y_R}{Q_R} = \frac{\tan \delta_R + \mu_R}{1 - \mu_R \tan \delta_R}, \quad (5)$$

where Y_L and Y_R indicate the left and right wheel–rail lateral forces, respectively; Q_L and Q_R are the wheel–rail vertical forces, respectively; $\tan \delta_L$ and $\tan \delta_R$ are the left and right wheel–rail contact angles, respectively; and μ_L and μ_R are the wheel–rail friction coefficients on the left and right sides, respectively.

If the left wheel–rail contact angle reaches the maximum flange angle, then the left wheel is in the critical condition of the flange climbing derailment. The roll angle of the wheelset is relatively small compared with the contact angle in the flange wheel. The wheel–rail vertical contact force applied at the left and right contact points can be expressed as

$$Q_i = Q\left(1 \mp \frac{\Delta Q}{Q}\right)(i = \mathrm{L}, \mathrm{R}), \quad (6)$$

where Q is the nominal wheel–rail vertical force, ΔQ is the wheel unloading force, and the ratio of the wheel unloading force to the nominal wheel–rail vertical force is defined as the wheel unloading ratio $\Delta Q/Q$. The negative sign in Eq. (6) is related to the left flange climbing wheel and the positive sign to the right non-flanging wheel. In addition, the H force can be written as

$$H = Y_L - Y_R. \quad (7)$$

Based on Eqs. (6) and (7), the wheelset derailment coefficient can then be written as

$$\frac{H}{Q} = \frac{Y_L}{Q_L} \cdot (1 - \frac{\Delta Q}{Q}) - \frac{Y_R}{Q_R} \cdot (1 + \frac{\Delta Q}{Q}), \quad (8)$$

where the wheelset derailment coefficient is defined as the ratio of the H force to the nominal wheel–rail vertical force.

In this indirect method, however, the wheelset is treated as a rigid body, which means that the left and right wheel–rail lateral forces cannot be separated. To overcome this disadvantage, a derailment evaluation based on the wheelset derailment coefficient and the wheel unloading ratio is derived in this section. By substituting Eqs. (4) and (5) into Eq. (8), we obtain the derailment criterion based on the wheelset derailment coefficient as below:

$$\frac{H}{Q}+\left[\frac{\tan\delta_L-\mu_L}{1+\mu_L\tan\delta_L}+\frac{\tan\delta_R+\mu_R}{1-\mu_R\tan\delta_R}\right]\cdot\frac{\Delta Q}{Q}=\frac{\tan\delta_L-\mu_L}{1+\mu_L\tan\delta_L}-\frac{\tan\delta_R+\mu_R}{1-\mu_R\tan\delta_R}. \quad (9)$$

Given the other wheel–rail contact parameters in Eq. (9), the wheelset derailment coefficient and wheel unloading ratio can be determined by the indirect method for wheel–rail force measurement. Based on the critical state of derailment, the boundaries of the safety domain can be determined, see Fig. 2. The black oblique line of the safety domain is determined from Eq. (9) with the contact angle of 70° and the friction coefficient of 0.3. If the flange climbing wheel is loaded with the wheel unloading ratio of −1.0, then no-flanging wheel is lifted from the rail. Finally, a triangular safety domain against derailment is determined. The vehicle can be assessed to be safe if the wheelset derailment coefficient and wheel unloading ratio are located in the left area of the black oblique line. It is seen that the safety area of the unloading flange wheel is smaller than that of the loaded wheel, which means that the flange climbing wheel possesses a higher derailment risk if the wheel is unloaded.

Besides, the wheel unloading ratio is only applicable to the situation that the wheel axle lateral force is zero or very small and should be used together with the derailment coefficient. The derailment evaluation by wheelset derailment coefficient *H/Q* is similar to the wheel axle force limit which is applied to evaluate the wheel–rail interactions rather than the flange climbing.

Furthermore, the '*H* force criteria' in the Standard GB5599-85 can be derived by substituting Eq. (6) into Eq. (9):

$$\frac{H+0.24\cdot Q_R}{Q_L}\leq 1.0. \quad (10)$$

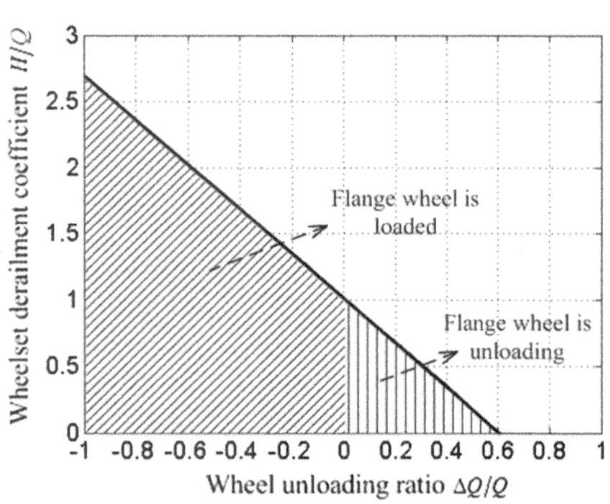

Fig. 2 Union safety domain based on wheelset derailment coefficient and wheel unloading ratio

In general, the new derailment criterion is equivalent to the traditional flange climbing derailment evaluation method and can be transfer to an equivalent form to the traditional method.

3 Applications

3.1 Safety evaluation of high-speed train under cross winds

Aiming at the high-speed train subjected to cross winds, a dynamic test using indirect method for wheel–rail force measurement was conducted in the Lanzhou-Xinjiang high-speed railway line. The environmental wind speed in the so-called 'hundred-mile wind area' between Urumqi and Hami was as high as 33 m/s. The axle load of the tested vehicle was 13.4 t and the normal speed of the vehicle was 250 km/h. Accelerometers, laser transducers, and instrumented wheelset were installed on the test vehicle, as shown in Fig. 3. The goal of this test was to identify the *H* force and the wheel vertical forces Q_i using this indirect method, by measuring the accelerations on the axle box and the displacements of the primary suspension. Comparisons of the identified forces with the measured results from instrumented wheelset were introduced in Ref [22], which showed that the identified wheel vertical forces had perfect correlation with the measured results from the instrumented wheelset.

The running safety of the high-speed train under cross winds is affected by many factors, e.g., the running speed, wind characteristic, car body shape, etc. The system response and wheel–rail forces of the vehicle subjected to strong cross winds are quite complicated. Figure 4 gives the measured wheel–rail interaction forces in some particular locations using this indirect method. The vertical forces show that the wheel facing to the wind (windward side) is loaded while the wheel away from the wind (leeward side) unloaded. Meanwhile, the *H* force synchronously increases with the wheel–rail vertical forces. The combination of the increasing lateral force and decreasing vertical force could result in a derailment risk.

For the traditional safety domain using the single-wheel derailment coefficient and wheel unloading ratio, the safety domain is presented as the rectangle type. The scatter diagram is plotted using the absolute values of *Y/Q* and $\Delta Q/Q$. For the new-derived derailment criterion, the safety domain using the wheelset derailment coefficient and wheel unloading ratio is presented as the triangular type. If the flange climbing wheel is unloaded (positive), the wheel is at high risk of derailment. Conversely, the flange climbing wheel is relatively safe if the wheel is loaded (negative). The safety assessment based on the proposed

Fig. 3 Illustration of primary structures and test instruments of the tested high-speed vehicle

criterion is shown in Fig. 5. We can see that the safety indices of the head car subjected to cross winds that exceed the allowable limit in some locations, while the measured indices for the tail car under cross winds are within the safety limit.

3.2 Safety evaluation of metro vehicle with hunting motions

The indirect method for wheel–rail force measurement was also used to conduct a dynamic test for a metro vehicle. The axle load of the tested vehicle was 8.57 t and the normal speed of the vehicle was nearly 100 km/h. In some locations of the track, the wheel–rail contact conicity of the worn wheel is very high, which may probably cause hunting motions of the vehicle system. The test instruments including accelerometers and laser transducers are shown in Fig. 6. The goal of this test was to identify the *H* force and the wheel vertical forces Q_i using this indirect method, by measuring the accelerations on the axle box and the displacements of the primary suspension. The safety evaluation of the metro vehicle with hunting motions is then carried out.

The running safety of the metro vehicle with hunting motions is affected by many factors, such as the running speed, wheel–rail contact relations, and track excitations. Figure 7 gives the measured wheel–rail interaction forces based on this indirect method. We can see that a very large *H* force appears in some certain locations of the track and the waveform behaves as harmonic vibration, indicating that hunting instability occurs to the vehicle.

The scatter diagram for the safety assessment against derailment of the metro vehicle is shown in Fig. 8. For the derived derailment criterion above, the safety domain using wheelset derailment coefficient and wheel unloading ratio is presented as the triangular type. We can see that the safety indices of the metro vehicle under unstable conditions exceed the allowable limit, while that under stable conditions is within the safety limit. Thus, the occurrence of hunting motions to the metro vehicle has great influence on the derailment safety. Further measures, e.g., wheel re-profiling to reduce the contact conicity, should be taken to solve this problem.

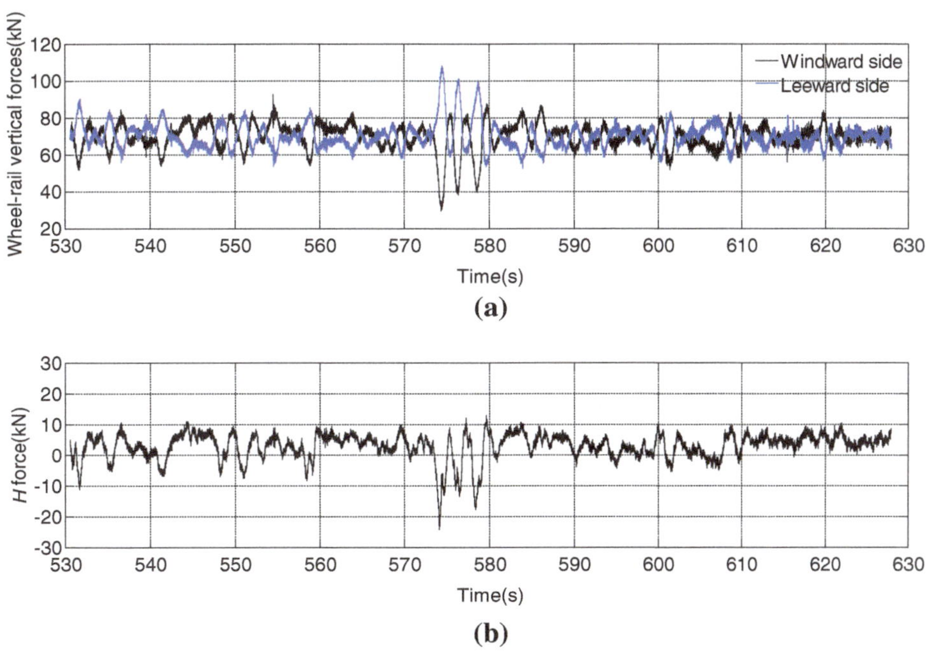

Fig. 4 Measured wheel–rail interaction forces of high-speed vehicle under cross winds. **a** Wheel–rail vertical forces. **b** *H* force

3.3 Safety evaluation of low-floor vehicle with resilient wheels

The traditional instrumented wheelset is not applicable to low-floor light rail vehicles (LRVs) with resilient wheels because the deformation of the resilient wheel is influenced by the rubber pad. An alternative method is to adopt the indirect method for wheel–rail force measurement. The forces exerted on the resilient wheelset are shown in Fig. 9. The idea of the indirect method for wheel–rail force measurement is also to establish force and moment equilibrium equations by analyzing the forces exerted on the resilient wheelset. The unknown forces H and Q_i can then be obtained from inertial forces and suspension forces, which can be determined by measuring the accelerations on the axle box and the relative displacements of the primary suspension. Besides, the interaction forces of the rubber (Y_{ei} and Q_{ei}) are difficult to measure because of the rotating contact position. The elastic rubber is herein treated as the primary suspension while the force is taken as the internal force. In principle, the methodology for the wheel–rail force measurement based on Eqs. (1), (2) and (3) is still applicable.

In this section, the simulation package Simpack is used to develop the low-floor vehicle model in which nonlinear wheel–rail kinematic constraints, nonlinear wheel–rail creep forces, and nonlinear suspensions are taken into account. The accelerations on the axle box, the displacements of the primary suspension, and the wheel–rail forces are generated from this model. To verify the indirect method theoretically, the wheel–rail forces determined from Eqs. (1), (2) and (3) are compared with the forces calculated by Simpack. The running speed for simulations is 55 km/h on the curved track with a radius of 300 m, and the measured track irregularities are adopted in the simulations. As can be seen from Fig. 10, this method can provide good results for the H force and the wheel–rail vertical forces Q_i, and is suitable for application. The small difference in the vertical force is caused by neglect of the rolling motion of the wheelset as well as the shift of wheel–rail contact points. The good agreement between the identified and actual wheel–rail forces implies that the output data can be accurately obtained from the simulations. However, further feasibility tests on LRVs are necessary for the validation of the indirect method, since the actual running environments are more complicated than those considered in the simulations.

Actually, the simulated results can not reflect completely the real responses as there are a lot of undesired signals in practice. Unfortunately, for low-floor vehicles with small-radium resilient wheelset, there is no instrumented wheelset available, so the indirect method cannot be verified experimentally by measurements on an actual instrumented wheelset. Alternatively, the indirect method can be verified theoretically using Simpack.

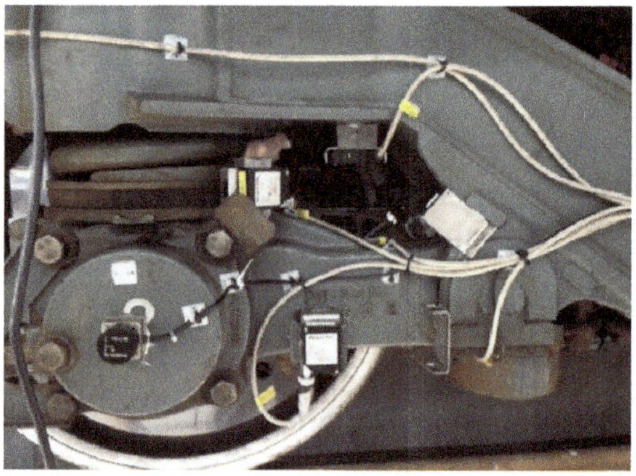

Fig. 6 Illustration of primary structures and test instruments of the tested metro vehicle

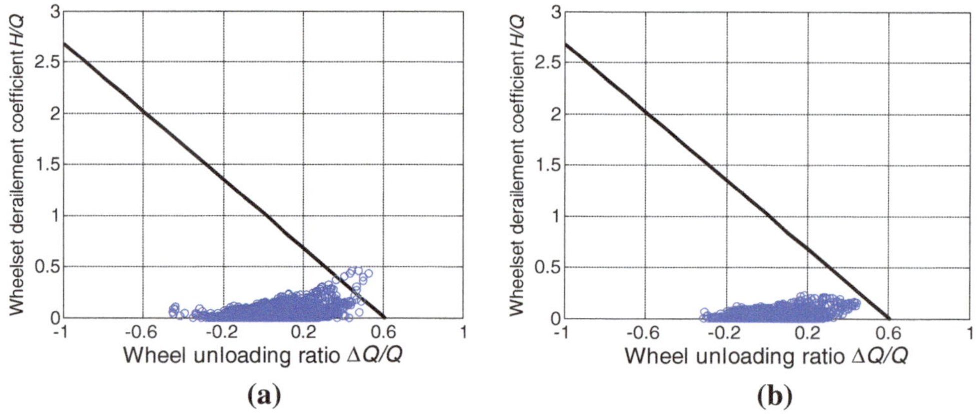

Fig. 5 Safety assessment against derailment for high-speed train under cross winds. **a** Head car. **b** Tail car

3.4 Identification of high-frequency wheel–rail forces

The working mechanism of the instrumented wheelset is to measure the deformation of the wheel hub or axle. However, the deformation of the wheelset in high frequencies (e.g., caused by rail defects and wheel out-of-roundness) may come from the shift of wheel–rail contact points or the elastic deformation itself. In addition, the measurement of high-frequency force puts forward higher requirements on the signal transition system. Therefore, it is necessary to discuss the feasibility of applying this indirect method to measure the high-frequency forces. Under the interaction of short-wavelength track excitations, the acceleration generated on the wheelset is very high, while the displacement of the primary spring is very small and can be ignorable. Thus, Eqs. (1)–(3) can be further written as

$$H = -m_w a_y, \tag{11}$$

$$Q_1 = G/2 + m_w a_z/2 + H \cdot r_0/l_c, \tag{12}$$

and

$$Q_2 = G/2 + m_w a_z/2 - H \cdot r_0/l_c. \tag{13}$$

It should be noted that this method is only applicable to the rigid body system in which the elastic vibration of the wheelset is not considered. To verify the indirect method theoretically, the wheel–rail forces determined from Eqs. (10)–(12) are compared with those calculated by Simpack. The running speed for the simulations is 50 km/h for the wheel in flat condition with a wavelength of 30 mm.

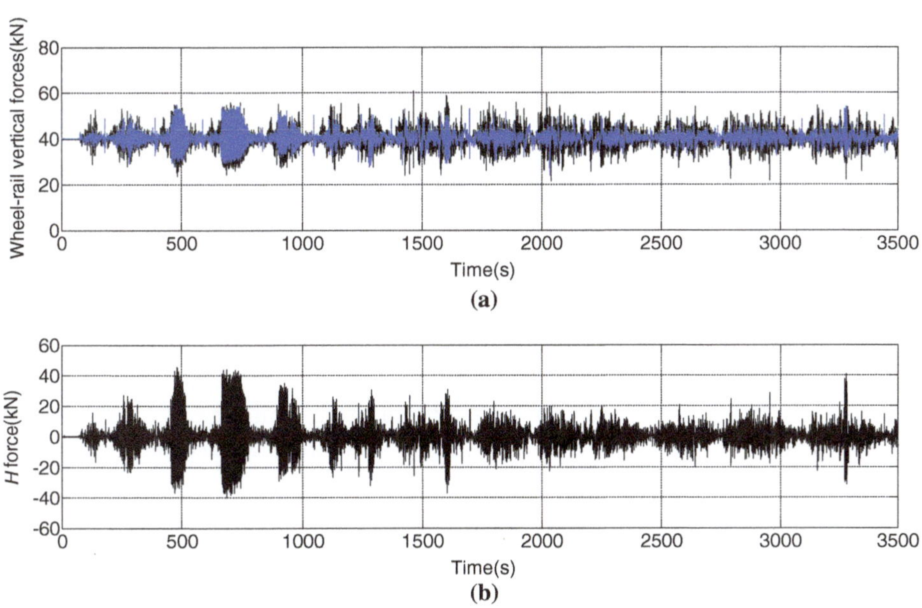

Fig. 7 Measured wheel–rail interaction forces of the metro vehicle with hunting motions. **a** Wheel–rail vertical forces. **b** *H* force

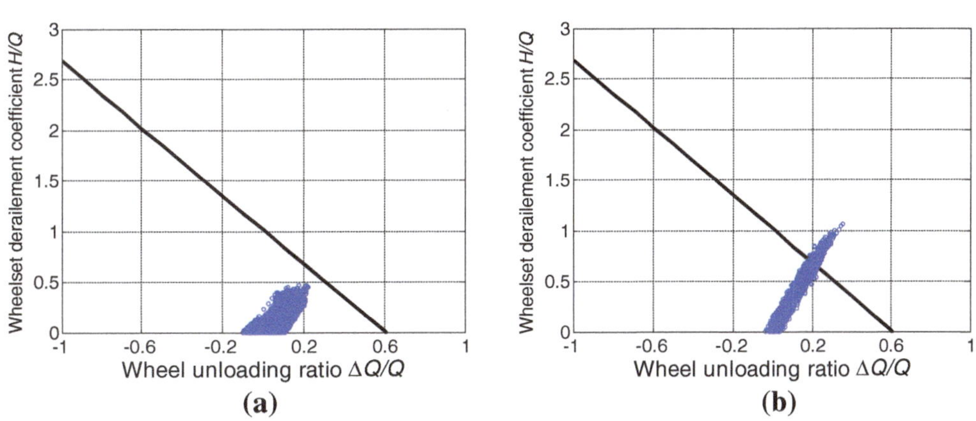

Fig. 8 Safetyassessment against derailment for metro vehicle with hunting motions. **a** Under stable condition. **b** Under unstable condition

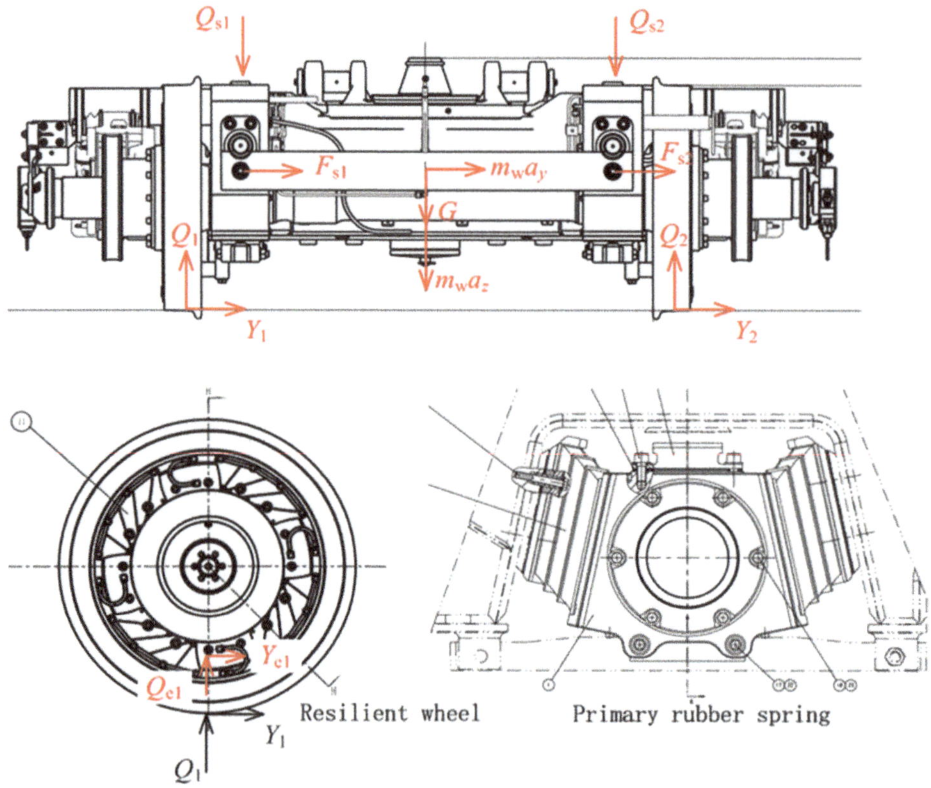

Fig. 9 Analysis of forces exerted on resilient wheelset for LRV

The comparison of the identified wheel–rail vertical force with the output value of Simpack is shown in Fig. 11, from which we can see that this method gives good results in the time history.

In fact, the accelerations on the axle box due to elastic vibrations are apparently higher than that due to rigid motions. In future applications, the measured accelerations are necessary to be processed with a band pass filter so as to eliminate the influence of high-frequency elastic vibrations. The cutoff frequency of the filter will be determined by the characteristic frequency due to the wheel out-of-roundness or rail corrugation. For example, the characteristic frequency due to rail corrugation can be calculated by the wavelength and train speed.

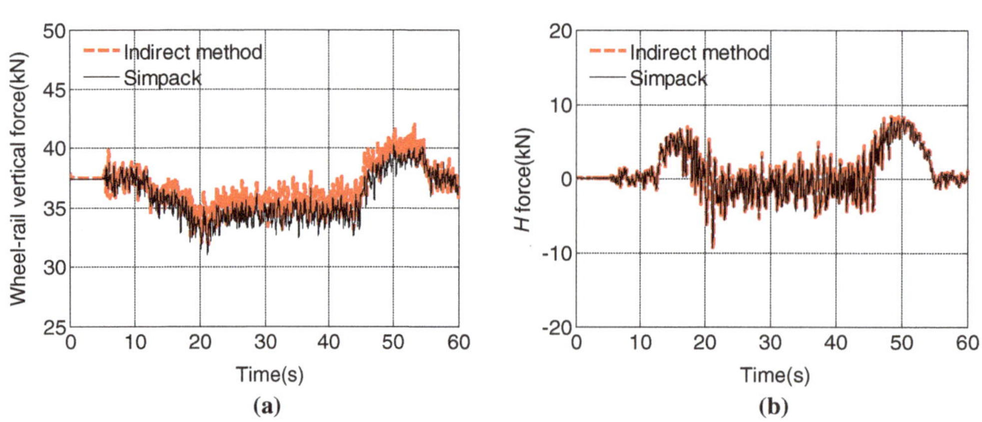

Fig. 10 Theoretical verifications by simulations on a curved track with radius of 300 m. **a** Wheel–rail vertical forces. **b** *H* force

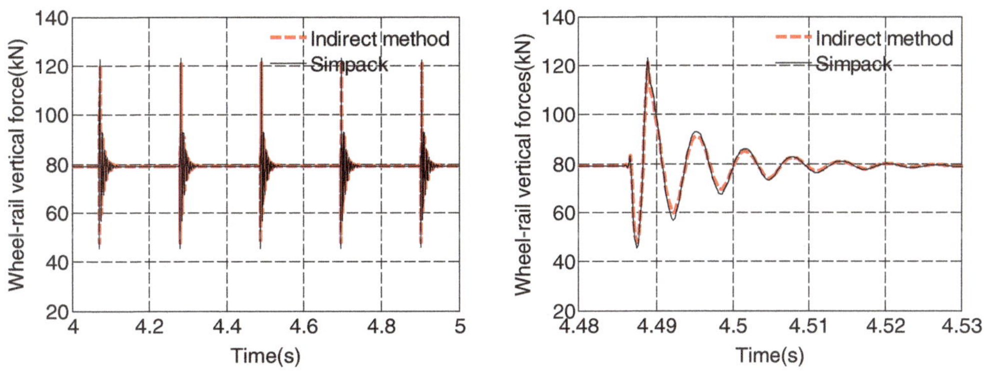

Fig. 11 Theoretical verifications by simulations for the wheel in flat condition. **a** Wheel-rail vertical forces. **b** Partial enlargement of one waveform

4 Conclusions

An improved indirect method for wheel–rail force measurement is proposed to evaluate the derailment safety of railway vehicles in various scenarios. From the applications, the following conclusions can be drawn:

(1) A union safety domain against derailment based on the wheelset derailment coefficient and the wheel unloading ratio is presented. The derailment safety can be evaluated by judging whether the scatter points are within the boundary limit of the triangular safety domain.

(2) Aiming at the high-speed train subjected to cross winds, the dynamic test using indirect method for wheel–rail force measurement was conducted. The results show that the wheel facing to the wind is loaded while the wheel away from the wind unloaded. The combination of the increasing lateral force and decreasing vertical force could result in derailment risk. The safety indices of the head car subjected to cross winds exceed the allowable limit in some locations, while the measured indices for the tail car under cross winds are within the safety limit.

(3) The dynamic test using the indirect method for wheel–rail force measurement for a metro vehicle was also conducted. It is found that the H force increases significantly and hunting instability occurs in certain sections of the track. Similarly, the safety indices for the metro vehicle with hunting motions exceed the allowable limit, while that in the stable conditions, are within the safety limit.

(4) The forces exerted on a resilient wheelset of a light rail vehicle are analyzed. To verify the indirect method theoretically, the identified wheel–rail forces are compared with those calculated by Simpack. The results show that the H force and the wheel–rail vertical forces identified by the proposed indirect method agree well with the calculated values by Simpack.

(5) The feasibility of applying this indirect method to measure the high-frequency forces is also discussed. The identified wheel–rail vertical force is compared with the output value by Simpack, showing a good coincidence in the time history of wheel–rail vertical force. However, this method is only applicable to the rigid body system in which the elastic vibration of the wheelset is not considered. Therefore, the measured accelerations are necessary to be processed with a band pass filter so as to eliminate the influence of high-frequency elastic vibrations.

Acknowledgments This work was supported by the National Natural Science Foundation of China (Grant No. U1334206 and No. 51475388) and Science & Technology Development Project of China Railway Corporation (Grant No. J012-C).

References

1. Riggall G (2008) IWT4 goes into operation. Railw Gazette Int 164:455–456
2. Braghin F, Bruni S, Cervello S, Cigada A, Resta F (2003) A new method for the measure of the wheel–rail contact forces. In: Proceedings of the 6th international conference on contact mechanics and wear of rail/wheel systems (CM2003), Goteborg, 313–319

3. Matsumoto A, Sato Y, Ohno H, Shimizu M, Kurihara J et al (2012) Continuous observation of wheel/rail contact forces in curved track and theoretical considerations. Veh Syst Dyn 50(sup1):349–364
4. Kanehara H, Fujioka T (2002) Measuring rail/wheel contact points of running railway vehicles. Wear 253(1):275–283
5. Ham YS, Lee DH, Kwon SJ, You WH, Oh TY (2009) Continuous measurement of interaction forces between wheel and rail. Int J Precis Eng Manuf 10(1):35–39
6. Elkins JA, Carter A (1993) Testing and analysis techniques for safety assessment of rail vehicles: the state-of-the-art. Veh Syst Dyn 22(3–4):185–208
7. Papini S, Pugi L, Rindi A, Meli E (2013) An integrated approach for the optimization of wheel–rail contact force measurement systems. J Mod Transp 21(2):95–102
8. GB5599-85 (1985) Railway vehicles-specification for evaluation the dynamic performance and accreditation test, National Bureau of Standard, Beijing **(in Chinese)**
9. Genaro G, Rade DA (1998). Input force identification in the time domain. In: SPIE proceedings series, Society of Photo-Optical Instrumentation Engineers, pp. 124–129
10. Parloo E, Verboven P, Guillaume P, Van Overmeire M (2003) Force identification by means of in-operation modal models. J Sound Vib 262(1):161–173
11. Xia F, Cole C, Wolfs P (2007) An inverse railway wagon model and its applications. Veh Syst Dyn 45(6):583–605
12. Xia F, Cole C, Wolfs P (2008) Grey box-based inverse wagon model to predict wheel–rail contact forces from measured wagon body responses. Veh Syst Dyn 46(S1):469–479
13. Uhl T (2007) The inverse identification problem and its technical application. Arch Appl Mech 77(5):325–337
14. Zhu T, Xiao S, Yang G, Ma W, Zhang Z (2014) An inverse dynamics method for railway vehicle systems. Transport 29(1):107–114
15. Mehrpouya M, Ahmadian H (2009) Estimation of applied forces on railway vehicle wheelsets from measured vehicle responses. Int J Veh Struct Syst 1(4):104–110
16. Tanaka H, Furukawa A (2008) The estimation method of wheel load and lateral force using the axlebox acceleration. WCRR, Seoul
17. Gao C, Lu H, Ren L (2012) Test on the wheel/rail forces of 70 % low-floor tram. Urban mass transit 15(10):43–45 **(in Chinese)**
18. Wang W, Wang Y, Sun S, Liang S (2015) Long-term load spectrum test of high speed train bogie. J Southwest Jiaotong Univ 50(1):84–89 **(in Chinese)**
19. Matsumoto A, Sato Y, Ohno H, Tomeoka M, Matsumoto K, Kurihara K, Nakai T (2008) A new measuring method of wheel–rail contact forces and related considerations. Wear 265(9):1518–1525
20. Wei L, Zeng J, Wu P, Gao H (2014) Indirect method for wheel–rail force measurement and derailment evaluation. Veh Syst Dyn 52(12):1622–1641
21. Wei L, Zeng J, Wu P, Gao H, Wang Q (2015) Derailment safety evaluation for railway vehicles based on wheelset model. J China Railw Soc 37(9):25–31 **(in Chinese)**
22. Wu P, Wei L, Zeng J, Gao H (2015) Safety assessment of high speed trains under cross winds. In: The 24th international symposium on dynamics of vehicles on roads and tacks (ISVSD 2015), Graz
23. Zeng J, Guan Q (2008) Study on flange climb derailment criteria of a railway wheelset. Veh Syst Dyn 46(3):239–251

18

A comparative study of automotive trip characteristics between older drivers and others among densely inhabited district and other areas

Yanning Zhao[1] · Toshiyuki Yamamoto[2] · Takayuki Morikawa[3]

Abstract This paper examines older driver's automotive trip (abbreviation: trip) characteristics which include trip frequency, trip length, destination distribution, and non-home-based (NHB) trips. A two-month experiment of 108 participants was carried out to collect GPS tracking data in Aichi Prefecture, Japan. To identify the effect of living area, a comparative analysis between older drivers and others is conducted in densely inhabited district (DID, i.e., urban) and other areas (non-DID, i.e., suburban, rural, etc), separately. The present study found that there was no significant difference between the trip characteristics of older drivers and others who were living in DID. Thus, we suggest that the education of safety driving and the recommendation of public transportation should be given to DID-living older drivers. However, the results of non-DID reflected that older drivers' trip frequency, trip length, destination, and NHB trips rate were shorter and lower than others'. This implies that electric vehicles may be suitable for promotion among older drivers in suburban and rural area. Furthermore, the regression analysis confirmed that "older driver" was a significant independent variable on trip frequency, trip length, and NHB trips, and there were interaction effects between "older driver" and "living areas" on all trip characteristics.

✉ Yanning Zhao
yzhaoatnu@gmail.com

1 Department of Civil Engineering, Nagoya University, Furo-cho, Chikusa-ku, Nagoya 464-8603, Japan

2 Institute of Materials and Systems for Sustainability, Nagoya University, C1-3(651), Furo-cho, Chikusa-ku, Nagoya 464-8603, Japan

3 Institute of Innovation for Future Society, Nagoya University, Chikusa-ku, Nagoya 464-8603, Japan

Keywords Older driver · Travel behavior · Living area · Trip frequency · Trip length · Destination distribution · Non-home-based (NHB) trips

1 Introduction

According to United Nations World Population Prospects (UNWPP), Japan has entered the period of a super-aged society [1]. With the highest proportion of older adults in the world, the number of older drivers in this country reached 16 million in 2015, and is expected to surpass 20 million by 2030 [2]. Meanwhile, the consequential increase of older driver-related accidents has also been an inevitable problem. For instance, although the number of traffic accidents in Tokyo has reduced by more than 50 % in the last decade, the ratio of older driver in all accidents nearly doubled from 10.9 % to 20.4 %.

Increasingly, researchers began to concern the aging problem in the car society. Many studies reported that physical and mental changes that often come with aging can affect older adult's travel behaviors. By a 5-year survey-based study on the relationship between health problems and risk ratio among older drivers, MacLeod et al. [3] suggested to provide more driving training and assistance to older adults. Delhomme et al. [4] found that older drivers have some driving characteristics, such as driving at a steady speed, conservative use of the accelerator, gear shifting, etc. Ross et al. [5] indicated that most older adults tend to overrate their driving ability. In their research, 85.14 % of older participants believed themselves as either good or excellent drivers regardless of their actual previous citation or crash rates. Meanwhile, aging may also change people's living and activity areas. Tahara and Iwadare [6] concluded that "younger" elderly people tend to move out from metropolitan areas, as opposed to "older" elderly people who have a higher tendency to move into metropolitan areas. Benekohal et al. [7] summarized the effects of aging on older drivers' travel characteristics, which included travel frequency, travel time, trip length, etc. It suggested that all these indicators decrease significantly as the age of the driver increases. Moreover, Rothe [8] associated the high accident ratio of elderly drivers with their prominently frequent around-home driving.

Home-based (HB) and non-home-based (NHB) trips, which play an important role in trip generation, are also important to understand driver's trip characteristics. By HB and NHB trips, the complex driving activities can be converted into number of trips [9]. Because trips must be generated by certain purposes, like from home to work, home to shop, or school to home, HB and NHB reflected drivers' driving intention and travel convenience. Rothe's statistics [8] showed that the main trip purpose of middle-aged group is commuting and social, while that of elderly group is personal and family. On the other hand, one consensus is that living area could affect driver's NHB rate. Victor and Ponnuswamy [10] estimated that NHB trips might constitute 10 %–15 % of the total trips in most urban areas. Through an improved calculation model, Schultz and Allen [11] confirmed that NHB trips accounts for 25 %–30 % of urban-living driver's travel. Moreover, Gonzalez-Ayala's report [12] concluded that the NHB rate in urban area is lower than that in rural area. About this phenomenon, Hildebrand et al. [13] explained that rural-living participants tend to chain more activities together because they live further away from their desired destinations. However, it should be noted that Hildebrand's study only focused on older drivers who were living in rural area, and hence, the comparative analysis between older drivers and other age groups was not conducted.

Another feature of these studies is that most of them are mainly based on the questionnaire survey. Although self-report study is widely used, it is difficult to avoid participant's subjective effects, such as exaggeration, concealment, bias, etc. [14]. Ross and Dodson [5] even recommended to abandon survey-based study on older drivers because their feeling is likely not related to actual driving proficiency. Considering the limited reliability of self-report from older drivers, some studies analyzed trip characteristics by the GPS-based travel diaries which were collected by on-vehicle device [15, 16]. However, they still lack the contrastive study of older drivers and others among urban and rural areas.

Although some studies [17, 18] focused on the consequences of aging-related driving habits change and suggested to improve transportation alternatives, it often lacked specific suggestions on transportation planning for older adults living in different areas. For instance, public transportation has many advantages in urban area, but its demand, cost, and efficiency may decrease in suburban area. Even in the same living area, whether older adults driving less than others is still unclear. On the other hand, the development of electric vehicles (EVs) is deeply affecting transportation planning in current years [19]. However, although some automobile companies have viewed older adults as potential users, the feature caused by living area has not been considered yet. Associated with some travel behaviors of older adults, transportation infrastructure planning models and related policy issues about EVs may be developed or improved.

The aim of this study was to evaluate the older adult's trip characteristics by comparing with other age groups both in urban and rural area. The research objects included trip frequency, trip length, destination distribution, and NHB trips. It should be noted that the trip in this paper was limited to automotive trip which did not consist of bus, walk, or any other transportation mode. A 2011's official survey [20] suggested that automotive trip accounted for about 67 % of all trips in the areas of this study. It could be further estimated that automotive trip occupied more than 83.75 % in the trips of this study's participants who were holding driving licenses and vehicles [20, 21]. According to some research [22], it can be inferred that the automotive trip's rate in suburban/rural is even higher than that in urban area.

Further aims of this study were to confirm whether the indicator of "older driver" could affect trip characteristics; and to identify the interaction impact factors with older driver on travel behavior. Except for the basic information such as gender, age, and address, this study mainly relied on probe vehicle (PV) data instead of questionnaire survey. The results may contribute to the improvement of driving assistant, driving training, and transportation plan for older drivers.

2 Data and terminology

121 Drivers were recruited in the data-collection experiment, which was carried out mainly in Aichi Prefecture, Japan from April 1 to May 31, 2013. Before the experiment, they were permitted to install data-collection devices in their vehicles and then fill in basic information survey. Due to machine faults or operation mistakes, there are 108 valid experiment subjects out of 121 participants. The following three broad categories of data were assembled in this study.

(1) Driver's characteristics (e.g., name, gender, age, address, etc.)

(2) PV data (e.g., location, speed, date and time, engine on/off, etc.)
(3) Region classification by population density (e.g., densely inhabited district or nondensely inhabited district, etc.)

Following the introduction of data, the terminology in this study will be given.

2.1 Driver's characteristics

Driver's characteristics were obtained by questionnaire survey before the experiments. Participants were asked 18 background questions. According to age (mean = 40.3; SD = 11), gender, and employment status, experimental subjects are categorized into different groups as shown in Table 1. The information of employment status is used in the judgment of commute trip which is engaged by only employee, student, or self-employed person. It must be admitted that older drivers (10, 9.3 %) and nonworkers (11, 10.2 %) make up a relatively small proportion of the whole sample. However, in order to ensure that the long experiment of data collection was carried out smoothly, the sample has to bias on workers and students who are below 60 years old. Table 1 also lists the population and proportion of the study area by each age group [23]. It should be noted that although female comprises a relatively higher proportion of older adults, they are fewer in this study because of the lower rate of female license-holding (30.02 %) in all older drivers [24].

2.2 PV data

The PV data were collected by in-vehicle devices which were equipped in each experimental car, separately. After starting the device, data were constantly sent from the moving car to the nearby base stations or receiving nodes. There are 78 channels which record the types of number, string, date, etc. Some of the data describe the road positions or environments, such as GPS (latitude, longitude, altitude, etc.) and road ID. Other data present vehicle's moving information, such as fuel consumption (0.1 km/L), running speed (km/h), and engine speed (rpm). Data are

Table 1 Sample distribution in this study and population distribution of the study area

Cases	Sample distribution in this study		Population distribution of the study area	
	Numbers	Percentage (%)	Population	Percentage (%)
Total	108		318,020	
Age				
20's	8	7.4	52,995	16.7
30's	27	25.0	58,911	18.5
40's	38	35.2	64,395	20.3
50's	25	23.1	48,307	15.3
60's and 70's	10	9.3	93,412	29.4
Male	67	62.0		
20's	5	62.5	30,567	57.7
30's	17	63.0	32,259	54.8
40's	23	60.5	34,205	53.1
50's	13	52.0	25,233	52.2
60's and 70's	9	90.0	46,461	49.7
Worker/student	97	89.8		
Worker/student in 20's	7	87.5		
Worker/student in 30's	24	88.9		
Worker/student in 40's	36	94.7		
Worker/student in 50's	22	88.0		
Worker/student in 60's & 70's	8	80.0		
DID-living				
DID-living in 20's	4	50.0		
DID-living in 30's	16	59.3		
DID-living in 40's	22	57.9		
DID-living in 50's	11	44.0		
DID-living in 60's and 70's	3	30.0		

eliminated, while engine is off or GPS data are invalid (e.g., driving in tunnel, on mountain road, or underneath the viaduct). It should be noted that once GPS signal is missed (e.g., after entering tunnel), localization estimation algorithm (LEA) embedded in the in-vehicle device will be operated automatically. Based on this algorithm, the vehicle is assumed in the extension line of the last-recorded direction at a constant speed which maintains the last-recorded speed. Considering the low credibility, the estimated records are also deleted in this study.

2.3 Region classification by population density

Based on the 2010's population census conducted by Statistics Bureau of Japan [25], densely inhabited districts (DIDs) are designated in units of census basic unit blocks, and census enumeration districts if there are several census enumeration districts in a census basic unit block, and should coincide with the following criteria.

1. A district containing basic unit blocks, etc., with a population density of 4,000 or more per square kilometer, such districts being adjacent to each other in a municipality
2. A district consisting of the above adjacent basic unit blocks, etc., whose population is 5,000 or more at the time of the Population Census of Japan

Based on the definition above, DIDs could be viewed as urban areas, while non-DIDs represents rural areas, suburban areas, etc. Then, the address of each participant could be located into DID or non-DID areas.

2.4 Terminology

Following words and concepts should be clearly defined before discussing the methods and results of this research.

- Older driver: driver who is 60 years old or older.
- Trip: automotive trip (abbreviation: trip) which does not consist of bus, walk, or any other transportation mode.
- Commute trip: trip from home to office/school. Discretionary trip: otherwise.
- Peak periods (rush time): 7:00–10:00 and 15:00–20:00 on weekdays (Fig. 1).
  Off-peak periods: otherwise.
- HB trip: a trip which has at least one end of the trip at the home of the trip maker.
  NHB trip: otherwise.

3 Methods

3.1 Trip frequency and length

Participants are classified by DID-living or non-DID-living, which is hypothesized to affect trip frequency and length. The data of some frequently driving participants may have an excessively high influence on the results. To avoid this condition, after each driver's trips per day are calculated, trip frequencies are further averaged by living area and age groups, respectively. About trips length, similar data process is operated. For the comparative analysis between older driver and others both in DID and non-DID, t test is used to evaluate whether there is significant difference among two age groups.

Since commute trips usually comprise a certain route in constant time, discretionary trips in off-peak periods are calculated (the same in the analysis of destination distribution and NHB rate) in this research. Each driver's home can be located by comparing their address information and the PV data of the first most frequently visited destination. For driver who is working or is a student, the second most

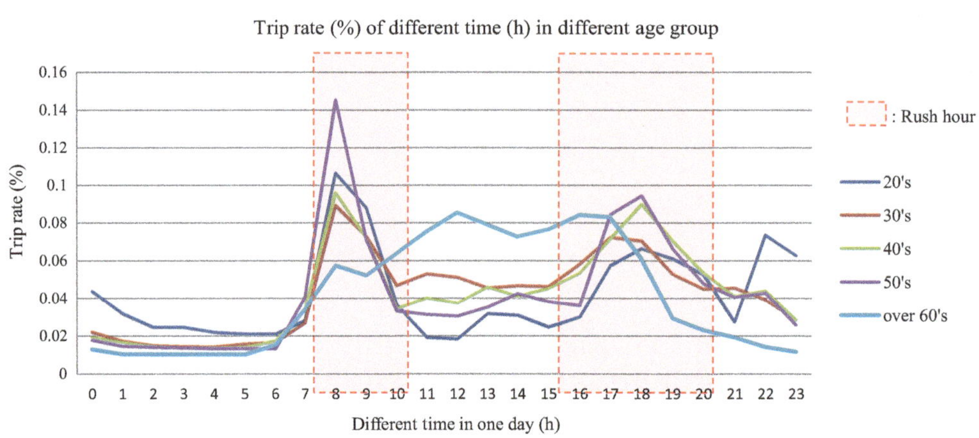

Fig. 1 Sample distribution on time of day of trip

frequently visited destination (only lower than home) on weekdays is considered as his/her commute destination.

3.2 Destination distribution

Different from trip length, the analysis of destination distribution is expected to help understand older drivers' driving range around home. Participants are also classified by age groups and living areas. Regardless of the origin, the distance from home to destination can be measured by GPS records. Considering that most trips might be distributed in a region not far away from home, this study further investigates the destination distribution in a radius of 5 km away from home.

3.3 NHB trips

By comparing the locations of trip's end points (origin and destination) and driver's home, each trip can be broadly classified as HB trips and NHB trips. Because the two age groups are mutually interdependent and have different sample sizes, *Z*-test is used to analyze the statistical significance between older drivers and others.

Regression model is used to identify the impact factors on all trip characteristics of this study. Because each GPS data point could be considered as a single sample case, and multiple samples with unequal sizes were collected from each person, a random effects regression model is constructed to consider unobserved heterogeneity among drivers [26, 27]. The model consists of 1 constant term (β_0) and 11 binary independent variables ($X_1, X_2,\ldots, X_{11}$): one is gender, five are age, the other five are working/at school, living area, weekday, rush time, and commute trip. It should be noted that because the trip frequency is classified by person and day rather than trip or hour, it is described by Poisson's regression model which is usually used for multilayer data. Meanwhile, for the same reason, the factors of rush time and commute trip are not included in the regression model of trip frequency. Furthermore, the interaction terms between the variable "Age (60 or older)" and the latter five factors are also added into the model. Through significance test, each variable's effect on trip characteristics will be examined. Considering that there may exist an association between the interaction variables, the association is measured by a Pearson correlation coefficient (also referred to as "phi coefficient" for two binary variables) [28]. If two variables interact, they may or may not be associated. However, if two interaction variables are associated (correlation between them is high), one variable should be eliminated from the regression model because the interaction effect is deduced to be caused by the association. Furthermore, elasticity information reflects the responsiveness of dependent variable to an independent variable change, and it depends on where the calculation starts [29]. If the elasticity is less than 1, we can claim that the dependent variable is inelastic, and vice versa. Thus, the elasticity at each age group is calculated to confirm the responsiveness of dependent variables to age.

4 Results

4.1 Trip frequency and length

The average daily trip frequency and average trip length are shown in Tables 2 and 3, respectively. Sample size means the amount of trips (the same in destination distribution and NHB rate). It reflected that in DID area, both indicators of older drivers are almost the same as those of other participants. In contrast, the older drivers in non-DID area drive significantly less and shorter than participants who are younger than 60.

Considering the traffic environment in urban area, such as intersection of multiple directions, complex road network, a large number of pedestrians and nearly unpredictable traffic congestion, driving is often not a recommended transportation mode not only for older drivers but also for younger drivers [30]. On the other hand, urban areas usually provide more travel options which include bus, metro, taxi, and car-sharing. One possible reason why DID-living older drivers do not drive less or

Table 2 Average daily trip frequency (trips/day)

	DID (e.g., urban)		Non-DID (e.g., suburban, rural)	
	Older drivers	Others	Older drivers	Others
Sample size	229	4,967	951	5,373
Mean	1.20	1.14	0.46	0.90
Std	0.99	0.54	0.19	0.61
t Stat	0.86		−42.49*	

* $P < 0.05$

Table 3 Average trip length (km/trip)

	DID		Non-DID	
	Older drivers	Others	Older drivers	Others
Sample size	229	4,967	951	5,373
Mean	18.75	17.70	12.70	19.45
Std	10.70	8.43	6.40	8.24
t Stat	1.47		−28.58*	

* $P < 0.05$

shorter than others is that the former have more free time to avoid rush hours. The relatively smooth traffic flow in off-peak period may increase their driving intention. Meanwhile, unlike moving from other region, aging is a gradual process which could not lead older adults to abandon their habit of driving immediately. As noted in the section of introduction, some physiological changes with aging are more easily to cause traffic accident. For the older drivers who are living in the DID area, the promotion of public transportation mode and the training of safety driving are therefore needed.

In some previous survey-based studies [31–33], most of rural-living older drivers claimed that giving up driving, so-called driving cessation, meant isolation for them. However, the data in this research show that older drivers living far from urban areas tend to reduce travel by driving. It can be inferred that older driver tends to buy a larger quantity of goods while shopping and hence drives less than others. Another possible reason is that the reduction of tour, entertainment, and socializing with aging may directly cut down both trip frequency and length. Meanwhile, in the long journey, middle-age people in family may be required to drive, and older adults are pleased to be passengers.

4.2 Destination distribution

Unlike trip length, destination distribution reveals the travel range around drivers' home. Regardless of origins, all trips can be imaged as distributed in concentric circles. As trip frequency and length, Table 4 shows that there is no significant difference between two age groups in DID area. Meanwhile, non-DID-living older drivers' travel range is smaller than others.

As noted in Sect. 4.1, more training and education should be given to DID-living older drivers for the prevention of accidents. On the other hand, since most of older drivers who are living in non-DID area inclined to drive near home, it should be further investigated and explained with the result of NHB trips in the next section.

4.3 NHB trips

NHB trips reflect the process of trips generation. Because driver, usually as trip planner, attempts to quantify the relationship between multiple activities and travel, a relatively high NHB rate means efficient travel plan. As shown in Table 5, the NHB rates of both age groups in DID area remains within the range between 25 % and 30 %, which is

Table 4 Destination distribution (km)

	DID		Non-DID	
	Older drivers	Others	Older drivers	Others
All trips				
Sample size	229	4,967	951	5,373
Mean	16.61	14.22	9.30	15.14
Std	18.79	22.36	15.36	16.37
t Stat	1.86		−10.70*	
Trips with the destination in 5 km range from home				
Sample size	101	1,980	580	2,840
Mean	2.43	2.94	2.00	2.78
Std	1.37	1.26	1.09	2.46
t Stat	−3.67		−12.00*	

* $P < 0.05$

Table 5 HB and NHB trips rate

	DID		Non-DID	
	HB trips	NHB trips	HB trips	NHB trips
Older drivers	161 (70.31 %)	68 (29.69 %)	764 (80.34 %)	187 (19.66 %)
Others	3,538 (71.23 %)	1,429 (28.77 %)	3,356 (62.46 %)	2,017 (37.54 %)
z Stat	−0.30		10.66*	

* $P < 0.05$

Table 6 Regression analysis of trip frequency, trip length, destination distribution, and NHB trips

Independent variable	Description	Possible values	Trip frequency (Possion model)		Trip length (log model)		Destination distribution (log model)		NHB trip (logit model)	
			Coef. (β)	*t*-stat	Coef. (β)	*t*-stat	Coef. (β)	*t*-stat	Coef. (β)	*t*-stat
β_0	Constant term		−0.52	−1.99*	2.08	7.03**	2.81	8.90**	−1.46	−4.51**
Driver attributes										
X_1	Gender	1. Male, 0. Female	0.03	0.22	0.27	1.95	0.21	1.84	0.27	1.56
X_2	Age (20's)	1. 20's, 0. Others	−0.19[a]	−0.80	0.29[a]	1.24	0.36	1.49	0.03[a]	0.10
X_3	Age (30's)	1. 30's, 0. Others	−0.07[a]	−0.43	0.23[a]	1.47	0.29[a]	1.31	−0.07	−0.33
X_4	Age (40's)	1. 40's, 0. Others	Base		Base		Base		Base	
X_5	Age (50's)	1. 50's, 0. Others	−0.02[a]	−0.13	0.20	1.24	0.23	1.17	−0.04[a]	−0.19
X_6	Age (60's or older)	1. 60's or older, 0. Others	**−1.12**[a]	**−2.59****	**−0.81**[a]	**−2.6***	−0.70[a]	−0.96	**0.57**[a]	**1.87***
X_7	Working/at school	1. Working or at school, 0.Others	0.15	0.57	0.26	0.98	0.24	0.71	0.81	2.56*
X_8	Living area	1. DID, 0. Non-DID	0.40	3.05**	0.06	0.49	0.39	1.66*	−0.07	−0.44
Day										
X_9	Weekday	1. Weekday, 0. Weekend	0.02	0.74	0.54	5.87**	0.94	2.34*	0.21	2.89**
Time										
X_{10}	Rush time	1. Yes, 0. No			0.91	8.49**	1.13	6.35**	0.27	3.97**
Commute trip										
X_{11}	Commute trip	1. Yes, 0. No			−1.40	−10.42**	−1.22	−8.77**	−21.77	−0.02
Interaction effects between age and living area										
X_{12}	X6 × X7	1. Yes, 0. No	−0.33	−0.61	0.97	0.90	0.65	0.83	−0.88	−1.09
X_{13}	X6 × X8	1. Yes, 0. No	**1.41**	**2.43***	**0.49**	**3.46***	**0.96**	**2.92***	**0.22**	**3.29****
X_{14}	X6 × X9	1. Yes, 0. No	**0.43**	**3.30****	−0.15	−0.21	−0.23	−0.38	0.03	0.08
X_{15}	X6 × X10	1. Yes, 0. No			0.20	0.28	0.23	0.25	−0.45	−1.73
X_{16}	X6 × X11	1. Yes, 0. No			0.05	0.04	0.03	0.05	0.29	0.00
Number of observations			6,588		16,320 (trips amount)		16,320 (trips amount)		16,320 (trips amount)	
Number of groups			108		108		108		108	
Log likelihood			−10,189		−17,187		−14,272		−3,086	
R^2			0.106		0.095		0.092		0.484	

* $P < 0.05$

** $P < 0.01$

[a] Elastic (greater than 1) at this independent variable

consistent with Schultz's model [11]. However, in the non-DID area, the NHB rate of older drivers is significant lower than that of others. Moreover, older drivers' NHB rate does not corroborate Gonzalez-Ayala's conclusion [12] that NHB rate in urban areas was lower than that in rural areas. It should be noted that "*z* Stat" in Table 5 is the result of Z-test.

With the smaller driving range from home (Sect. 4.2), it can be inferred that older adults in non-DID do not prefer continuous driving to different positions far away from home. It reveals some older drivers' life style: driving for single purpose around home, such as farming, social events, shopping in convenient store, etc; relying on other drivers or other transportation modes while going to a further destination.

4.4 Regression analysis

The result of regression analysis is shown in Table 6. If one older driver-related independent variable significantly affects the research object, the number will be in bold.

Trip frequency, length, and NHB trip, except for the destination distribution, are influenced significantly by older driver (X_6) even considering the interaction effect. As analyzed in the previous section, the interaction between older driver and living area is a significant influence factor on each travel behavior. Moreover, about the trip frequency, older driver also has a positive interaction effect (X_{14}) with the impact factor of weekday/weekend, which implies that older adults driving on weekday can increase the trip frequency. It accords with some reports [34, 35] and may be caused by their expectation of avoiding the traffic congestion during the weekends. As mentioned in Sect. 3, correlation between interaction variables is measured by Pearson correlation test. The correlation coefficients are -0.4922 (X_6 and X_7), -0.1751 (X_6 and X_8), 0.0566 (X_6 and X_9), -0.0186 (X_6 and X_{10}), and -0.0244 (X_6 and X_{11}), respectively. And all of them are nonsignificant at the 0.05 level. Thus, we can conclude that the interaction effects are not caused by the correlation between independent variables. The regression analysis also demonstrates that living area (X_8) plays an important role in the trip characteristics, which confirms that it is necessary to divide the participants of this research into DID-living and non-DID-living groups. Furthermore, the elasticity information is also computed and indicated by superscript "IE (inelastic)" or "E (elastic)." It shows that the responsiveness of each dependent variable to "Age (60 or older)" has high elasticity, which means that the change of the former is strongly influenced by the latter.

5 Conclusions

Through a two-month experiment of 108 participants in Aichi Prefecture, Japan, the present study primarily focused on older driver's automotive trip characteristics which included trip frequency, trip length, destination distribution, and NHB trips. To confirm whether living areas (DID and non-DID) were relevant for older drivers' trip characteristics, participants of this study were divided not only into age groups but also into living area groups.

At first, this study found that there was no significant difference between the trip characteristics of older drivers and others who were living in DID. Many reports have warned that some physiological and psychological changes with aging are easily to cause driving mistakes. The other unavoidable fact is that the complex traffic environment in urban area is one important reason for traffic accidents. Thus, we suggest that the education of safety driving should be given to DID-living older drivers. On the other hand, more convenient public transportation modes in DID, such as bus, metro, and taxi, should be further promoted for older adults.

Unlike the results in DID, trip characteristics had great difference between two age groups in non-DID. The results showed that older drivers' trip frequency, trip length, destination, and NHB trips rate were shorter or lower than others. The first three indicators reflected that older drivers living in non-DID tended to drive less and travel in a relatively small range, while NHB trips demonstrated that most non-DID-living older drivers had a tendency to travel for fewer purposes. It should be noted that some previous studies claimed that the NHB rate of rural-living drivers was higher than that of urban-living drivers because the former lived further away from their desired destinations. However, in this study, this phenomenon was verified to exist only in the participants who were younger than 60. It reflected that aging might have greater effect on non-DID-living older drivers than living area. All these characteristics suggested that EVs might be suitable for promotion among older drivers in rural area, because EVs' weaknesses of charging space/time and endurance ability are relatively easily to be accepted, while their advantages of cleaner and greener could be developed.

Furthermore, to identify the impact factors on all trip characteristics of this study, regression analysis was conducted. The results showed that "older drivers" was a significant independent factor on trip frequency, trip length, and NHB trips. Moreover, it had interaction effects with "living areas" on all trip characteristics during this study. It also confirmed the necessity of grouping not only by age but also by living area in this study.

However, the main limitations of this study were its small sample size (number of older drivers = 10) and limited geographic coverage (Aichi Prefecture, Japan). Therefore, caution should be exercised while extrapolating the results to general population. On the other hand, some hypotheses about the reasons of the phenomenon in this study may not be solid because they were assumed only based on PV data but lacking questionnaire survey. For instance, we can deduce that older drivers are living a retired life in rural areas, and therefore the need to travel is less, but in urban areas this age group can still be active. Thus, further survey-based research is needed to find out participant's attitude to driving.

Future research will seek to analyze older drivers' bio-information which could not be estimated by PV data. Biological data-collection device will be used in the experiment, and the data are expected to respect the correspondence between older drivers' psychological characteristics and their travel behaviors. A larger group of elderly individuals is planned to be recruited in the further experiment. Similarly, more participants from other age groups will be tracked for the comparison of the results.

Acknowledgments This research is partially supported by the Center of Innovation Program from Japan Science and Technology Agency, JST.

References

1. Population Division, United Nations (2015) World population prospects: 2015 revision. http://esa.un.org/unpd/wpp/. Accessed 15 Dec 2015
2. License Division, Traffic Bureau, National Police Agency, Japan (2014) Driving license statistic ver. 2014. https://www.npa.go.jp/toukei/menkyo/index.htm. Accessed 15 Dec 2015
3. MacLeod KE, Satariano WA, Ragland DR (2014) The impact of health problems on driving status among older adults. J Transp Health 1(2):86–94
4. Delhomme P, Cristea M, Paran F (2013) Self-reported frequency and perceived difficulty of adopting eco-friendly driving behavior according to gender, age, and environmental concern. Transp Res Part D 20:55–58
5. Ross LA, Dodson JE, Edwards JD et al (2012) Self-rated driving and driving safety in older adults. Accid Anal Prev 48:523–527
6. Tahara Y, Iwadare M (1999) Where the elderly move to: a review and a study of elderly migration flows in Japan. Komaba Stud Hum Geogr 13:1–53
7. Benekohal RF, Michaels RM, Shim E et al (1994) Effects of aging on older drivers' travel characteristics. Transp Res Rec 1438(HS-042 017):91–98
8. Rothe JP (1990) The safety of elderly drivers: yesterday's young in today's traffic. Transaction Publishers, New Brunswick, p 120
9. Verma A, Ramanayya TV (2014) Public transport planning and management in developing countries. CRC Press, Boca Raton, pp 116–117
10. Victor DJ, Ponnuswamy S (2012) Urban transportation: planning, operation and management. Tata McGraw-Hill Education, Noida, p 23
11. Schultz G, Allen W (1996) Improved modeling of non-home-based trips. Transp Res Rec 1556:22–26
12. Gonzalez-Ayala S (1999) Estudio Integral de Transporte (III)/multimodal transportation study: development of travel demand and mobile source emissions models for base year 1996, Juarez. https://www3.epa.gov/ttn/catc/dir1/emis-06.pdf. Accessed 20 Oct 2015
13. Hildebrand E, Gordon M, Hanson T (2004) Understanding the travel behaviour of the rural elderly. In: Proceedings of the 39th annual conference of the Canadian transportation research forum: revolutions in transportation, Calgary, Canada, pp 1183–2770
14. McDonald JD (2008) Measuring personality constructs: the advantages and disadvantages of self-reports, informant reports and behavioural assessments. Enquire 1(1):1–18
15. Hanson T, Hildebrand E (2011) Revealed choice of a new generation: travel behavior of older drivers in rural New Brunswick, Canada. In: Transportation research board 90th annual meeting, Washington, DC
16. Hanson T, Hildebrand E (2011) Experiences with GPS travel diaries in rural older driver research. In: Transportation research board 90th annual meeting, Washington, DC
17. Harrison A, Ragland D (2003) Consequences of driving reduction or cessation for older adults. Transp Res Rec 1843:96–104
18. Mollenkopf H, Marcellini F, Ruoppila I et al (2005) Enhancing mobility in later life: personal coping, environmental resources and technical support; the out-of-home mobility of older adults in urban and rural regions of five European countries. IOS Press, Amsterdam
19. Kempton W, Letendre SE (1997) Electric vehicles as a new power source for electric utilities. Transp Res Part D 2(3):157–175
20. Ministry of Land, Infrastructure and Transport Japan (2011) A brief report of the 5th investigation on person trip in Chūkyō Metropolitan Area. http://www.cbr.mlit.go.jp/kikaku/chukyo-pt/persontrip/pdf/gaiyou05_02.pdf. Accessed 15 Oct 2015
21. Japan Automobile Manufacturers Association Inc (2011) Survey on automobile market trend, 2011. http://www.jama.or.jp/lib/invest_analysis/four-wheeled.html. Accessed 15 Oct 2015
22. Ogawa K (2003) An analysis of household car ownership in a local city and its suburban area. In: Research meeting on civil engineering planning, Japan, 2003. https://www.jsce.or.jp/library/open/proc/maglist2/00039/200311_no28/pdf/268.pdf. Accessed 15 Oct 2015
23. Home Page of Toyota City Government (2015) http://www.city.toyota.aichi.jp/shisei/tokei/1008302.html. Accessed 15 Oct 2015
24. License Division, Traffic Bureau, National Police Agency, Japan (2014) Driving license statistic ver. 2014. https://www.npa.go.jp/toukei/menkyo/index.htm. Accessed 15 Oct 2015
25. Statistics Bureau Japan (2015) What is a densely inhabited district? http://www.stat.go.jp/english/data/chiri/did/1-1.htm. Accessed 1 July 2015
26. Wooldridge JM (2005) Simple solutions to the initial conditions problem in dynamic, nonlinear panel data models with unobserved heterogeneity. J Appl Econom 20(1):39–54
27. Jaccard J (2003) Interaction effects in multiple regression, 2nd edn. Sage, California, pp 16–43
28. Warrens MJ (2008) Similarity coefficients for binary data. Proefschriftmaken.nl, Oisterwijk 2008
29. Gary S (2004) Simple linear regression. New York University, Stern School of Business. http://pages.stern.nyu.edu/~wgreene/

Statistics/SimpleLinearRegressionCollection.pdf. Accessed 15 Jul 2015
30. Japan National Tourism Organization (2015) Driving a car. http://www.seejapan.co.uk/jnto_consumer/plan-your-trip/transportation/driving-a-car.aspx. Accessed 15 Oct 2015
31. Hassan H, King M, Watt K (2015) The perspectives of older drivers on the impact of feedback on their driving behaviours: a qualitative study. Transp Res Part F 28:25–39
32. Johnson JE (1998) Older rural adults and the decision to stop driving: the influence of family and friends. J Community Health Nurs 15(4):205–216
33. Adler G, Rottunda S (2006) Older adults' perspectives on driving cessation. J Aging Stud 20(3):227–235
34. Maryland Department of Transportation, USA (2015) Older driver safety in Maryland. http://www.mva.maryland.gov/safety/_docs/FY15_Older_ProgramAreaBriefCombinedFinal1.pdf. Accessed 15 Oct 2015
35. National Institute on Aging, USA (2015) How aging affects driving. http://nihseniorhealth.gov/olderdrivers/howagingaffectsdriving/01.html. Accessed 15 Oct 2015

19

Key Technologies of the Independent and Innovative Communications-Based Train Control Systems in China

Bo Liu[1] · Chunhai Gao[1]

Abstract In the past 10 years, the independent and innovative communications-based train control system has experienced breakthroughs in key technologies by conducting pilot experiments, field tests, and demonstration projects. For the first time, the design theories and technical aspects of moving block systems, which cover the entire life cycle of fail-safe systems, are set up. The models and algorithms are proposed to control the safe operation of trains. The safety technologies for short-interval train sequencing are adopted to realize the 90 s headway between consecutive trains. The optimized automatic train operation strategies are used to save energy and allow trains to safely reach their destinations. The highly dependable and bi-directional train-to-ground communication technologies are introduced for reliable transmission of safety related train-to-ground information in a complex environment which is compatible with multiple media, such as free waves, leaky waveguides, and leaky feeders. The simulation, tests, and verification technologies based on the minimum system and test cases were adopted to conduct the following: principle demonstration, functional tests, integrated tests, field failure data replay, analysis and disposal, system upgrade, and maintenance.

Keywords Communications-based train control (CBTC) · Moving block · Automatic train operation (ATO) · Safe train operation

✉ Bo Liu
18811446007@163.com

Chunhai Gao
jun.chen@bj-tct.com

[1] Beijing Traffic Control Technology CO, Ltd, Beijing, China

Editor: Xuesong Zhou

1 Background

Communications-based train control (CBTC) is the most advanced technology in the field of urban rail traffic signal control. Since 2004, CBTC has been adopted in newly constructed and upgraded old subway lines in Beijing, Shanghai, Guangzhou, and other cities [1]. China's first set of CBTC systems with completely independent intellectual property rights was implemented in the Yizhuang and Changping lines of the Beijing subway on December 30, 2010. The Yizhuang line is currently the most complete line in terms of functionality, and had the best performance in the world at the time of its opening. Based on the primary evaluation, an additional 6000-km line will be built into China's subway systems by 2020. The implementation of CBTC systems made China the fourth Country, after Germany, France, and Canada, to own the key technologies and to successfully apply them in practical engineering practices. The successful opening and operation of the Yizhuang line is an indication that the independent research, development, and engineering application of the domestic signal system have made significant progress [2].

The development of self-dependent innovative CBTC systems proposed in this paper takes 10 years and can be divided into five stages as follows [3]:

- The first stage: professors and scholars put forward the scientific problem and conduct theoretical analysis.
- The second stage: universities complete research on key technologies.
- The third phase: operation users conduct field and pilot tests.
- The fourth stage: construction users develop demonstration projects.

- The fifth stage: enterprises lead the application of research results to real-world implementation.

Our CBTC system can achieve the same design intervals and safety integrity levels (SIL) as those of the foreign signal companies that are experienced in signal systems development. However, the cost is more than 20 % lower than other companies. Table 1 shows comparison between domestic and foreign CBTC systems.

As seen in the table above, except CBTC system supplied by HITACHI, our CBTC system as well as other foreign CBTC systems can achieve the design intervals that is within 90 s. Meanwhile, our CBTC system follows the safety standard which is obeyed by most foreign CBTC systems and also is the highest standard in the industry.

2 Key Technologies

2.1 System Design and Equipment Development

The developed design theories and technologies of moving block with high security and high reliability cover the entire life cycle. We studied the rail transit CBTC system integration technology and method. According to the overall goal of the system and operation requirements, we use integrated method to combine the system-related discipline theory, method, technology, and demand, in turn, making an overall analysis of structure, environment, and function of the system.

1. System design for moving block

The moving block system design theory was established to realize the security tracking of the train group in a "single file" style and 90 s minimum interval [4]. It includes multi-particle models for train traction and braking, precise positioning model for multi-sensor fusion of single-train complex line, minimum distance information searching, early warning model, automatic collaborative energy-saving control model for train groups, strategies, and methods for bi-directional transmission of security control information between large-scale train groups and stations, etc.

2. Critical life-cycle safety system design

According to CEN-ELEC international standards, the system design and development process, which covers the entire life cycle, must be based on risk analysis and be able to trace risks back [5]. After the process was established, CBTC risk source identification component was completed. A reliability and security analysis model of a security module with interval and speed control was built. Nearly 100,000 test cases were done in the comprehensive test platform.

3. A secure computer platform investigated and developed independently

Based on the theory of fault tolerant computers, a fault tolerant and safety management (FTSM) mechanism was proposed. This mechanism can prevent known or unknown software and hardware faults. Design principles of space diversity and the time diversity (time triggered software scheduling mechanism) were adopted to meet the characteristics of deterministic fault security. Figure 1 shows a security framework adopted in Yichuan Line of the Beijing subway.

As seen in Fig. 1, this security framework is hierarchical. The layer of software is divided into three parts, application software, security platform software, and real-time multitasking operating system from top to down. There comes the layer of hardware. It contains double FTSM hardware, real-time multitasking operating system, and FTSM communication bus. On the bottom, dual redundancy high-speed-switched Ethernet is used to connect hardware.

4. Technical equipment

Technical equipments are developed based on integrated design, 2-level scheduling model, and three types of control. Research on equipment development with rank-free conversion and cooperative control was completed. The integrated design, which includes multiple components ATS (Automatic Train Supervision), ATP (Automatic Train Protection), ATO (Automatic Train Operation), and CI (Computer Interlocking), was proposed. Then the

Table 1 Comparison of domestic and foreign CBTC

Signal system	Supplier	Design interval	Safety standards
Introduced CBTC	THALES	Within 90 s	SIL4, highest standard
	SIEMENS	Within 90 s	SIL4, highest standard
	ALSTOM	Within 90 s	SIL4, highest standard
	ANSALDO	Within 90 s	SIL4, highest standard
	Bombardier	Within 90 s	SIL4, highest standard
	HITACHI	120 s	Japanese standard
Independent and innovative CBTC	TCT	Within 90 s	SIL4, highest standard

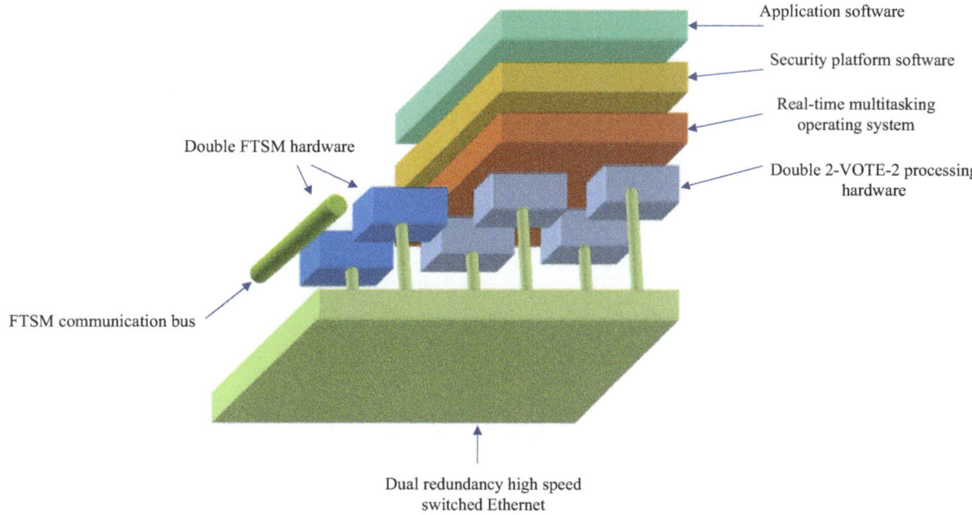

Fig. 1 Security framework of Yizhuang Line of Beijing Subway

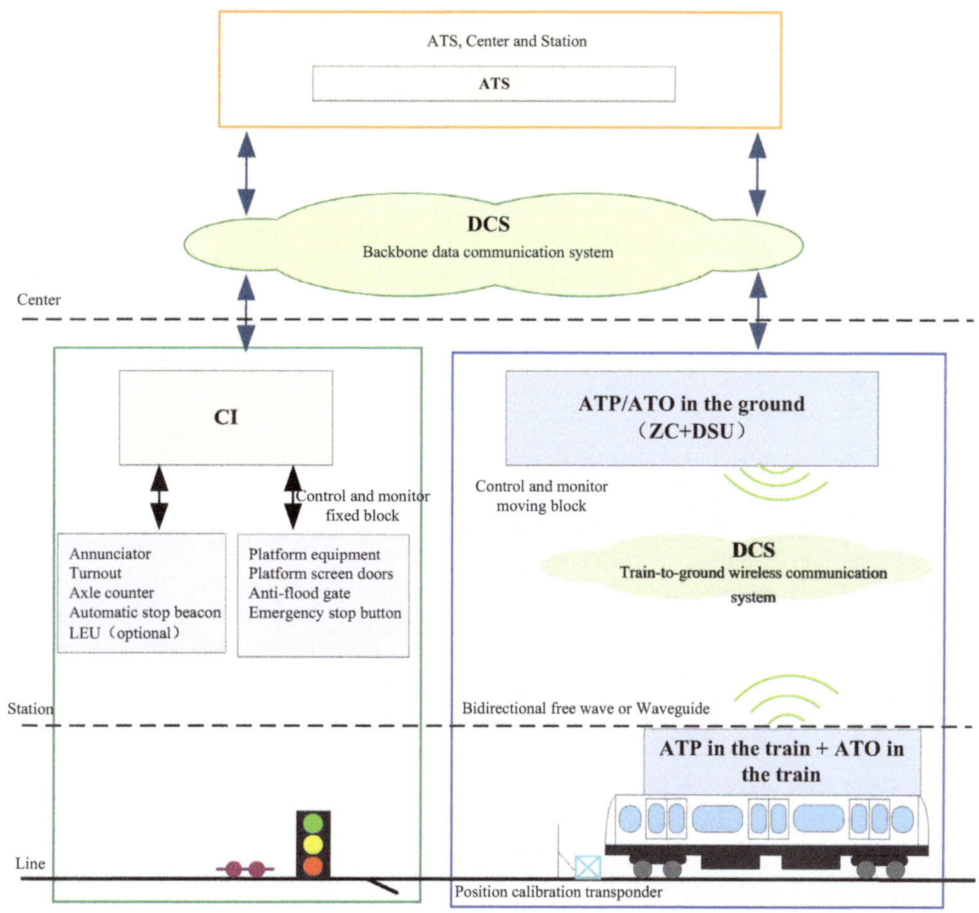

Fig. 2 Integrated design, 2-level scheduling model, and 3 types of control

complete set of technical equipment was developed. The complex coordination of 3-level asynchronous, non-intrusive, multi-vehicle-mixed operations of the mobile block, fixed block, and the interlocking of the station was constructed (Fig. 2).

As seen in Fig. 2, ATS is on the top level. It is always put in the command center and stations. Its main function is to monitor the train running state, use the software to realize the automatic networking, communications, and train operation management. There comes the DCS (Backbone data communication system) [6]. It takes the responsibility of data transmission between ATS and other blocks below. CI plays a role in controlling and monitoring the fixed block which is composed of annunciators, turnouts, axle counters platform equipment. ATP's main function is to achieve the train interval control and speed control, and then ensure driving safety. In order to do this, there is a channel for the information transmission between the on-board ATP equipment and ground ATP equipment, which we called Train-to-ground wireless communication system here. ATO in the ground and ATO in the train work together to fulfill automatic driving between stations, speed adjustment, and precise stop inside the stations.

2.2 Safety Protection and Automatic Driving

The safety protection technology for highly dense train traffic, which connects the vehicle with the ground, was proposed initially at the domestic level. The problems of unmanned operation and safety tracing for high speed, high density, complex lines, and adverse weather conditions were addressed. This ensures the safety of the train operations around the clock. Optimal automatic driving technology was created for the first time at the domestic level. Automatic adjustment of operation time tables in the complex environment of multi-objective optimization was used, and the adjustment error is less than 5 s. Energy-saving optimization control of complex lines was realized [7]. Precise stopping of trains at stations was achieved, and the parking precision was less than 0.3 m. A few key achievements are listed below:

1. Optimal automatic driving technology was adopted.
2. Train dynamics model, which combines single- and multi-particle models and then realizes train over-speed protection control, was established (Fig. 3). If the speed of train reaches to v_4, it strikes the common braking line, so the system will activate the common braking model. Seriously, if the speed reaches to v_0, it strikes the emergency braking line, so the system will activate the emergency braking model to cut down the speed.
3. The train group safety control strategy was adopted.
4. Optimal automatic driving technology was applied.

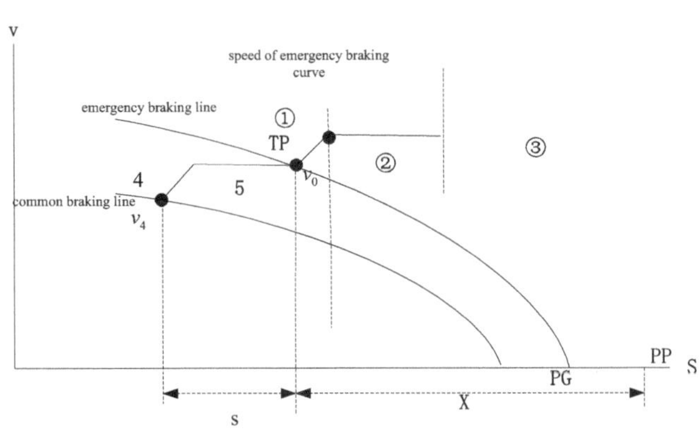

Fig. 3 Train dynamics model which combines simple particles and multi particles

2.3 Bi-directional Large Capacity Transmission Between the Vehicle and Ground

Three transmission modes, namely free wireless wave, leaky waveguide, and leaky feeder, were adopted for the first time in the self-dependent innovative CBTC system. The train operation control information realizes bi-directional transmission. As a result, the engineering adaptability and flexibility of the CBTC system are greatly improved. The information transmission synchronization correction algorithm was proposed. CBTC security transfer protocol was developed. The CBTC information redundancy transmission network was developed independently, providing a reliable and safe data channel between each subsystem of the CBTC system. We studied all the way of WLAN that are used for railway traffic control system at present, including IEEE802.11, IEEE802.11b. IEEE802.11g [8]. Meanwhile, we also studied the leaky waveguide. Finally, we made experiments of IEEE802.11g train–ground transmission mode in the field.

2.4 Simulation, Testing, and Verification

1. Technology of a semi-physical system with the hardware in the loop

Based on the event-driven technology, the CBTC simulation and testing platform was developed [9]. The minimum system with minimization hardware and maximization function was established. Based on visual identification characteristics of rail transit virtual intelligent scene of generation technology, flexible virtual reality, and the mutual control were proposed, and the comprehensive test of visual and high coverage was operated.

2. Test sequence based on fault injection and script verification

Using a formal method, the non-formal description of test specification was described and the test sequence was optimized [10]. The test case sets of the CBTC system was formed using data-driven testing technologies. The test sequence toolset based on engineering data was set up through the method of hierarchical decision tree analysis. Subsequently, test sequences under different fault scenarios were generated automatically.

3. Recording, playback, and evaluation technologies

Logging efficiency and capacity were improved greatly by log record mode based on encoding storage. The recorded data, which uses log format standardization, can be played again in the test platform, providing the basis for the dynamic analysis of the operation status of equipment evaluation and replication site problems [11].

2.5 Other Innovative Results

In the process of the implementation of the CBTC system, the structure of the signal system was optimized and functional limitation was surpassed by large numbers of research and in-depth analysis of the international standards and systems [12]. The innovative results are as follows: the standard system equipment configuration principles, signal equipment layout design principles, design principles of CBTC backup mode, platform screen door linkage function of intermittent type, interface design principles, design and on-site verification of signal system with 2-min tracking interval running ability, the guide tube optimization scheme, wireless network security testing, and driving process fault handling function optimization [13–15].

3 Conclusion

Independent innovation of the CBTC system can encourage us to keep up with the latest technological development of the international train operation control systems. Core technologies of the CBTC system, such as bi-directional transmission technology between the vehicle and ground, precise positioning of the train, automatic safety protection, and automatic driving, are mastered [16]. CBTC for urban rail transit systems in China were developed in full accordance with the CEN-ELEC international security standards, thereby reaching the highest level internationally.

The successful application of independent innovation CBTC systems breaks the long-term monopoly of the core technology of the subway signal in our country, bringing significant economic and social benefits. This innovation brings down the price of imported system by 20 % and reduces the cost of our own CBTC by 10 %. With the development of the core technologies and industrial chain, the independent CBTC system will be the most cost-effective system, providing a solid technical foundation for the construction and operation of urban rail transit in China.

Acknowledgments This work is supported by the Beijing municipal science and technology plan project (Project Number: D15110000581 5001) and the Beijing urban rail operation control system engineering technology research center.

References

1. Fraszczyk A, Brown P, Duan S (2015) Public perception of driverless trains. Urban Rail Transit 1(2):78–86
2. Batty P, Palacin R (2015) The circumvention of barriers to urban rail energy efficiency. Urban Rail Transit 1(2):71–77. doi:10.1007/s40864-015-0014-9
3. An J, Yang D, Kang Q, Wang L, Wu Q (2008) Security radio data transmission mode based on turbo codes for the CBTC system. Intelligent control and automation, 2008. WCICA 2008. 7th world congress on 25–27 June 2008, pp 6663–6668
4. Fitzmanrice M (2006) Use of 2.4 GHz frequency band for communication based train control data communication system. In: Rail conference, 2006. Proceeding of the 2006 IEEE/ASME joint, 4–6 April 2006, pp 263–267
5. Gao C, Tang T, Yao F (2006) Research and design of vehicle equipment ATP based on CENELEC. J China Railw Soc 28(1):99–107
6. Tang T, Yao F, Gao C (2005) Research on security assessment and authentication system of rail traffic signal system. Urban Rapid Rail Transit 17(6):28–32
7. Huang Y, Tang T (2008) Research on virtual simulation technology in metro train operation simulation system. J Syst Simul 20(12):3208–3211
8. Li Z, Bin N (2010) Design of CBTC train-ground communication system based on 802.11 g IEEE standard. China Railw Sci 31(5):119–124
9. Ma L, Gao B (2004) Design and implementation of a high security and fault tolerant control computer. China Railw Sci 8
10. Wang H (2010) Analysis method and application of functional hazard analysis based on model. Trans Beijing Inst Technol 7:849–854
11. Pascoe RD, Eichorn TN (2009) What is communication-based train control? IEEE J Mag 4
12. Zhao X, Tang T, Yan F (2009) A functional safety analysis approach for analyzing CBTC system. In: IEEE conference publications, 2009, vol 3
13. Chen L, Ning B, Xu T-H (2007) Research on modeling and simulation of vehicle-on-board automatic train protection subsystem of communication based train control system. In: IEEE conference publications, 2007, vol 2
14. Schifers C, Hans G IEEE standard for communication-based train control (CBTC) performance and functional requirements
15. Peng Z, Li G, Wang H, Wu Q (2008) A wireless transmission mechanism of the wireless CBTC system and performance analysis. Audio, language and image processing, 2008. ICALIP 2008. International Conference on 7–9 July 2008, pp 443–448
16. Zhang Z, Wang CQ, Zhang W (2015) Status analysis and development suggestions on signaling system of Beijing rail transit. Urban Rail Transit 1(1):1–12

20

A technical review on persistent scatterer interferometry

Hongguo Jia[1,2,3,4] · Luyao Liu[1]

Abstract This article focuses on reviewing the technologies of persistent scatterer interferometry (PSI), which has been often used to monitor the deformation of Earth surface. Three critical steps in the implementation of PSI were introduced, i.e., (1) detection of persistent scatterer (PS), (2) construction of PS network, and (3) PSI modeling and solution. Finally, the main problems and outlooks on the PSI technique are discussed and given.

Keywords Review · Persistent scatterer interferometry (PSI) · PS detection · PS network construction · PSI modeling and solution

1 Introduction

Since 1980s, interferometry synthetic aperture radar (InSAR) technique has been utilized to monitor the deformation of the Earth's surface [1] and applied to detecting hazards, e.g., earthquakes [2], ice motion [3], volcanism [4], landslides [5], and groundwater flow [6]. InSAR technique is able to detect motion phenomena on ground surface for large areas and has high accuracy of 3 to 24 cm and spatial resolution (up to 1 m). However, the implementation of InSAR technique is often influenced by three main factors: (1) signal interference due to atmosphere condition; (2) temporal decorrelation due to the changes of the scattering characteristics at different times; and (3) geometric decorrelation due to different imaging geometries arising from the far distances between the satellite repeat orbits. These factors may result in the reduction in the accuracy of deformation monitoring.

To solve these problems, much research work has been devoted to the time series analyses, i.e., the utilization of images collected at different times for the same areas. Furthermore, persistent scatterer interferometry synthetic aperture radar (PSInSAR) approach [7, 8] was developed to conduct the deformation monitoring. These PS points have quite constant scattering properties over time and the reflection dominance within a pixel cell so as to reduce the temporal decorrelation [9, 10] and the geometry decorrelation. In addition, the signal interference can be estimated and removed using the series of images acquired at different times. Later, the small baseline, i.e., the small distances among either the satellite positions or different acquisition times, is introduced to reduce the geometric and temporal decorrelation. This method is also time-series analysis approach and is called the small baseline subset (SBAS) [11, 12].

Recently, many researchers have focused on the improvement of PSInSAR technique, and significant progress has been achieved. In this article, we focus on the past development and the state of the art of this technique. This article is organized as follows: Sects. 2, 3. and 4 review the three steps involved in PSInSAR methods, i.e., PS detection, network formation, and resolution modeling,

✉ Hongguo Jia
lemon_gg@hotmail.com

1 Faculty of Geosciences and Environmental Engineering, Southwest Jiaotong University, Chengdu, Sichuan, China

2 State Key Laboratory of Geodesy and Earth's Dynamics, Wuhan, Hubei, China

3 Collaborative Innovation Center for Rail Transport Safety, Ministry of Education of the People's Republic of China, Southwest Jiaotong University, Chengdu, Sichuan, China

4 State-Province Joint Engineering Laboratory of Spatial Information Technology for High-Speed Railway Safety, Southwest Jiaotong University, Chengdu, Sichuan, China

respectively; in Sect. 5, conclusions are made and outlooks on PSInSAR technique are presented.

2 Persistent scatterer detection

A series of SAR images used in PSInSAR contain billions of image pixels. For such large amount of data, it is necessary to firstly select the most promising scatter locations, i.e., the persistent scatterer candidates (PSCs), to reduce the data in order to ensure the implementation efficiency. The selected PSCs are commonly divided into two classes with different levels. The first level of PSCs is utilized to form a network covering the target area. The second level of PSCs is for the densification of the PS distribution. If PS density is still low over a target area, artificial corner reflectors can be used to provide strong responses in the SAR image so as to achieve a good interferometric phases for the estimation of the deformation and assess the performance of DInSAR and PSI [13–15]. Technical aspects on artificial corner reflectors in DInSAR are described in Ye et al. [16] and Crosetto et al. [17]. This technique has been applied to PSI [18, 19].

The selection of PSCs is based on the amplitude information because of the wrap of original interferometric phases and an amount of phase contributors:

$$\Phi = \varphi_{\text{flat}} + \varphi_{\text{topo}} + \varphi_{\text{def}} + \varphi_{\text{atm}} + \varphi_{\text{noise}}, \tag{1}$$

where φ_{flat}, φ_{topo}, φ_{def}, φ_{atm}, and φ_{noise} are the phase components about reference surface, topography, deformation, atmospheric delay, and noise, respectively. The detection of PSs depends on the phase stability or coherence of PSs. As the calibrated amplitude stability has a statistical relation to phase stability [8], if a scatterer has a large amplitude so as to dominate the resolution cell, the clutter has little influence on the phase. Based on this phenomenon, pixels with high-phase stability can be selected as PSCs.

Currently, there are three basic methods to detect PSs, i.e., the methods based on normalized amplitude dispersion, phase stability, and correlation. Many attempts have been made to improve these methods [20] or design a hybrid strategy to incorporate three basic methods [21]. In the following part, three typical methods are introduced, respectively.

2.1 Method based on amplitude dispersion

This method is used to detect the PSs according to the analysis of the amplitudes of pixels in time series. The amplitude dispersion D_A can be written as [8]

$$D_A = \frac{\sigma_A}{m_A}, \tag{2}$$

where σ_A and m_A are the standard deviation and the mean of the backscattering intensity. In general, the scatterers with low-amplitude dispersions also have low-phase dispersions, while high-amplitude dispersions do not mean high-phase dispersions. Hence, a pixel with the amplitude less than a preset threshold can be regarded as a PS point. In this method, the phase information has been not considered. The method has the merits of high efficiency, easy implementation, and does not need to consider the relationship among adjacent pixels. Thus, it is suitable to the detection of single PS point. Ferretti [8] pointed out that the pixel with the amplitude from 0.25 to 0.4 can be selected as PS. Generally, the minimum number of 25 SAR images is needed in the detection of PS points in order to ensure the statistical significance and the quality of PSs [22].

2.2 Method based on phase stability

As PSs have the characteristic of the stable backscatter in images collected at different times, the phase stability is often considered as a criterion to select PSs. The phase stability can be assessed according to the signal-to-clutter ratio (SCR) [23]. If the average SCR of a pixel is more than a given threshold, the pixel can be regarded as a PS point. The form of SCR can be written as

$$r_{\text{sc}} = \frac{s^2}{c^2}, \tag{3}$$

where s is the amplitude and c represents the clutter. Under the assumption of the equal clutter for a pixel and its surrounding pixels, the SCR for each pixel can be estimated by the phase standard variance σ_ϕ [8] and the relation among pixels can be written as

$$\sigma_\phi = \frac{1}{\sqrt{2r_{\text{sc}}}}. \tag{4}$$

If a pixel has the phase standard deviation less than a certain threshold, it can be regarded as a PS. However, if adjacent pixels contain point scatterers [23], the clutter may be overestimated, which may result in the rejection of suitable PSC. In addition, a lot of computation time is needed in the implementation of this method.

2.3 Method based on correlation

If there are N images for a target region, N-1 interferometric pairs can be formed by a master image selected from these images and a slave image. As the backscatter of PSs is almost consistent in the images collected at different times, higher correlation for PSs should be kept in interferometric pairs. If the correlation in a pixel is more than a certain threshold, it can be regarded as PSC. This method can greatly reduce the data in the selection of PSs. In practice, for a pixel, the correlation coefficient γ is estimated by the pixel and its surrounding pixels in a given window size:

$$\gamma = \frac{\left|\sum_{i=1}^{m}\sum_{j=1}^{n} M(i,j)S^{*}(i,j)\right|}{\sqrt{\sum_{i=1}^{m}\sum_{j=1}^{n}|M(i,j)|^{2}\sum_{i=1}^{m}\sum_{j=1}^{n}|S(i,j)|^{2}}}, \tag{5}$$

where M and S are the local information sets of pixels on two SAR images of an interferometric pair; $*$ is the complex conjugation operator.

However, the calculation of the correlation coefficient greatly depends on the selection of the window size. Large size may result in the reliability reduction of PS detection, while small size results in the reliability reduction of the correlation coefficient estimation. In addition, the estimation of the correlation coefficient greatly depends on the quality of master image.

3 PS network construction

After PSC selection, the first-order PSCs are utilized to form a reference network for the full target area. This network is the basic for estimation of the atmospheric and orbital phase screen (APS) and lower-order PSCs can be added to densify this network.

For PS networking, there is a common algorithm based on Delaunay triangulation [24]. Figure 1 shows the examples of the constructed network. In Delaunay network, the PSCs are connected based on the Delaunay triangulation principle. However, the number of arcs in such networks may be low, which results in lower accuracy for the deformation estimation. Furthermore, if an arc is longer than the maximum correlation length of the atmosphere, it will be eliminated, which further reduce the redundancy of the Delaunay network. In order to enhance the redundancy of Delaunay network, Kampes [24] proposed the multi-order networking algorithm. The implementation of this algorithm consists of four steps: (1) setting the minimal number of connecting arcs at each PSC; (2) equally dividing the area around each PSC into sub-areas; (3) selecting the closest PSCs around a PSC in all directions in a circular order; and (4) ignoring the sub-area when the distance between a PSC and its closest PSCs in the direction is too long. This procedure is iteratively implemented until the minimal number of connections at each PSC can be reached. By this algorithm, the PSCs at the outer sides of the crops, rivers, and grassland should be well connected. Once the network is constructed, the differential phase observations at each arc are calculated.

Later, Liu et al. [25] proposed a freely connected network (FCN) algorithm which is shown in Fig. 2. In this network, if the distance between two PSs is less than a given threshold, they are connected as an arc:

$$S\left(x_i, y_j, x_i, y_j\right) = \sqrt{f_r^2 \cdot (x_i - x_j)^2 + f_a^2 \cdot (y_i - y_j)^2} \le S_0, \tag{6}$$

where (i, j) are the pixel coordinates in an image; f_r and f_a are the scaling factors in range and azimuth directions, respectively, which are used to transfer the pixel distance to the geometric distance; and S_0 is a given threshold for the minimal distance. S_0 is determined by the atmospheric gradients. If the atmospheric delay has faster spatial changes, a lower threshold will be selected. FCN can reliably estimate subsidence rates and elevation errors of PSs than the algorithms based on Delaunay triangulation. However, it costs much more computation time.

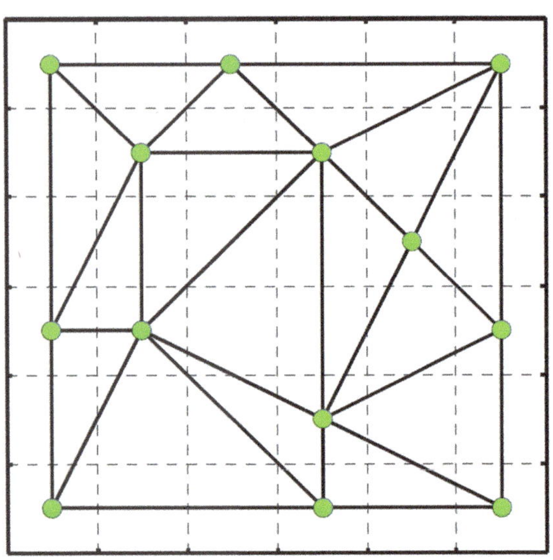

Fig. 1 Delaunay triangulation (Liu et al. [25])

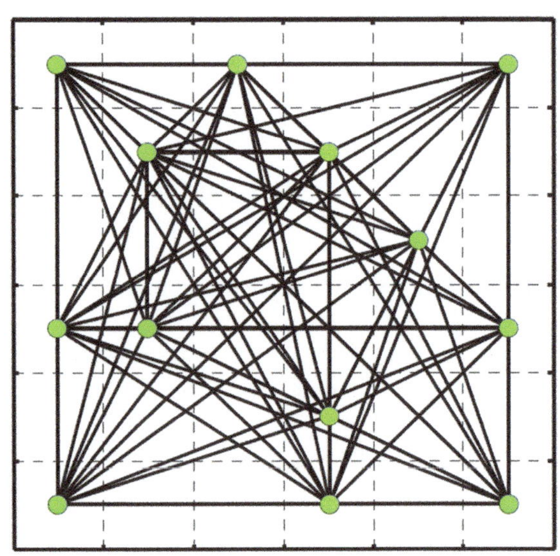

Fig. 2 A freely connected network (FCN) (Liu et al. [25])

Zhang et al. [26] developed an algorithm by constructing a local Delaunay triangulation network. In this algorithm, the temporarily coherent point (TCP) which remains reflectivity only for some time spans is used, and the TCPs in a regular patch are connected and some regular patches are placed over the image. This algorithm can ensure the extensive TCP connections without the phase ambiguity. In addition, the constructed network by this algorithm has the connections with small lengths so as to reduce the atmospheric artifact.

4 PSInSAR modeling and solution

DInSAR phase measurements are wrapped in the range of $-\pi$ to π. Because of the ambiguous nature of the observations, one of the main limitations of PSI is the incapability of measuring "fast" deformation phenomena. As the ambiguous nature can be eliminated by the network neighborhood differencing, the phase ambiguity can be estimated based on the arc connecting two PSs, instead of two subsequent acquisitions at a single PS. In fact, if the differential deformation phase between two subsequent acquisitions exceeds the range of $-\pi$ to π, the actual deformation cannot be explicitly retrieved.

After removing the non-deformation phase components, for DInSAR observations, the differential phase of π corresponds to a maximum differential deformation of $\lambda/4$ over the revisit interval. For ENVISAT ASAR, TerraSAR-X, Sentinel-1, and ALOS PALSAR, the wavelengths of which are 56, 31, 56, and 236 mm, respectively, and the revisiting time are 35, 11, 12, and 46 days, respectively, the maximum measurable differential deformation rates between two PSs are 147, 257, 426, and 468 mm/year, respectively. Note that these are theoretical values ignoring the phase noises and data processing errors. The noise level of data and the methods adopted to obtain the phase ambiguities greatly influence the actual deformation rates. In addition, the deformation estimation by PSI depends also on the spatial pattern and available PS density of the deformation phenomenon.

If the deformations are greater than the maximum phase ambiguity limit, one effective method for the estimation is SAR offset-tracking (OT) using the amplitude information, i.e., complementing the measurements by tracking very fast deformation. The OT method exploits the SAR amplitude information of at least two SAR images, covering the same area, acquired at different times. The main procedure includes pixel offset estimation, systematic offset removal, and offset field conversion (i.e., from offset in pixel to deformation in m or mm). OT can estimate 2D displacements, the across-track (range) and along-track (azimuth), of a given ground target. The processing utilizes the image matching techniques. Although OT is not affected by the ambiguous nature of the interferometric approaches, it has the disadvantage of low sensitivity to deformation and the coarse accuracy. Theoretically, the accuracy of OT method is 1/10th or 1/20th of the SAR pixel (Fialko et al. [27] and Strozzi et al. [28]). Taking TerraSAR-X as an example, its range pixel spacing and azimuth pixel spacing are around 2.0 m and 1.9 m, respectively, and then the absolute value of the OT measurements accuracy in range and azimuth direction will be in the ranges of [0.1 0.2] m and [0.095 0.19] m, respectively, which is much lower than the measurement accuracies by DInSAR and PSI.

Another key factor of PSI is the deformation model for phase modeling and parameter estimation, and a linear model is often used. If the deformation phenomenon fits well with the linear model, the PSI estimation has desirable accuracy. However, for non-linear deformation phenomenon, the utilization of the linear model may significantly reduce the accuracy of the deformation estimation. Lately, some PSI approaches have been developed by building other deformation models or hybrid models, e.g., polynomial and periodic models. However, all these approaches are under the assumption of the spatial smoothness of the deformation phenomena, which means that all current approaches are limited by the ambiguous nature of the interferometric phases.

For the deformation estimation by PSI, the thermal expansion is also a critical factor [29–31]. Further research on this issue has been extended from single PS to single object, e.g., buildings and bridges. For modeling and estimating thermal expansion, there are three typical approaches. The first one is to extract the thermal expansion from the total observed displacement [31]. However, in this approach, the thermal expansion is not explicitly considered in the modeling such that the estimated deformation is greatly affected by large distortions, especially when there are less SAR images in short periods [32]. In the second approach, the thermal expansion has been explicitly considered as a parameter in the model, e.g., Gernhardt et al. [33], Monserrat et al. [34], Fornaro et al. [35], and Zhang et al. [36]. The third approach aims to extract the thermal expansion coefficient by thermal maps [32, 34].

5 Conclusions and outlook

PSI technique has been widely applied to the measurement of displacements and is able to achieve a high accuracy. Many researchers have focused on this technique and many improvements for PSI have been achieved. In addition, some active SAR missions and the plans have been conducted by agencies to further improve the accuracy and reliability of this technique.

Currently, main tasks on PSI technique include the following aspects:

(1) Unbiased estimation of thresholds for cohesive points.
(2) In current research on the detection of DS points, typical approaches are the region growing and sample statistical test, e.g., Kolmogorov–Smirnov (KS) test, Anderson–Darling (AD) test, Baumgartner–Weiss–Schindler (BWS) test, and detect the homologous points around the detection center according to the amplitude or intensity. However, as DS consists of its adjacent points with similar scatter characteristics, it is necessary to conduct a further research on the detection of these adjacent points in a clustering manner.
(3) For the network construction of PSs, the main problem is how to reduce the possibility of the ambiguity in the arcs in the network. In addition, current methods do not consider the elevation and scatter intensity in the process of the network construction.
(4) Current methods reduce the atmosphere delay generally using external data or modeling, while the characteristics of SAR data, e.g., the distribution characteristic, are not used.
(5) In the PSI modeling, the estimation of parameters, e.g., for the temperature and atmosphere delay, should be more precise by considering the actual influences of these factors.

Acknowledgments This project was supported by State Key Laboratory of Geodesy and Earth's Dynamics (SKLGED2015-5-1-E), the Fundamental Research Funds for the Central Universities (2682015CX015), the National Natural Science Foundation of China (41474003), and Program for Changjiang Scholars and Innovative Research Team in University (IRT13092).

References

1. Gabriel AK, Goldstein RM, Zebker HA (1989) Mapping small elevation changes over large areas: differential radar interferometry. J Geophys Res 94(B7):9183–9191
2. Massonnet D, Rossi M, Carmona C, Adagna F, Peltzer G, Feigl K, Rabaute T (1993) The displacement field of the Landers earthquake mapped by radar interferometry. Nature 364(8):138–142
3. Goldstein RM, Engelhardt H, Kamp B, Frolich RM (1993) Satellite radar interferometry for monitoring ice sheet motion: application to an antarctic ice stream. Science 262:1525–1530
4. Massonnet D, Briole P, Arnaud A (1995) Deflation of Mount Etna monitored by spaceborne radar interferometry. Nature 375:567–570
5. Fruneau B, Achache J, Delacourt C (1996) Observation and modelling of the Saint-Etienne-de-Tińee landslide. Tectonophysics 265:181–190
6. Amelung F, Galloway DL, Bell JW, Zebker HA, Laczniak RJ (1999) Sensing the ups and downs of Las Vegas: InSAR reveals structural control of land subsidence and aquifer-system deformation. Geology 27(6):483–486
7. Ferretti A, Prati C, Rocca F (2000) Nonlinear subsidence rate estimation using permanent scatterers in differential SAR interferometry. IEEE Trans Geosci Remote Sens 38(5):2202–2212
8. Ferretti A, Prati C, Rocca F (2001) Permanent scatterers in SAR interferometry. IEEE Trans Geosci Remote Sens 39(1):8–20
9. Usai S (1997) The use of man-made features for long time scale insar. In: International geoscience and remote sensing symposium, Singapore, 3–8 Aug 1997, p 1542–1544
10. Usai S, Hanssen R (1997) Long time scale INSAR by means of high coherence features. In: Third ERS symposium—space at the service of our environment, Florence, 17–21 March 1997, pp 225–228
11. Berardino P, Fornaro G, Lanari R, Sansosti E (2002) A new algorithm for surface deformation monitoring based on small baseline differential SAR interferograms. IEEE Trans Geosci Remote Sens 40(11):2375–2383
12. Mora O, Mallorqui JJ, Broquetas A (2003) Linear and nonlinear terrain deformation maps from a reduced set of interferometric SAR images. IEEE Trans Geosci Remote Sens 41(10):2243–2253
13. Ferretti A, Monti-Guarnieri A, Prati C et al (2007) InSAR principles-guidelines for SAR interferometry processing and interpretation. ESA Publications, Noordwijk
14. Marinkovic P, Ketelaar G, van Leijen F et al (2007) InSAR quality control: analysis of five years of corner reflector time series[C]//fifth international workshop on ERS/Envisat SAR interferometry,'FRINGE07', Frascati, 26–30 Nov 2007
15. Quin G, Loreaux P (2013) Submillimeter accuracy of multipass corner reflector monitoring by PS technique. Geosci Remote Sens IEEE Trans 51(3):1775–1783
16. Ye X, Kaufmann H, Guo XF (2004) Landslide monitoring in the Three Gorges area using D-InSAR and corner reflectors. Photogramm Eng Remote Sens 70(10):1167–1172
17. Crosetto M, Gili JA, Monserrat O et al (2013) Interferometric SAR monitoring of the Vallcebre landslide (Spain) using corner reflectors. Nat Hazards Earth Syst Sci 13(4):923–933
18. Teatini P, Tosi L, Strozzi T (2012) Comment on "Recent subsidence of the Venice Lagoon from continuous GPS and interferometric synthetic aperture radar" by Y. Bock, S. Wdowinski, A. Ferretti, F. Novali, and A. Fumagalli. Geochem Geophys Geosyst 13(7):1–13
19. Van Leijen F, Hanssen R (2008) Ground water management and its consequences in Delft, the Netherlands as observed by persistent scatterer interferometry. In: Fifth international workshop on ERS/Envisat SAR interferometry, 'FRINGE07', Frascati, 26–30 Nov 2007, p 6
20. Jiang M, Ding X, Hanssen RF et al (2015) Fast statistically homogeneous pixel selection for covariance matrix estimation for multitemporal InSAR. Geosci Remote Sens IEEE Trans 53(3):1213–1224
21. Jiang M, Ding X, Li Z (2014) Hybrid approach for unbiased coherence estimation for multitemporal InSAR. Geosci Remote Sens IEEE Trans 52(5):2459–2473
22. Colesanti C, Ferretti A, Novali F, Prati C, Rocca F (2003) SAR monitoring of progressive and seasonal ground deformation using the Permanent Scatterers Technique. IEEE TGRS 41(7):1685–1701

23. Adam N, Kampes BM, Eineder M (2004) Development of a scientific persistent scatterer system: modifications for mixed ERS/ENVISAT time series. In: ENVISAT and ERS symposium, Salzburg, 6–10 September 2004, p 9
24. Kampes BM (2006) Radar interferometry: persistent scatterer technique. Springer, Dordrecht
25. Liu G, Jia H, Zhang R et al (2012) Ultrashort-baseline persistent scatterer radar interferometry for subsidence detection. ISPRS Ann Photogramm Remote Sens Sp Inform Sci 1:41–48
26. Zhang L, Lu Z, Ding X et al (2012) Mapping ground surface deformation using temporarily coherent point SAR interferometry: application to Los Angeles Basin. Remote Sens Environ 117:429–439
27. Fialko Y, Simons M, Agnew D (2001) The complete (3-D) surface displacement field in the epicentral area of the 1999 M_W7. 1 Hector Mine earthquake, California, from space geodetic observations. Geophys Res Lett 28(16):3063–3066
28. Strozzi T, Luckman A, Murray T, Wegmüller U, Werner CL (2002) Glacier motion estimation using SAR offset-tracking procedures. IEEE TGRS 40(11):2384–2391
29. Ferretti A, Perissin D, Prati C, Rocca F (2005) On the physical nature of SAR permanent scatterers. In: URSI commission F symposium on microwave remote sensing of the earth, oceans, ice and atmosphere, Ispra, 20–21 April 2005
30. Perissin D, Rocca F (2006) High-accuracy urban DEM using permanent scatterers. IEEE TGRS 44(11):3338–3347
31. Crosetto M, Monserrat O, Agudo M (2008) Validation of existing processing chains in terrafirma stage 2; process analysis report–part 2: IG inter-comparison. Technical report, Institut de Geom'atica. European Space Agency TerraFirma report, ESRIN contract No. 19366/05/I-E
32. Crosetto M, Monserrat O, Cuevas-González M et al (2015) Measuring thermal expansion using X-band persistent scatterer interferometry. ISPRS J Photogramm Remote Sens 100:84–91
33. Gernhardt S, Adam N, Eineder M, Bamler R (2010) Potential of very high resolution SAR for persistent scatterer interferometry in urban areas. Ann GIS 16(2):103–111
34. Monserrat O, Crosetto M, Cuevas M, Crippa B (2011) The thermal expansion component of persistent scatterer interferometry observations. IEEE Geosci Remote Sens Lett 8:864–868
35. Fornaro G, Reale D, Verde S (2013) Bridge thermal dilation monitoring with millimeter sensitivity via multidimensional SAR imaging. IEEE Geosci Remote Sens Lett 10:677–681
36. Zhang L, Ding X, Lu Z et al (2014) A novel multitemporal InSAR model for joint estimation of deformation rates and orbital errors. Geosci Remote Sens IEEE Trans 52(6):3529–3540

21

Concurrent Optimization of Rail Transit Alignments and Station Locations

Xiaorong Lai[1] · Paul Schonfeld[2]

Abstract Designing urban rail transit systems is a complex problem, which involves the determination of station locations, track geometry, and various other system characteristics. Most of the previous rail transit route optimization studies have focused on the alignment design between predetermined stations, whereas a practical design process has to account for the complex interactions among railway alignments and station locations. This paper proposes a methodology for concurrently optimizing station locations and the rail transit alignment connecting those stations, by accommodating multiple system objectives, satisfying various design constraints, and integrating the analysis models with a geographical information system database. The methodology incorporates demand and station costs in the evaluation framework and employs a genetic algorithm for optimizing the decision variables for station locations, station types, and track alignments. It is expected that transit planners may greatly benefit from the proposed methodology, with which they can conveniently and efficiently optimize candidate alternatives. The Baltimore Red Line is used as a case study to demonstrate how the model can find very good solutions in regions with complex geography.

✉ Xiaorong Lai
xlai@rkk.com

Paul Schonfeld
pschon@umd.edu

1 Rummel, Klepper & Kahl, LLP, 81 Mosher Street, Baltimore, MD 21217, USA

2 Department of Civil and Environmental Engineering, University of Maryland, College Park, MD 20742, USA

Editor: Xuesong Zhou

Keywords Rail transit · Concurrent optimization · Station location · Track alignment

1 Introduction

As a high-capacity transportation mode that is free from surface road congestion, urban rail transit has witnessed significant expansion in many cities during recent decades despite the substantial investment needed for its construction and maintenance. Designing urban rail transit systems is a complex problem, which involves the determination of station locations, track geometry, and various other system characteristics. Currently, the design of urban rail transit systems is mostly approached empirically with a trial-and-error process: planners develop alternatives subject to design specifications and local conditions, and then evaluate these alternatives based on project budget as well as system performance criteria. Such a design process is time-consuming and cannot guarantee that its results are even close to optimal.

In response, various optimization models have been proposed to establish alignments that meet various geometry requirements. The early work focused especially on modeling techniques, such as calculus of variations [1, 2], numerical search [3, 4], linear programming [5, 6], network optimization [7–9], and dynamic programming [3, 8, 10, 11]. Recent studies advanced the methodology by removing unrealistic assumptions and incorporating real-world constraints. Researchers at the University of Maryland have proposed a series of GA-based alignment optimization models. Jong [12] and Jong and Schonfeld [13] first demonstrated the concept in highway alignment optimization, featuring its comprehensive cost function and its consistency with engineering practice in the generated alignments. Jha [14], Jha and Schonfeld [15] and Jha et al. [16] extended the model by integrating a GIS to better accommodate the complex topological and environmental features. Kim [17] and Kim et al. [18] developed methods for incorporating the cost of major structures, i.e., bridges and tunnels. Kang et al. [19] further improved the GA-based solution algorithm by introducing the Feasible Gates approach.

Jha et al. [20] extended the previous highway models to railway alignment optimization. Lai and Schonfeld [21] presented a practical rail transit alignment optimization method to account for vehicle dynamics, which aims to balance the initial construction cost with the operation and user costs recurring throughout the system's life cycle. Kim et al. [22, 23] focused on vertical alignments between rail transit stations that exploited gravity to help accelerate and decelerate trains. Most of these publications on optimizing urban rail transit designs focused on the alignment optimization between two or more predetermined stations, whereas the selection of station locations may actually be even more challenging. Planners sometimes have to identify the potential station locations and select the best set among these locations, while accounting for various geometric, topological, environmental, and financial constraints.

This paper proposes a methodology for concurrently optimizing station locations and the rail transit alignment connecting these stations, by accommodating multiple system objectives, formulating various design constraints, and integrating the analysis models with a geographical information system (GIS) database. The methodology incorporates demand and station costs in the evaluation framework and employs a genetic algorithm (GA) for optimizing the decision variables for station locations, station types, and track alignments. It is expected that transit planners may greatly benefit from the proposed methodology, with which they can conveniently and efficiently optimize alternatives.

2 Methodology

Figure 1 illustrates the framework of the proposed methodology for concurrently optimizing rail transit alignment and station locations, and also for helping planners select the type of each station.

Following the engineering practice, this paper models the 3-dimensional rail track alignment with three separate components: the selection of stations, including station sequence and station type, the horizontal alignment that defines the track' path on the *XY* plane, and the vertical alignment that defines the elevation along the horizontal alignment. The model also incorporates various geometry constraints derived from engineering practice [24], using the cutting-plane concept [12] to define the Point of Intersection. The simulation of vehicle dynamics is based on the essential train dynamics equations in Hay [25], which divide the train operation between each pair of neighboring stations into acceleration, cruising, and braking stages, and then calculates train energy consumption and travel time along a given track alignment with an iterative process [21].

The remaining sections will detail three other key steps, i.e., the generation of the candidate station pool, the mode choice model, and the calculation of the cost function.

2.1 Generation of the Candidate Pool of Potential Rail Transit Stations

The proposed concurrent optimization model directly addresses the tradeoffs between ridership and cost in selecting station locations and generating the alignment in between. As it is impractical to check every point in the study area as potential station sites, this section presents procedures that apply quantified constraints to screen the study area and build a candidate pool of possible station locations. This candidate pool is then used as an input in the concurrent station and alignment optimization model to identify the best station locations.

Based on engineering practice, the candidate station locations typically should satisfy the following general requirements:

1. Stations cannot be located within infeasible areas (e.g., lakes or rivers), environmentally sensitive areas (e.g., wetland or residence of protected species), or historically sensitive areas (e.g., churches or cemeteries).
2. Stations should have the potential to attract considerable ridership, which can be realized in three ways. First, the catchment areas of stations could cover a sufficient number of households or employment locations. However, the size of the catchment area is related to the selected station type: walking-based stations have a shorter radius compared to park-and-ride stations. Secondly, stations may be located at existing activity centers or transfer centers of railway or bus transit systems. Finally, areas having the potential to support future growth at higher densities, such as land reserved for future transit-oriented development, may also be good candidate station locations.
3. To attract more ridership, stations should have good accessibility for their target population. Park-and-ride stations should have easy access to the existing road network and preferably be near the roadways carrying significant traffic volumes; walking-based stations should have good accessibility for pedestrians.
4. Stations should avoid locations that could incur extremely high cost, such as extensively developed

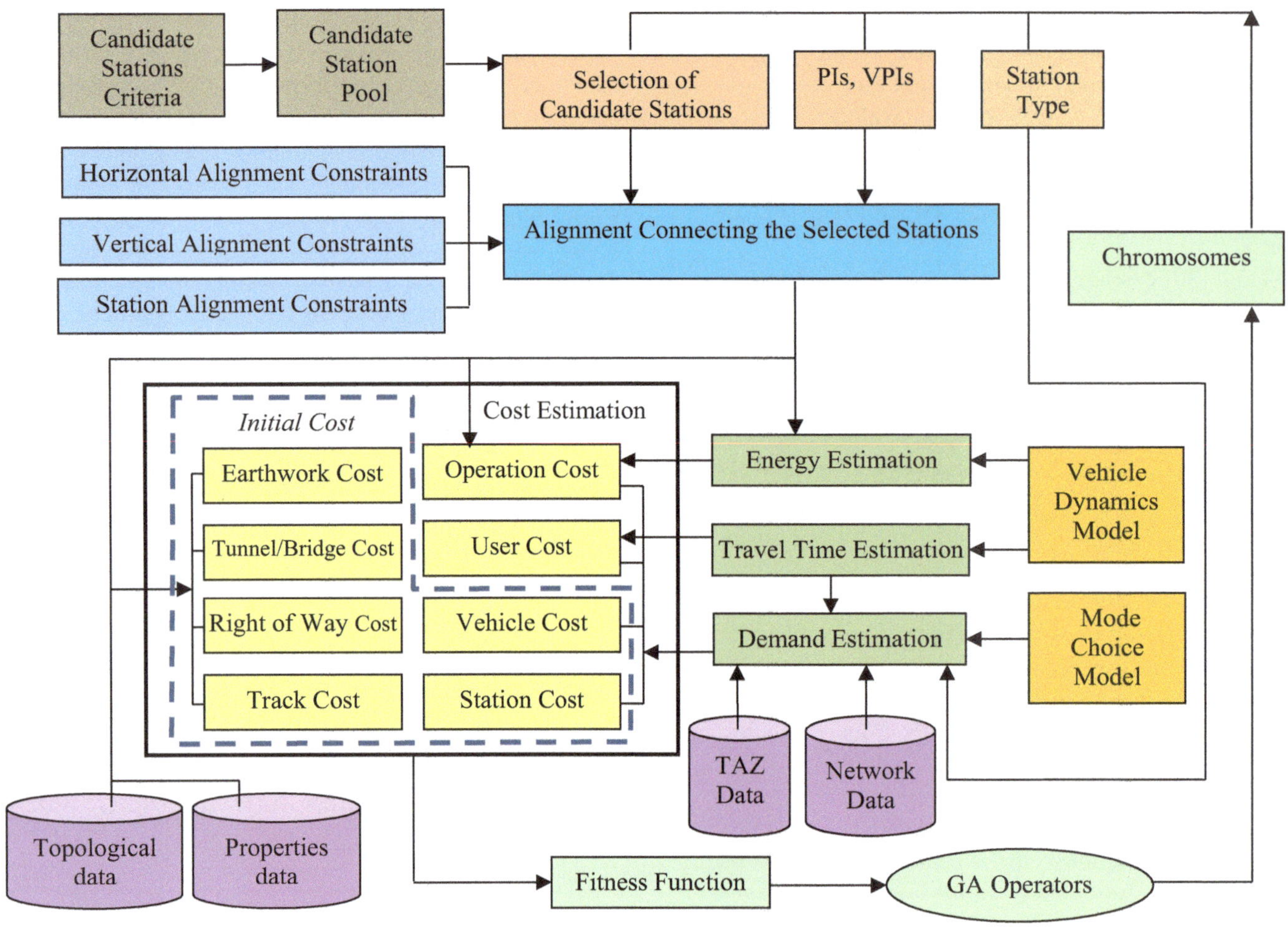

Fig. 1 Framework of the concurrent optimization model

neighborhoods with expensive right-of-way cost. The park-and-ride stations should not be too close to downtown; otherwise commuters would be unlikely to use rail transit for very short trips.

In accordance with these selection criteria, the following procedures are used here to generate the candidate pool of possible rail transit stations:

Step 1 Create a layer of grids inside the study area, Ω_S, with attributes:
S^i_w, feasibility of grid i to be pedestrian-orientated station (0—infeasible; 1—feasible)
S^i_p, feasibility of grid i to be park-and-ride station (0—infeasible; 1—feasible)
Set $S^i_w = 0$, $S^i_p = 0$

Step 2 Create a layer Ω_i for infeasible areas, which combines wetlands, historic districts, historically sensitive areas, and topography features such as rivers, lakes, and valleys

Step 3 Overlay the two layers from Step 1 and Step 2 to create the feasible grid layer

$$\Omega_f = \Omega_s \bigcap \bar{\Omega}_1 \tag{1}$$

Step 4 Find all grids with the number of households within walking distance above a threshold

For each $G_i \in \Omega_f$,

$$\text{set } S^i_w = 1 \text{ if } \left(\sum_j H_j | R^i_j \le R_w \right) \ge H_w \bigcap \bar{\Omega}_1, \tag{2}$$

where H_j is the number of households in block j, from census data, R^i_j is the distance from the center of census block j to G_i, R_w is the walking distance, H_w is the pre-specified threshold number of households within walking distance

Step 5 Find all grids with the number of households within driving distance above a threshold

For each $G_i \in \Omega_f$,

$$\text{set } S_p^i = 1 \text{if} \left(\sum_j H_j | R_j^i \le R_p \right) \ge H_p, \tag{3}$$

where R_p is the radius for park-and-ride stations, H_p is the pre-specified threshold number of households within driving distance

Step 6 Find all grids with the number jobs within walking distance above a threshold

For each $G_i \in \Omega_f$,

$$\text{set } S_w^i = 1 \text{ if } \left(\sum_j E_j | R_j^i \le R_w \right) \ge E_w \tag{4}$$

where E_j is the employment number of block j, from census data, E_w is the pre-specified threshold number of jobs within walking distance

Step 7 Find the grids close to rail transfer or bus stations

For each $G_i \in \Omega_f$, set $S_w^i = 1$ if count of rail or bus stations which satisfy $D_j^i \le R_w$ exceeds N_s,

where D_j^i is the distance from the bus/rail station j to G_i, N_s is pre-specified threshold number of bus/rail stations

Step 8 Find the grids without extremely high right-of-way (ROW) cost

For each $G_i \in \Omega_f$,

$$\text{set } S_w^i = 0 \text{ if } \left(\sum_j C_j | L_j^i \le L_w \right) \ge C_w,$$

$$\text{set } S_p^i = 0 \text{ if } \left(\sum_j C_j | L_j^i \le L_p \right) \ge C_p, \tag{5}$$

where C_j is the ROW cost for property j, L_j^i is the distance from property j to G_i, L_w is the impact distance for pedestrian-orientated station, C_w is the maximal allowed ROW cost for pedestrian-orientated station, L_p is the impact distance for park-and-ride station, C_p is the maximal allowed ROW cost for park-and-ride station

Step 9 Obtain the layer of annual average daily traffic (AADT) polyline features and select those with AADT above a user input value to form a new layer Ω_A. Find grids that are far away from these features. For each $G_i \in \Omega_f$, set $S_p^i = 0$ if $B_i \bigcap \Omega_A = \emptyset$, where B_i is a buffer area around grid i with a buffer radius R_A

Step 10 Find grids that are close to the downtown center. For each $G_i \in \Omega_f$ Set $S_p^i = 0$ if $K_i \le K_d$, where K_i is the distance from grid i to the downtown center, K_d is a pre-specified threshold distance from the downtown center

Step 11 Let $\Omega_s = \left\{ G_i \in \Omega_f | S_w^i + S_p^i > 0 \right\}$ be the candidate pool. Sort Ω_s, based on the distance from the starting terminal $\Omega_s = \{G_1, G_1, \ldots, G_i, G_{i+1}, \ldots, |L_i^0 \le L_{i+1}^0\}$, where L_i^0 is the distance from the starting terminal to G_i

After the proposed procedures screen the study area and generate the candidate pool of potential station locations, the concurrent optimization model encodes the decision variable for the selection of potential station site G_i as X_i, where $X_i = 1$, if G_i is selected; $X_i = 0$, if G_i is not selected. The selection of stations must satisfy the following constraints:

(a) Minimum number of stations N_L

$$\sum_i X_i \ge N_L \tag{6}$$

(b) Maximum number of stations N_U

$$\sum_i X_i \le N_U \tag{7}$$

(c) Minimum spacing between any two selected stations

$$\forall X_i = 1, X_j = 1,\ i \ne j:\ Y_{i,j} \ge Y_{\min}, \tag{8}$$

where $Y_{i,j}$ is the distance along the alignment from Station i to Station j, $Y_{\min}$ is the minimum spacing required between stations

(d) Maximum spacing between any two selected stations

$$\forall X_i = 1, X_j = 1,\ i \ne j:\ Y_{i,j} \le Y_{\max}, \tag{9}$$

where $Y_{\max}$ is the maximum spacing required between stations

(e) Minimum distance to depart from the starting terminal

$$\forall X_i = 1,\, X_j = 1, j > i:\; L_j^0 - L_i^0 \geq L_0, \tag{10}$$

where L_i^0 is the distance from the starting terminal to Station i, L_0 is the minimum spacing required to depart from the starting terminal

(f) Minimum distance to approach the end terminal

$$\forall X_i = 1, X_j = 1, j > i:\; L_i^1 - L_j^1 \geq L_1, \tag{11}$$

where L_i^1 is the distance from Station i to the end terminal, L_1 is the minimum spacing required to approach the end terminal

These decision variables and constraints are then incorporated into a GA-based solution heuristic for generating alignments that connect the selected stations.

2.2 Forecast of Rail Transit Demand

Existing station location models represented ridership attraction either by the number of rail transit users calculated with simple mode choice models, or by the alignment coverage estimated as line coverage or station coverage [26]. Such methods are quite simplified compared to various transit ridership forecasting models that are used in rail transit planning studies. This paper incorporates in the proposed concurrent optimization framework a discrete choice model, which is a widely accepted transit ridership forecasting model in real-world practice.

2.2.1 Choice Modeling for Rail Travel Demand Forecast

Discrete choice models model the travelers' choice among different transportation modes. The choice modeling is based on the random utility theory, which assumes that the decisions maker's preference for a discrete alternative is captured by a value called a utility, and his/her choice is reflected in the choice set with the highest utility. Choice models can be aggregate or disaggregate, according to the type of input data. The aggregate approach directly models the aggregate share of all decision makers choosing each alternative as a function of the characteristics of the alternatives and socio-demographic attributes of the group. The disaggregate approach recognizes that aggregate behavior is the result of numerous individual decisions and to model individual choice responses as a function of the characteristics of the alternatives available to and socio-demographic attributes of each individual. In this paper, it is assumed that the total trips matrix is known from external regional demand forecasting models, and thus aggregate choice models are employed. The models use the trip matrix as input and split the matrix into separated matrices, one for each mode.

Depending on the logit structure for the alternatives in the study area, the proposed concurrent optimization model employs two types of choice models in its rail ridership forecasting module: a multinomial logit choice model for pedestrian-oriented stations and a nested logit choice model for Park-and-Ride facilities.

2.2.2 Multinomial Logit Choice Model for Pedestrian-Oriented Stations

The multinomial logit choice (MNL) model is the most widely used discrete choice model, as its formula for the choice probabilities has a closed form and is readily interpretable. MNL relies on the assumption of independence of irrelevant alternatives (IIA). The basic utility U_m for choosing alternative m in MNL model is

$$U_m = V_m + \varepsilon_m, \tag{12}$$

where V_m is the representation of utility using observed variables, ε_m is the unknown part which is treated as random.

The MNL model is obtained by assuming that each ε_m is an independently identically distributed extreme value. The relation of the logit probability to representative utility is sigmoid, or S-shaped. This shape has implications for the impact of changes in explanatory variables. If the representative utility of an alternative is very low or high compared with other alternatives, a small change in the utility of the alternative has little effect on the probability of it being chosen. The point at which the increase in representative utility has the greatest effect on the probability of it being chosen is when the probability is close to 0.5, meaning a 50–50 chance of the alternative being chosen. In this case, a small improvement tips the balance in people's choices, inducing a large change in probability. For pedestrian-oriented stations, the structure of the MNL model is shown in Fig. 2.

The probability of taking mode m between OD pair ij is given as

$$P_{ijm} = \frac{e^{V_{ijm}}}{\sum_m e^{V_{ijm}}}. \tag{13}$$

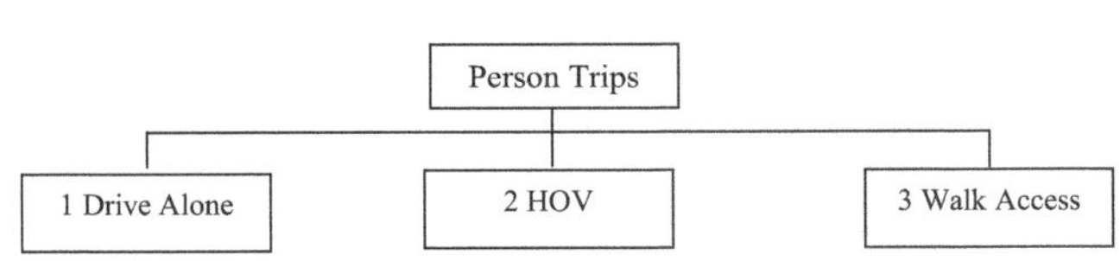

Fig. 2 Multinomial logit choice model for pedestrian-oriented stations

Here U_{ijm} is the utility of mode m between OD pair ij for a representative traveler. Representative utility is usually specified to be linear in parameters $V_{ijm} = \beta' x_{ijm}$, where x_{ijm} is a vector of observed variables relating to alternative m. With this specification, the logit probabilities become

$$P_{ijm} = \frac{e^{\beta' x_{ijm}}}{\sum_m e^{\beta' x_{ijm}}}. \tag{14}$$

2.2.3 *Nested Logit Choice Model for Park-and-Ride Stations*

The nested logit model (NLM), also known as generalized extreme value (GEV) model, allows partial relaxation of IIA property. It is useful when the unobserved portions of utility for some alternatives are correlated and IIA does not hold. An NLM is considered when the set of alternatives can be partitioned into subsets, called nests, so that the following properties hold:

1. For any two alternatives that are in the same nest, the ratio of probabilities is independent of the attributes or existence of all other alternatives. That is, IIA holds within each nest.
2. For any two alternatives in different nests, the ratio of probabilities can depend on the attributes of other alternatives in the two nests. IIA does not hold in general for alternatives in different nests

In the nested logit model, the utility is expressed as

$$U_m = W_k + Y_m + \varepsilon_m \tag{15}$$

$$V_m = W_k + Y_m. \tag{16}$$

Here the observed component of utility can be decomposed into two parts. The part W_k is constant for all alternatives within a nest and depends only on variables that describe nest k. These variables differ over nests but not over alternatives within each nest. The part Y_m depends on variables that describe alternative m and varies over alternatives within a nest k. For park-and-ride stations, the structure of the nested logit model is shown in Fig. 3.

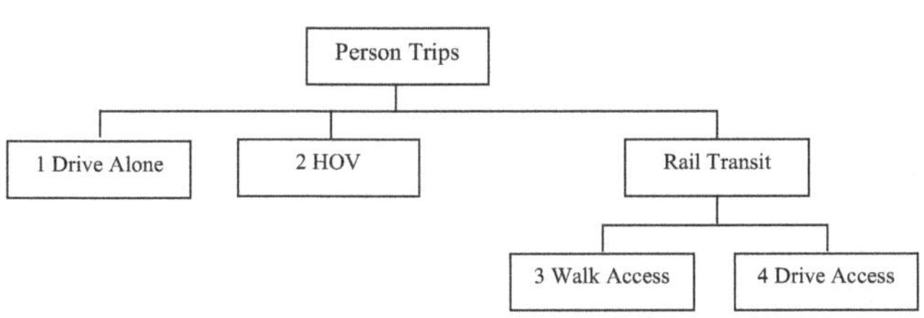

Fig. 3 Nested logit model for park-and-ride stations

The probability of taking mode m between OD pair ij is given as the product of two standard logit probabilities. The probability of choosing alternative $m \in B_k$, P_{ijm}, is the product of two probabilities:

- The probability that an alternative within nest B_k is chosen, P_{ijB_k}, which is the marginal probability of choosing an alternative in nest B_k
- The probability that then alternative m is chosen given that an alternative within B_k is chosen, $P_{ijm|B_k}$, which can be obtained by using MNL model

$$P_{ijm} = P_{ijm|B_k} \times P_{ijB_k} \tag{17}$$

$$P_{ijm|B_k} = \frac{e^{Y_{ijm}/\lambda_k}}{\sum_{l \in B_k} e^{Y_{ijl}/\lambda_k}} \tag{18}$$

$$P_{ijB_k} = \frac{e^{W_k + \lambda_k I_{ijk}}}{\sum_{l=1}^{K} e^{W_l + \lambda_l I_{ijl}}} \tag{19}$$

$$I_{ijk} = \ln\left(\sum_{m \in B_k} e^{Y_{ijm}/\lambda_k}\right). \tag{20}$$

The choice of nest is a marginal probability, also called the upper model. The choice of alternative within the nest is a conditional probability, also called the lower model. The quantity I_{ijk}, which is called the inclusive value or inclusive utility of nest k, links the upper and lower models by bringing information from the lower model into the upper model. The coefficient λ_k of I_{ijk} in the upper model is called the log-sum coefficient. It indicates the degree of independence among the unobserved portions of utility for alternatives in nest B_k. A lower λ_k indicates less independence (more correlation).

2.3 Estimation of System Cost

The proposed concurrent optimization model uses the total net cost C_{net} as the fitness function, which is a function of initial costs C_c, operation cost saving S_O, and user cost saving S_U.

$$C_{net} = C_c - S_O - S_U. \tag{21}$$

2.3.1 *Initial Cost*

The initial cost of the alignment is the capital cost, which includes earthwork costs C_E, bridge costs C_B, tunnel costs C_T, right-of-way costs C_R, track costs C_L, train vehicle costs C_V, and station costs C_s:

$$C_c = C_E + C_B + C_T + C_R + C_L + C_V + C_s. \tag{22}$$

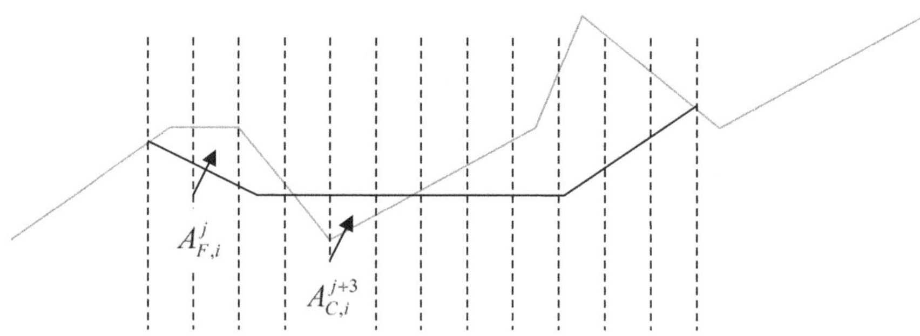

Fig. 4 Earthwork of a typical cut/fill section

Following engineering practice, this paper applies the typical track cross sections, as shown in Fig. 4, at an equal spacing L_{CS} along the horizontal alignment on the corresponding elevation from the vertical alignment. It then uses the elevation data from the GIS database to characterize the cut/fill sections that contribute earthwork costs and bridge/tunnel sections that contribute to structure costs. The total earthwork cost for a typical cut/fill section is calculated as follows:

$$\begin{aligned} E_N &= \sum_i E_{C,i} s_e - \sum_i E_{F,i} \\ C_E &= K_C \sum_i E_{C,i} + K_F \sum_i F_{F,i} + K_l \max(E_N, 0) \cdot \\ &\quad - k_b \min(E_N, 0) \end{aligned} \tag{23}$$

Here E_N is the net earthwork, C_E is the total earthwork cost, s_e is earth shrinkage factor, K_C is unit cutting cost, K_F is unit filling cost, and K_l and k_b are unit transportation cost for, respectively, moving earth to a landfill and from a borrow pit. The cut volume $E_{C,i}$ and fill volume $E_{F,i}$ are calculated for each cross section i by stratifying each section with very small intervals and calculating the cut area $A^j_{C,i}$ and fill area $A^j_{F,i}$ between the proposed and the existing ground for each stratum j.

$$\begin{aligned} E_{C,i} &= 0.5 \times \left(\sum_j A^j_{C,i} + \sum_j A^j_{C,i+1}\right) \times L_{CS} \\ E_{F,i} &= 0.5 \times \left(\sum_j A^j_{F,i} + \sum_j A^j_{F,i+1}\right) \times L_{CS} \end{aligned} \tag{24}$$

The total structure cost includes both bridge cost and tunnel cost. For each bridge i, an enumeration method is used to find the optimal span length L_{Bi} that minimizes the sum of superstructure C^U_{Bi} and substructure costs C^L_{Bi}. The cost calculation is based on the predefined bridge width, the bridge length identified in the first step, and the pier height that depends on the vertical alignment and the ground elevation extracted from the GIS database. The cost for each tunnel i depends on the predefined unit cost for tunnel excavation, the area of tunnel cross sections and the tunnel length. To calculate the right-of-way cost, this paper first generates the right-of-way band along the horizontal alignment by connecting the edge points of each cross section. For a cut/fill section, the edge points are outside the edge of earthwork with a buffer width. For bridges, the edge points are outside the bridge cross section with a buffer width. Tunnel sections require no right-of-way. The track cost depends only on the track length and a unit track installation cost. The vehicle costs C_V are the product of the number of trains N_T needed in the fleet, the number of cars per train N_c, and the cost per car K_V in millions:

$$C_V = N_T \times N_c \times K_V. \tag{25}$$

Assuming a fixed headway H in train schedule, the number of needed trains N_T can be calculated as the round trip travel time divided by the headway:

$$N_T = \frac{T_R}{H}. \tag{26}$$

The above formulations of train vehicle costs indicate that lower travel times require fewer trains, and thus decrease the cost of purchasing vehicles. The proposed concurrent optimization model assumes that station cost includes two parts: a fixed station cost that is independent of station locations, and a location-based station cost. The fixed station cost includes the cost for station facilities and for parking facilities. Assuming the station site to have a rectangular shape with user specified length and width, the fixed cost for station facilities varies only with the construction type, i.e., at-grade, elevated, or underground, which depends on the elevation difference between the proposed station and the existing ground. Assuming the cost of parking facilities is linear with respect to the park-and-ride demands, their fixed cost is calculated based on a preset unit cost per parking space. The location-based station cost includes the ROW cost and the earthwork cost. The ROW cost and earthwork cost for at-grade stations and parking facilities can be obtained from a GIS.

2.3.2 Operation Cost Saving

The operating cost includes energy costs and other operation and maintenance cost. In each time period p, the number of train trips needed N_p is calculated based on the estimated rail transit ridership.

$$N_{\mathrm{p}} = \frac{\sum_i \sum_j D_{ijp}}{N_{\mathrm{c}} D_{\mathrm{c}}}, \tag{27}$$

where D_c is the average number of passengers a train car can carry, N_c is the number of cars per train. Assuming the train service is only provided on workdays, the annual energy costs are

$$A_{\mathrm{e}} = (52 \times 5) \sum_p N_{\mathrm{p}} \times E_{\mathrm{R}} \times \frac{K_{\mathrm{e}}}{10^6}, \tag{28}$$

where E_R is the round-trip energy consumption (kwh), K_e is the unit cost of energy ($/kwh). The railway operation and maintenance costs are

$$A_m = (52 \times 5) \sum_p \sum_i \sum_j D_{ijp} L_{i,j} \times \frac{K_O}{10^6}, \tag{29}$$

where $L_{i,j}$ is the travel distance from station i to station j (mile), K_O is the unit operation and maintenance cost for rail ($/passenger-mile). The auto operation cost for the park-and-ride trips is

$$A_{\mathrm{P}} = (52 \times 5) \sum_p \sum_i \sum_j D^{\mathrm{P}}_{ijp} L^{\mathrm{P}}_{i,j} \times \frac{K_{\mathrm{a}}}{10^6}, \tag{30}$$

where D^P_{ijp} is the number of park-and-ride trips from Traffic Analysis Zone (TAZ) i to j in period p, $L^P_{i,j}$ is the auto travel distance for D^P_{ijp} (mile), K_a is the unit operationcost for auto ($/passenger-mile). The original auto operation cost for the rail riders is

$$A_a = (52 \times 5) \sum_p \sum_i \sum_j D^r_{ijp} L^{\mathrm{a}}_{i,j} \times \frac{K_a}{10^6}, \tag{31}$$

where D^r_{ijp} is the number of trips from TAZ i to TAZ j using rail in period p, $L^{\mathrm{a}}_{i,j}$ is the auto travel distance for trip from TAZ i to TAZ j (mile). The annual operating cost saving is

$$A_o = A_a - \left(A_e + A_m + A_p\right). \tag{32}$$

Assuming an annual interest rate of r and a life cycle of n_{a} years, the present value of operating cost saving over the system's life cycle is

$$S_{\mathrm{o}} = A_{\mathrm{o}} \frac{(1+r)^{n_{\mathrm{a}}} - 1}{r(1+r)^{n_{\mathrm{a}}}}. \tag{33}$$

2.3.3 User Cost Saving

Similarly to the calculation of the operation cost saving, the annual user cost saving for railway riders is

$$A_U = (52 \times 5) \sum_p \sum_i \sum_j D^{\mathrm{r}}_{ijp} \times \left(T^{\mathrm{a}}_{i,j,p} \times \frac{1}{3600} \times \frac{K^{\mathrm{a}}_U}{10^6} - T^{\mathrm{r}}_{i,j,p} \times \frac{1}{3600} \times \frac{K^{\mathrm{r}}_U}{10^6}\right), \tag{34}$$

where $T^a_{i,j,p}$ is the travel time by auto from TAZ $i$to TAZ j in time period p (s), $T^{\mathrm{r}}_{i,j,p}$ is the travel time by rail from TAZ i to TAZ j in time period p(s), K^{a}_U is the unit user cost for auto ($/passenger-hour), K^{r}_U is the unit user cost for rail ($/passenger-hour).

The present value of user cost saving is

$$\mathrm{S}_U = \mathrm{A}_U \frac{(1+r)^{n_{\mathrm{a}}} - 1}{r(1+r)^{n_{\mathrm{a}}}}. \tag{35}$$

3 Case Study

Using Baltimore City as the study area, this case study aims to illustrate the data preparation procedures of the proposed concurrent station location and alignment optimization model, and to demonstrate its effectiveness compared to the sequential optimization methodology where stations are first selected and alignment is then designed between these selected stations.

3.1 Data Preparation for the Proposed Concurrent Optimization Model

3.1.1 Data for Candidate Stations

Following the aforementioned procedures, a grid layer inside the study area is created with user-specified grid size (1000 feet by 1000 feet in the case study). Then a series of GIS operations are applied to identify the grids for candidate pedestrian-oriented stations and park-and-ride stations, as shown below.

- Land use pattern
  Certain types of land use are excluded for railway alignment and stations, such as forest, river, wetlands, historical area, and some restricted area due to political or economic concerns. The grid layer generated from the previous step is overlaid with the land use layer in GIS. All the grids intersecting with those restricted zones are identified as infeasible grids for railway stations.
- Census block data
  Census data are obtained from the United States Census Bureau. Year 2000 census data for Baltimore City and Baltimore County are used in the case study, as shown in Fig. 5.

Fig. 5 Census block data

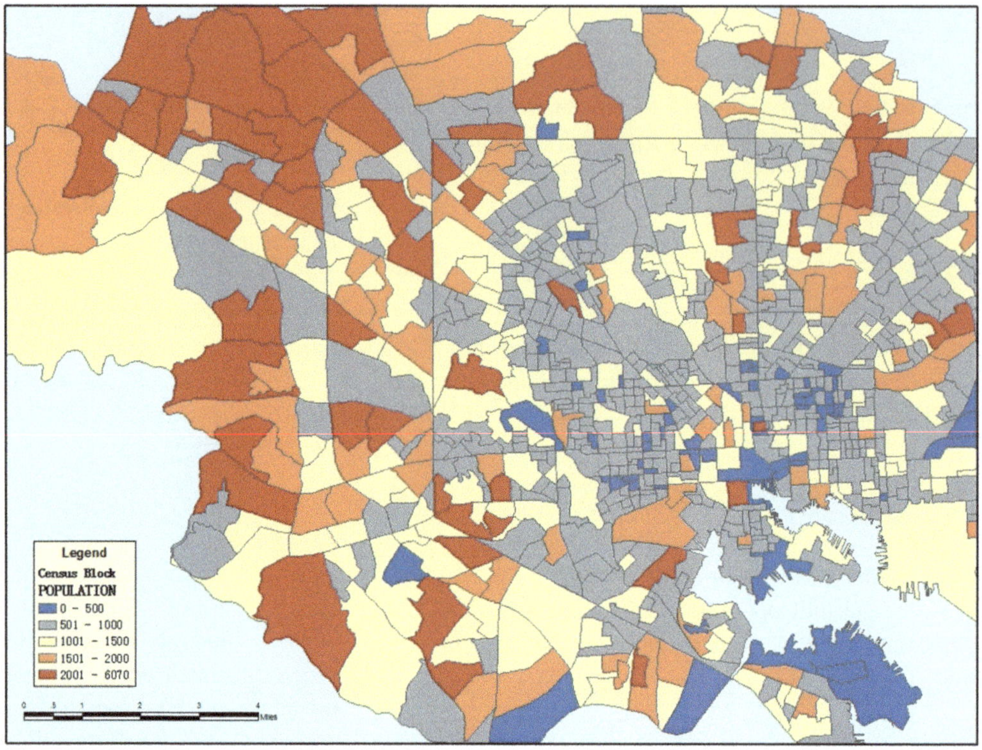

The stations grid cells which attract high population or high employment are considered as potential station locations. As noted earlier, pedestrian-oriented stations and park-and-ride stations have different catchment area dimensions and thresholds of population and employments.

- AADT
  Candidate park-and-ride stations need to meet the requirement of easy access to the existing road network. AADT line information obtained from the Maryland State Highway Administration is used to determine the accessibility of potential park-and-ride station locations: the roadways near the candidate location should carry significant traffic volumes, as shown in Fig. 6.
- Properties Data
  Based on the properties distribution, a 250 feet by 250 feet grid layer is created in the study area with a ROW cost value for each grid cell. The candidate station locations should avoid the high ROW cost grid cells.

After applying the proposed procedures for generating the candidate pool of potential rail transit stations, 52 candidate pedestrian-orientated station locations and 10 candidate park-and-ride station locations are found, as shown in Fig. 7.

3.1.2 Data for Estimating Railway Travel Demand

The proposed concurrent optimization method applies a nested logit mode choice model to calculate mode choice for personal trips and outputs the following four trip tables:

- Mode 1: Drive alone
- Mode 2: High occupancy vehicle
- Mode 3: Walk to rail
- Mode 4: Drive to rail.

The utility U_{ijm} is a function of the alternative characteristics and decision maker's characteristics, which includes the following variables:

- T_{ijm}^{IN}—Travel time in the vehicle or train from TAZ i to j for mode m
- T_{ijm}^{OUT}—Travel time outside of the vehicle or train from TAZ i to j for mode m
- T^{w}—Waiting time or headway of the train at the boarding station
- C_{ijm}—Cost of mode m (gas, parking, and ticket) from TAZ i to j
- S_{ijm}—Travel distance from TAZ i to j for mode m
- A_i—The number of autos per person in TAZ i
- B_j—Binary variable to check if the TAZ j is close to CBD
- E_j—Employment density of the TAZ j.

This study considers all the Traffic Analysis Zones (TAZs) within 1 mile of candidate Pedestrian-oriented stations and/or within 5 miles of candidate Park-and-Ride stations. The TAZ data are obtained from Baltimore Metropolitan Council (BMC) models [27]. TAZ data contain zone-related information, such as population,

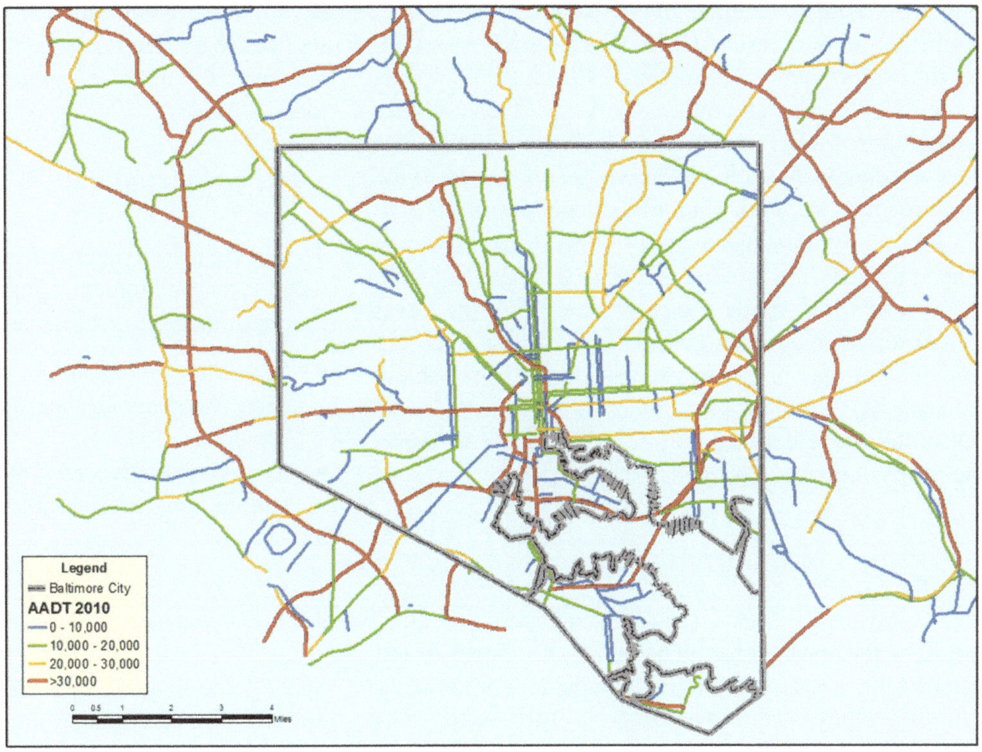

Fig. 6 AADT data

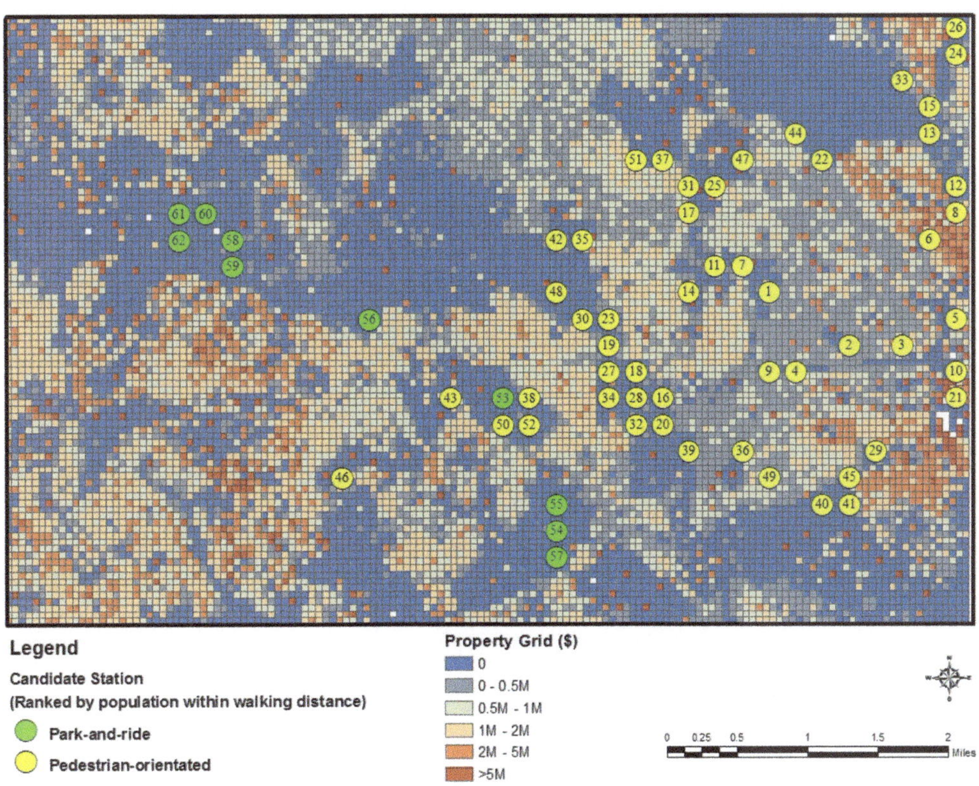

Fig. 7 Candidate station locations

employment density, income, number of autos per person, and whether or not inside CBD. The roadway network travel time information is also obtained from BMC models [27] for AM/PM peak and midday periods, which consider the congestion level for different time periods. All TAZ centers and candidate station locations are connected to the existing roadway network via artificial connectors. TAZ to TAZ, TAZ to Station, and Station to TAZ travel time and distance matrices are generated using GIS shortest path function prior to the mode choice process. Then the travel time and distance by train between any two stations are computed via an iterative process based on vehicle dynamics [21].

All of the above data are used to calculate the variables in the utility function. For mode m,

$$U_{ijm} = (a_m^0 + a_m^1 T_{ij2}^{\text{IN}} + a_m^2 T_{ij2}^{\text{OUT}} + a_m^3 T_{ij}^{\text{w}} + a_m^4 C_{ijm} + a_m^5 S_{ijm} + a_m^6 A_i + a_m^7 B_j + a_m^8 E_j), \quad (36)$$

where a_m^0 is the constant for mode m, a_m^i, $i = 1$ to 8, is the coefficient for the aforementioned 8 variables for mode m.

The case study adapts the values in Table 1 for constants, coefficients, and correlations from BMC model [27]. It is noted that the original model used different sets of parameters depending on the trip type (home-based, work-based, other-based) and income level (I, II, and III). For simplicity, this section only uses home-based trip and level II parameters in the mode choice modeling, which should be sufficient for examining the effectiveness of the proposed concurrent station and alignment optimization model.

After applying the above mode choice model to all OD pairs of TAZs in the study area, the total trip matrices in three time periods are split into 4 modes: drive alone, HOV, walk to rail, and drive to rail. The trip matrices for the latter two modes are used to compute the rail transit station to station demands, and are incorporated into the fitness calculation in the GA process.

3.2 Model Results

To test the effectiveness of the proposed concurrent station location and alignment optimization model, the following two optimization methods are examined:

- *Two-stage optimization* Locate stations first to maximize the demand, then find the alignment to minimize the cost.
- *Concurrent optimization* Concurrently optimize the station locations and alignment to minimize the system cost.

3.2.1 Comparison of the Two Optimization Methods

Table 2 presents the optimization results of the two-stage optimization and concurrent optimization.

Compared to the two-stage optimization, the concurrent optimization significantly reduces the total cost from 41.8 to 17.6 M. Concurrent optimization reduces the rail line's passenger trips by 21.8 % compared to the two-stage optimization. By compromising in the passenger attraction, concurrent optimization reduces the travel time from 11.8 to 10.8 min, shortens the track length from 6.7 to 5.8 miles, decreases the initial cost from 74.3 to 54.1 M, and also decreases the operation and user cost for about 4.0 M. The numerical results for this case demonstrate the advantage of concurrent optimization over the two-stage optimization.

Table 1 Nested logit model parameters

	Drive alone	HOV	Walk access	Drive access
m	1	2	3	4
a_m^0		−1.05989	1.29430	−0.55102
a_m^1	−0.0338	−0.0338	−0.0125	−0.0125
a_m^2	−0.0443	−0.0443	−0.0443	−0.0443
a_m^3 if $T_{ij}^w \leq 7.5$ min			−0.0291	−0.0291
a_m^3 if $T_{ij}^w > 7.5$ min			−0.0186	−0.0186
a_m^4	−0.1430	−0.1430	−0.0529	−0.0529
a_m^5		0.0991	0.3038	0.3038
a_m^6		−2.1822	−4.7928	−4.7928
a_m^7		0.3393	0.3393	0.3393
a_m^8			0.00003	0.00003
λ_m	1.0000	1.0000	0.7274	

Table 2 Comparison of two-stage optimization and concurrent optimization

	Two-stage	Concurrent	Difference (%)
Number of stations	5	5	0
Total daily passengers	4879	3814	−21.8
Travel time (min)	11.8	10.8	−8.5
Total length (miles)	6.7	5.8	13.4
Total initial cost (M)	74.3	54.1	−27.2
Operation and user cost saving (M)	32.5	36.5	12.3
Total cost (M)	41.8	17.6	−57.9

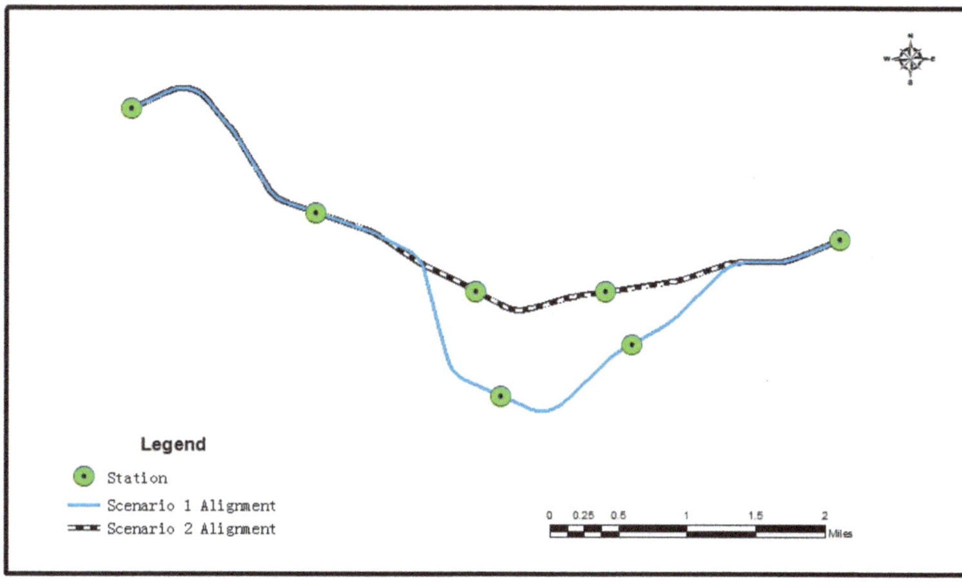

Fig. 8 Optimized station locations and alignments

Figure 8 presents the station locations and horizontal alignments generated from the two optimization methods. Both alignments have three intermediate stations. The first intermediate station is the same. For the second and the third intermediate stations, the two-step optimization selected two dispersed locations to attract more railway passengers, whereas the concurrent optimization selected two closer locations to shorten the alignment length and travel time, so as to decrease the system total cost. The alignments are similar at both ends for the two optimization methods, whereas the middle sections of the alignments are shifted to connect different selected stations.

3.2.2 Impact of Demand Variation on Optimization Results

This section examines how the proposed concurrent optimization model adjusts its station selection and alignment design with variations in demand distribution so as to minimize the total cost. The design scenario adjusts the total demands from/to the four TAZs of 86, 87, 88, and 90 from 12,869 to 51,476, as shown in Fig. 9. Figure 9 also compares the optimized station locations and alignments for the original and the adjusted demand distributions.

Two of the three stations selected for the original demand distribution are shifted to locate within the four TAZs with the adjusted demands. The two alignments start with the same segments at the western end of the study area, until they approach the first intermediate station. The solution algorithm then generates different tangent segments through the first intermediate station to adjust the alignment toward the two shifted station locations. Compared to the optimized station locations and alignment generated for the original demand distribution, the shifted station locations and alignment incur an initial cost increase of 7.7 M from 54.1 to 61.8 M. However, the savings in operation and user cost over the system's entire life cycle increase by 91.4 M for the shifted station locations, as they attract more than twice of the original

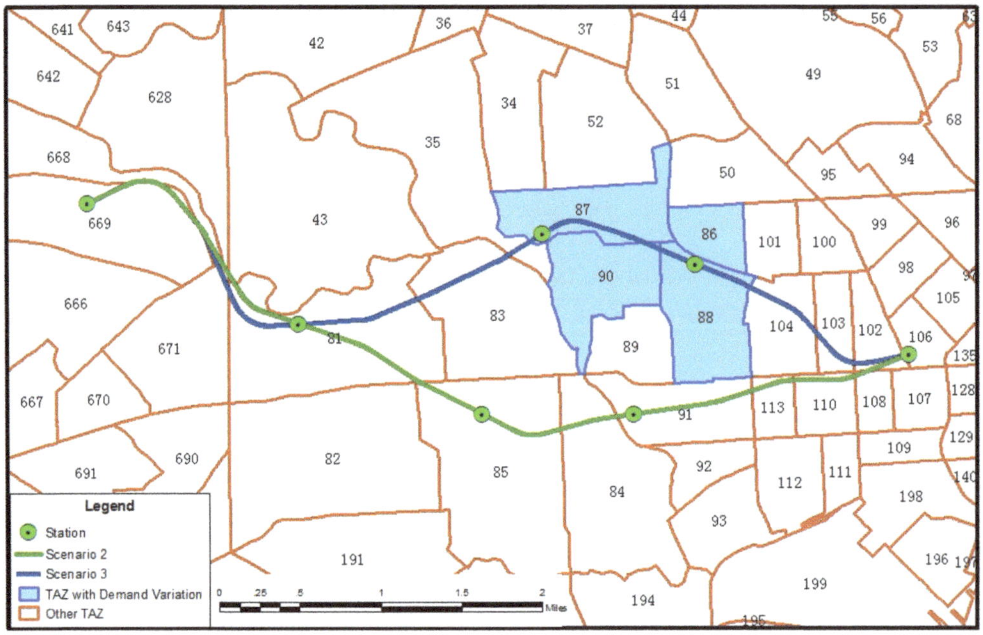

Fig. 9 Impact of demand variation on optimization results

demand by directly serving the TAZs with higher demand. The results show that the algorithm is effective in adapting the demand patterns and can concurrently optimize the station locations and alignment accordingly.

3.2.3 Statistical Test of Solution Goodness

This case study applies a statistical method from Jong and Schonfeld [13] to test the effectiveness of the proposed algorithm in finding near-optimal solutions. This procedure is a sampling process. 50,000 random solutions are generated, of which 23.7 % (11,870) are feasible. The average cost of the feasible solutions is 161.1 M, and standard deviation is 46.9 M. The least cost of any random solution is 41.5 M, which is very far above the best solution (17.6 M) found with our proposed algorithm.

We also use the Gamma and Normal distributions to fit the cost distribution from the feasible random solutions, as shown in Fig. 10, with R^2 values of 0.99 and 0.91, respectively.

The optimized cost from the proposed concurrent optimization model in scenario 2 is 17.6 M, which is better than 99.89 % of solutions in the fitted Normal distribution, and close to 100 % of solutions in the fitted Gamma distribution, but far better than any of the random feasible solutions. This indicates that neither the Normal nor Gamma distributions fit well the extreme low-cost end of the randomly generated solutions. This test shows that although we cannot guarantee a global optimum using the proposed genetic algorithm (or other such metaheuristic algorithms), the optimized solution is extremely good compared to the other solutions in the search space.

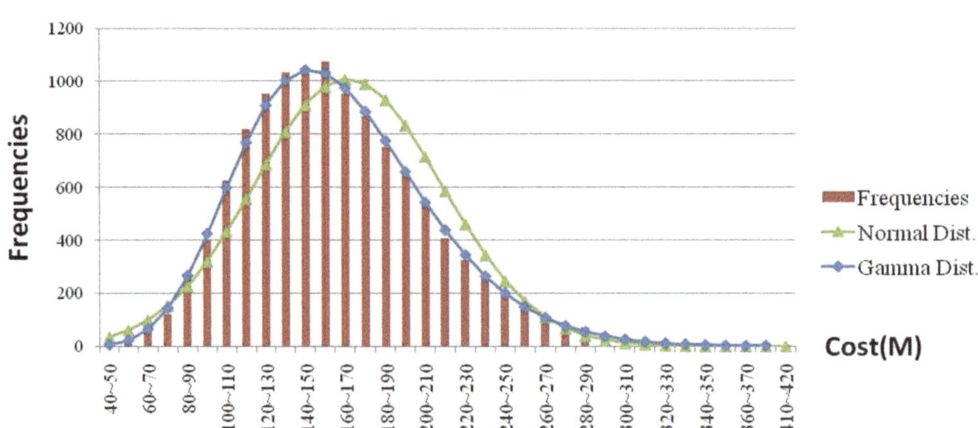

Fig. 10 Statistical test of solution goodness

4 Summary

This paper presents a concurrent railway station location and alignment optimization methodology. The methodology first constructs the candidate pool of potential rail transit stations based on the consideration of various site requirements regarding topological features, accessibility to the existing roadway network, and land availability. These candidates are then selected along with the alignment between each pair of neighboring stations using the concurrent optimization model to minimize the total system cost while satisfying station selection and track geometry constrains.

The proposed methodology demonstrates how massive amounts of geographic information can be processed and employed within an optimization framework in planning an urban rail transit system in real-world practice. In particular, it demonstrates how route alignments and station locations can be jointly optimized in continuous space (rather than on an abstracted graph) using whatever relevant geographic information is available. The optimization objectives and constraints used in this paper can be changed without drastically changing the modeling approach presented here. The comprehensive cost evaluation framework addresses the essential trade-off between minimizing capital cost and maximizing savings in operation and user costs by shifting trips from auto to rail. The embedded GA-based heuristic can concurrently optimize the decision variables for station locations, station types, and track alignments connecting each pair of neighboring stations. The case study demonstrates the applicability of the proposed concurrent optimization model and its advantages over the previously used two-stage optimization.

The proposed methodology is expected to help transit planners generate viable alternatives for a single rail transit line effectively and efficiently. Future research could consider the complex interactions between the rail transit line and the existing surface road network, provide the ability to repair infeasible alignments by relaxing some geometric constraints (e.g., by reducing design speeds), improve the computational performance with a distributed genetic algorithm operated on parallel processors, and address the problem of optimizing networks of rail transit lines with feeder bus lines.

Acknowledgments This research was partly funded by the USDOT University Transportation Center Program through the National Center for Strategic Transportation Policies, Investments, and Decisions at the University of Maryland.

References

1. Shaw JFB, Howard BE (1981) Comparison of two integration methods in transportation routing. Transp Res Rec 806:8–13
2. Shaw JFB, Howard BE (1982) Expressway route optimization by OCP. J Transp Eng 108(3):227–243
3. Goh CJ, Chew EP, Fwa TF (1988) Discrete and continuous models for computation of optimal vertical highway alignment. Transp Res Part B 22(6):399–409
4. Chew EP, Goh CJ, Fwa TF (1989) Simultaneous optimization of horizontal and vertical alignments for highways. Transp Res Part B 23(5):315–329
5. Moreb AA (1996) Linear programming model for finding optimal roadway grades that minimize earthwork cost. Eur J Oper Res 93(1):148–154
6. ReVelle CS, Whitlatch EE, Wright JR (1997) Civil and environmental systems engineering. Prentice Hall, New Jersey
7. Parker NA (1977) Rural highway route corridor selection. Transp Plan Technol 3:247–256
8. Trietsch D (1987) A family of methods for preliminary highway alignment. Transp Sci 21(1):17–25
9. Trietsch D (1987) Comprehensive design of highway networks. Transp Sci 21(1):26–35
10. Fwa TF (1989) Highway vertical alignment analysis by dynamic programming. Transp Res Rec 1239:1–9
11. Nicholson AJ, Elms DG, Williman A (1976) A variational approach to optimal route location. Highw Eng 23:22–25
12. Jong JC (1998) Optimizing highway alignments with genetic algorithms. Ph.D. Dissertation, University of Maryland, College Park
13. Jong JC, Schonfeld P (2003) An evolutionary model for simultaneously optimizing three-dimensional highway alignments. Transp Res Part B 37:107–128
14. Jha MK (2000) A geographic information systems-based model for highway design optimization. Ph.D. Dissertation, University of Maryland, College Park
15. Jha MK, Schonfeld P (2004) A highway alignment optimization model using geographic information systems. Transp Res Part A 38:455–481
16. Jha MK, Schonfeld P, Jong JC, Kim E (2006) Intelligent road design. WIT Press, Southhampton
17. Kim E (2001) Modeling intersections & other structures in highway alignment optimization. Ph.D. Dissertation, University of Maryland, College Park
18. Kim E, Jha MK, Schonfeld P (2004) Intersection construction cost functions for alignment optimization. J Transp Eng 130(2):194–203
19. Kang MW, Schonfeld P, Jong JC (2007) Highway alignment optimization through feasible gates. J Adv Transp 41(2):115–144
20. Jha MK, Schonfeld P, Samanta S (2007) Optimizing rail transit routes with genetic algorithms and GIS. J Urban Plan Dev 133(3):161–171
21. Lai X, Schonfeld P (2012) Optimization of rail transit alignments considering vehicle dynamics. Transp Res Rec 2275:77–87
22. Kim M, Schonfeld P, Kim E (2013) Comparison of vertical alignments for rail transit. J Transp Eng 139(2):230–238
23. Kim M, Schonfeld P, Kim E (2014) Simulation-based rail transit optimization model. Transp Res Rec 2374:143–153
24. Lai X, Schonfeld P (2010) Optimizing Rail Transit Alignments Connecting Several Major Stations. Presented at the 89th TRB Annual Meeting, Washington DC
25. Hay WW (1982) Railroad engineering. Wiley, New York

26. Laporte G, Mesa JA, Ortega FA (2000) Invited review: optimization methods for the planning of rapid transit systems. Eur J Oper Res 122:1–10

27. Baltimore Metropolitan Council (2004) Baltimore Region travel demand model for base year 2000 Task Report 04-01, January 2004

Permissions

All chapters in this book were first published in JMT, by Springer; hereby published with permission under the Creative Commons Attribution License or equivalent. Every chapter published in this book has been scrutinized by our experts. Their significance has been extensively debated. The topics covered herein carry significant findings which will fuel the growth of the discipline. They may even be implemented as practical applications or may be referred to as a beginning point for another development.

The contributors of this book come from diverse backgrounds, making this book a truly international effort. This book will bring forth new frontiers with its revolutionizing research information and detailed analysis of the nascent developments around the world.

We would like to thank all the contributing authors for lending their expertise to make the book truly unique. They have played a crucial role in the development of this book. Without their invaluable contributions this book wouldn't have been possible. They have made vital efforts to compile up to date information on the varied aspects of this subject to make this book a valuable addition to the collection of many professionals and students.

This book was conceptualized with the vision of imparting up-to-date information and advanced data in this field. To ensure the same, a matchless editorial board was set up. Every individual on the board went through rigorous rounds of assessment to prove their worth. After which they invested a large part of their time researching and compiling the most relevant data for our readers.

The editorial board has been involved in producing this book since its inception. They have spent rigorous hours researching and exploring the diverse topics which have resulted in the successful publishing of this book. They have passed on their knowledge of decades through this book. To expedite this challenging task, the publisher supported the team at every step. A small team of assistant editors was also appointed to further simplify the editing procedure and attain best results for the readers.

Apart from the editorial board, the designing team has also invested a significant amount of their time in understanding the subject and creating the most relevant covers. They scrutinized every image to scout for the most suitable representation of the subject and create an appropriate cover for the book.

The publishing team has been an ardent support to the editorial, designing and production team. Their endless efforts to recruit the best for this project, has resulted in the accomplishment of this book. They are a veteran in the field of academics and their pool of knowledge is as vast as their experience in printing. Their expertise and guidance has proved useful at every step. Their uncompromising quality standards have made this book an exceptional effort. Their encouragement from time to time has been an inspiration for everyone.

The publisher and the editorial board hope that this book will prove to be a valuable piece of knowledge for researchers, students, practitioners and scholars across the globe.

List of Contributors

A. A. Memon
IVES (International Vision for Engineering Solutions), Abu Dhabi, United Arab Emirates

M. Meng
TUM CREATE, 1 CREATE Way #10-02 CREATE Tower, Singapore 138602, Singapore

Y. D. Wong
School of Civil and Environmental Engineering, Centre for Infrastructure Systems, Nanyang Technological University, Singapore 639798, Singapore

S. H. Lam
Transportation Infrastructure Office, Macao S.A.R., China

Milan Janić
Faculty of Civil Engineering and Geosciences, Department of Transport & Planning, Delft University of Technology, Stevinweg 1, 2628 CN Delft, P.O. Box 5048, 2600, GA, Delft, The Netherlands

Taimur Usman
Department of Civil & Environmental Engineering, University of Waterloo, Waterloo, ON N2L 3G1, Canada

Liping Fu
Department of Civil & Environmental Engineering, University of Waterloo, Waterloo, ON N2L 3G1, Canada
School of Transportation and Logistics, Southwest Jiaotong University, Chengdu 610031, China

Luis F. Miranda-Moreno
Department of Civil Engineering & Applied Mechanics, McGill University, Montreal, QC H3A 2K6, Canada

Silvia Morales-Ivorra, Julia Irene Real, César Hernández and Laura Montalbán
University Institute for Multidisciplinary Mathematics, Polytechnic University of Valencia, 46022 Valencia, Spain

İlker Erkan and Hasan Hastemoglu
Faculty of Architecture, Suleyman Demirel University, 32260 Cunur, İsparta, Turkey

Guangning Wu, Wenfu Wei, Guoqiang Gao, Jie Wu and Yue Zhou
School of Electrical Engineering, Southwest Jiaotong University, Chengdu 610031, China

Naiwala P. Chandrasiri
The Kogakuin University, Tokyo, Japan

Kazunari Nawa
Toyota InfoTechnology Center, Co., Ltd., Tokyo, Japan

Akira Ishii
AISoftware Inc., Kobe, Japan

Qunzhan Li
School of Electrical Engineering, Southwest Jiaotong University, Chengdu 610031, China

Dazhi Sun and Anxi Jia
Texas A&M University-Kingsville, Kingsville, TX, USA

Xiaobo Liu
School of Transportation and Logistics, Southwest Jiaotong University, Chengdu 610031, People's Republic of China

Yi Deng
Parsons Transportation Group, 100 Broadway, New York, NY 10005, USA

Scott Le Vine
School of Transportation and Logistics, Southwest Jiaotong University, Chengdu 610031, People's Republic of China
State University of New York at New Paltz, New Paltz, NY 12561, USA
Imperial College London, South Kensington, London SW7 2AZ, England, UK

Sachin Kumar
Centre for Transportation Systems (CTRANS), Indian Institute of Technology Roorkee, Roorkee, Uttarakhand 247667, India

Durga Toshniwal
Computer Science & Engineering Department, Indian Institute of Technology Roorkee, Roorkee, Uttarakhand 247667, India

G. X. Chen, X. L. Cui and W. J. Qian
State Key Laboratory of Traction Power, Tribology Research Institute, Southwest Jiaotong University, Chengdu 610031, China

Ping Wang, Li Wang, Rong Chen, Jinhui Xu, Jinmang Xu and Mingyuan Gao
Key Laboratory of High-speed Railway Engineering, Ministry of Education, Southwest Jiaotong University, Chengdu 610031, China
School of Civil Engineering, Southwest Jiaotong University, Chengdu 610031, China

Yang Yue
Department of Road and Bridge Engineering, School of Transportation, Southeast University, Nanjing 210096, Jiangsu, China

Sida Luo and Tianming Luo
Department of Traffic Engineering, School of Transportation, Southeast University, Nanjing 210096, Jiangsu, China

Zhijie Dong
School of Civil and Environmental Engineering, Cornell University, Ithaca, NY 14853, USA

Meicheng Wang
Department of City and Regional Planning, Cornell University, Ithaca, NY 14853, USA

Xianfeng Yang
Department of Civil, Construction, & Environmental Engineering, San Diego State University, San Diego, CA 92182, USA

Yamur K. Al-Douri, Phillip Tretten and Ramin Karim
Division of Operation and Maintenance Engineering, Lulea° University of Technology, Luleå, Sweden

Jing Zeng, Lai Wei and Pingbo Wu
State Key Laboratory of Traction Power, Southwest Jiaotong University, Chengdu 610031, China

Yanning Zhao
Department of Civil Engineering, Nagoya University, Furo-cho, Chikusa-ku, Nagoya 464-8603, Japan

Toshiyuki Yamamoto
Institute of Materials and Systems for Sustainability, Nagoya University, C1-3(651), Furo-cho, Chikusa-ku, Nagoya 464-8603, Japan

Takayuki Morikawa
Institute of Innovation for Future Society, Nagoya University, Chikusa-ku, Nagoya 464-8603, Japan

Bo Liu and Chunhai Gao
Beijing Traffic Control Technology CO, Ltd, Beijing, China

Luyao Liu
Faculty of Geosciences and Environmental Engineering, Southwest Jiaotong University, Chengdu, Sichuan, China

Hongguo Jia
Faculty of Geosciences and Environmental Engineering, Southwest Jiaotong University, Chengdu, Sichuan, China
State Key Laboratory of Geodesy and Earth's Dynamics, Wuhan, Hubei, China Collaborative Innovation Center for Rail Transport Safety, Ministry of Education of the People's Republic of China, Southwest Jiaotong University, Chengdu, Sichuan, China
State-Province Joint Engineering Laboratory of Spatial Information Technology for High Speed Railway Safety, Southwest Jiaotong University, Chengdu, Sichuan, China

Xiaorong Lai
Rummel, Klepper & Kahl, LLP, 81 Mosher Street, Baltimore, MD 21217, USA

Paul Schonfeld
Department of Civil and Environmental Engineering, University of Maryland, College Park, MD 20742, USA

Index